T0402130

International Neurolaw

Tade Matthias Spranger
Editor

International Neurolaw

A Comparative Analysis

 Springer

Editor
Associate Professor Tade Matthias Spranger
BMBF Research Group ELSI at the Institute of Science and Ethics
University of Bonn
Bonner Talweg 57
53113 Bonn
Germany
spranger@iwe.uni-bonn.de

ISBN 978-3-642-21540-7   e-ISBN 978-3-642-21541-4
DOI 10.1007/978-3-642-21541-4
Springer Heidelberg Dordrecht London New York

Library of Congress Control Number: 2011939219

Printed on acid-free paper

Springer is part of Springer Science+Business Media (www.springer.com)

Preface

While the past few years have repeatedly been entitled as the "era of biotechnology", most recently one has to get the impression that at least the same degree of attention is being paid to the latest developments in the field of neurosciences. As in the fields of biotechnology and nanotechnology, neuroscientific research also opens a barely manageable range of possible applications, some of which are still related to a purely experimental setting, while others are already in practical use. The possible applications cover aspects as diverse as the development of mind reading machines, lie detection methods, or brain–computer interface applications for the improvement of disabled person's daily life.

It is by now nearly impossible to oversee the number of research projects dealing with the functionality of the brain – for instance concerning the organizational structure of the brain – or projects dealing with the topic neuro-economics or neuro-marketing. Massive efforts have also been taken in the field of prediction; for instance, some scientists consider it possible to predict a person's decision before he has ever told it. Hence, a huge practical interest is being paid to neuroscientific developments. This especially holds true for the usage of neuroscientific methods in court trials. In the USA, companies such as Cephos and NO Lie MRI canvass the usage of image-guided procedures – especially the functional magnetic resonance tomography – in different legal areas. Also other countries show a strong development of comparable methods: In India, two states with together about 160 million citizens use the so-called BEOS-Test (Brain Electrical Oscillations Signature), which is initially based on the electroencephalography (EEG). The activity measured by the EEG is evaluated by specific software; an interpretation of the data by the investigator is not planned. In 2008, the decision of an Indian court gained worldwide attention, basing its conviction essentially on the result of a BEOS-Test and convicting the concerned woman to a life-long sentence.

The above-mentioned procedures are connected to a wide variety of legal questions. These questions concern the frame conditions of scientific projects as well as the right approach toward the usage of generated findings. With regard to this utmost importance of the topic for latest developments, it is of special interest to compare the different legal systems and strategies which they hold at hand for dealing with those legal implications. Therefore, this volume contains several "country reports" from around the world, as well as reports of selected international organizations, in order to show the different legal approaches towards the topic. Each chapter aims to survey the relevant legal order's landscape both for the generation of neuroscientific knowledge (i.e. probands' rights, the relationship between research participant and researcher, the problem of incidental findings etc) and for the usage of neuroscientific knowledge.

This book aims to give a first and lasting impetus for further and – given the dimension of the issue – much-needed internationalization of the discussion. My special thanks go to my colleagues involved for their commitment, to the members of my research group at the University of Bonn for their substantial input, and last but not least, to Dr. Brigitte Reschke at Springer Publishing for supporting this project.

Bonn   Tade Matthias Spranger

Contents

Contributors

A. Aciduman Department of Medical Ethics and History of Medicine, Ankara University, Faculty of Medicine, Ankara, Turkey

B. Arda Department of Medical Ethics and History of Medicine, Ankara University, Faculty of Medicine, Ankara, Turkey

S. Bischof Forschungsgemeinschaft für Rechtswissenschaft der Universität St. Gallen, St. Gallen, Switzerland

P. Catley University of the West of England, Bristol, UK

T. Caulfield Faculty of Law, Law Centre, University of Alberta, Edmonton, Canada

L. Claydon University of the West of England, Bristol, UK

M.A.S. deFreitas University of São Paulo Law School, São Paulo, Brazil

G.-M. Gkotsi ETHOS, Interdisciplinary Ethics Platform, Quartier UNIL-Sorge, Batiment Amhipole, Lausanne, Switzerland

M. Henaghan Faculty of Law, University of Otago, Dunedin, New Zealand

M.J. Hilf Institut für Strafrecht, Strafprozessrecht und Kriminologie, Universität St. Gallen, St. Gallen, Switzerland

L. Houston University of Technology Sydney, Sydney, Australia

O.D. Jones MacArthur Foundation Research Network on Law and Neuroscience, Vanderbilt University, Nashville, TN, USA

K. Kai Center for Professional Legal Education and Research (CPLER), Waseda Law School, Tokyo, Japan

L. Klaming Tilburg Institute for Law, Technology, and Society, Tilburg University, Tilburg, The Netherlands

B.-J. Koops Tilburg Institute for Law, Technology, and Society, Tilburg University, Tilburg, The Netherlands

S. Lötjönen Docent in Medical and Bio Law, University of Helsinki, Helsinki, Finland

D. Macer RUSHSAP, UNESCO, Bangkok, Thailand

H.M. Prata Alameda Joaquim Eugênio de Lima, São Paulo, Brazil

C. Rödiger Institute of Science and Ethics, University of Bonn, Bonn, Germany

K. Rouch Faculty of Law, University of Otago, Dunedin, New Zealand

A. Santosuosso European Center for Law, Science and New Technologies, University of Pavia, Corso Strada Nuova, Italy

R.J. Schweizer Forschungsgemeinschaft für Rechtswissenschaft der Universität St. Gallen, St. Gallen, Switzerland

F.X. Shen Tulane University Law School and The Murphy Institute, New Orleans, LA, USA

S. Silvola University of Helsinki, Helsinki, Finland

Tade Matthias Spranger BMBF Research Group ELSI, Institute of Science and Ethics, University of Bonn, Bonn, Germany

K. Stöger Institut für Österreichisches, Europäisches und Vergleichendes Öffentliches Recht, Politikwissenschaft und Verwaltungslehre, Universität Graz, Graz, Austria

C. Toole Faculty of Law, Law Centre, University of Alberta, Edmonton, Canada

T. Vidalis National Bioethics Comission, Athens, Greece

A. Vierboom University of Technology Sydney, Sydney, Australia

H. Wegmann Institute of Science and Ethics, Bonn, Germany

A. Zarzeczny Faculty of Law, Law Centre, University of Alberta, Edmonton, Canada

Neurosciences and the Law: An Introduction

Tade Matthias Spranger

The scientific field called (modern) neurosciences covers a wide spectrum of most diverse branches of research and techniques. The scientific disciplines involved comprise inter alia, biology, medicine, chemistry, physics, psychology, mathematics, computer science, engineering, but also philosophy and – last but not least – law. The range of topics in the field of neurosciences now covers studies of the molecular, cellular, evolutionary, developmental, structural, functional, and medical aspects of the nervous system. Furthermore, the techniques used by have been developing rapidly, reaching from studies of individual cells to the imaging of sensory and motor skills.

Electroencephalography (EEG) is an approach to record the electrical activity of the brain provoked by the firing of neurons. Electrodes are attached to the scalp and connected to an electrical box that is in turn attached to an EEG machine. The EEG machine picks up small electrical signs during a short period of time (20–40 min) and reproduces them as a record (as "brain waves") on a paper or on a computer screen. EEG is a convenient and comparatively inexpensive way to establish a diagnosis, but its spatial resolution is limited. It can only record electrical activity of the excitatory and inhibitory postsynaptic potentials in the more superficial layers of the cortex. Thus, some areas of the brain have to be activated synchronously, otherwise changes could not be registered. Furthermore, some amounts of cortex, particularly in basal and mesial areas of the hemisphere, are not covered by standard electrode placement, so that spatial sampling in routine is incomplete.[1] Therefore, insufficient EEG information has to be completed by detailed spatial neuroimaging data.[2] Its clinical applications are in neurological disorders, notably

---

[1] Smith (2005), p. ii2.

[2] See Fattouch et al. (2007), pp. 170–173.

T.M. Spranger (✉)
Institute of Science and Ethics, Bonner Talweg 57, 53113 Bonn, Germany
e-mail: spranger@iwe.uni-bonn.de

T.M. Spranger (ed.), *International Neurolaw*,
DOI 10.1007/978-3-642-21541-4_1, © Springer-Verlag Berlin Heidelberg 2012

in epilepsy[3] and schizophrenia,[4] but EEG machines are also used for nonmedical purposes such as brain–computer interfaces (BCIs)[5] and lie detectors or in neuromarketing research.

Functional Magnetic Resonance Imaging (fMRI) is a variant of structural MRI, which is presumed to be the most important imaging advance since the introduction of X-rays by Conrad Röntgen in 1895.[6] It measures indirectly the neural activity by determining the blood oxygen level dependent (BOLD) signal,[7] that is the different levels of oxyhemoglobin and deoxyhemoglobin in the brain.[8] The main advantages of fMRI lie in its noninvasive and nondestructive nature, ever-increasing availability, high spatiotemporal resolution, and its capacity to represent the whole brain activity when subjects perform different tasks. Data have to be interpreted carefully, as their spatial and temporal accuracy is liable to both physical and biological constraints. fMRI is principally used to produce structural images of the central nervous system and of organs, but it has also the potential to generate information on the physicochemical state of tissues, their vascularization, and perfusion.[9] It plays a decisive role in brain disease research. For instance, fMRI is utilized to reduce the rate of misdiagnosis of a vegetative state, which is otherwise estimated to be up to 40%[10] and to find treatment options for psychopaths.[11] The fMRI scanner has already been used to address yes/no questions to patients with different disorders of consciousness[12] and also as a real brain–computer interface (BCI) that allows control of computers and other external devices by the modulation of neural activity.[13] In contrast to EEG-based BCI, neuroimaging-based BCI has the potential to achieve higher spatial resolution[14] and to access the whole brain.[15] fMRI is also used outside a clinical setting to examine the behavior of consumers in neuromarketing research.[16] Furthermore, the scanner is supposed to be suitable for interrogation of terror suspects.[17] Several companies have already discovered its commercial value and offer fMRI lie detectors. Although the accuracy is estimated

---

[3] Tatum et al. (2008), p. 283.

[4] Winterer and McCarley (2011), p. 311.

[5] See Saddique and Siddiqui (2009), pp. 550–554.

[6] Logothetis (2008), p. 869.

[7] Matthews et al. (2006), p. 733.

[8] Poldrack (2008), p. 223.

[9] Logothetis (2008), p. 869.

[10] Monti et al. (2010), p. 579.

[11] See Kiehl et al. (2004), pp. 297–312; Birbaumer et al. (2005), pp. 799–805.

[12] Monti et al. (2010), p. 579.

[13] Owen et al. (2009), p. 403.

[14] Sitaram et al. (2007), p. 2.

[15] Weiskopf et al. (2004), p. 966.

[16] See Häusel (2007), pp. 210–220; Ariely and Berns (2010), pp. 284–292.

[17] Spranger (2009a), p. 210.

to be up to 90% or more by some scientists,[18] the use of fMRI scanners as lie detectors in courtrooms is controversially discussed with a view to reliability and standards of scientific evidence.[19] There is wide consensus that fMRI is never going to be a mind reader.[20] On the other hand, fMRI-based lie detection is widely used in private settings or for private reasons, respectively, for instance in relationships.

Positron Emission Tomography (PET) is a noninvasive nuclear medicine imaging technique that produces three-dimensional images of functional brain and body activities. The general aim of PET imaging is to visualize, describe, and quantify biological processes at the cellular, subcellular, and molecular level. It is at relatively high cost, has radioactive expositions, and requires vascular puncture.[21] Positron-emitting probes are mainly introduced to determine the expression of indicative molecular targets at different stages of cancer progression.[22] EEG, MRI, and PET are frequently combined to improve diagnostic accuracy such in the case of epilepsy[23] or Alzheimer's disease whose results show an improvement of up to 10–20% if neuroscientific data sources are combined compared to the individual use of each technique.[24]

Single Photon Emission Computed Tomography (SPECT) is a three-dimensional nuclear medicine tomographic imaging technique using a gamma-emitting radioisotope, which is injected into the bloodstream. The photons emitted by the rapid radioactive decay are subsequently detected by a SPECT camera.[25] Artifacts, patient movements, and improper positioning can be sources for poor image quality and data misinterpretation.[26] SPECT has a limited resolution, which is why it is often used in combination with MRI.[27] Its clinical applications are mainly in brain imaging.[28]

Diffuse Optical Imaging/Diffuse Optical Tomography (DOI/DOT) is a noninvasive three-dimensional imaging technique that uses near-infrared (NIR) light to create images of the brain and body by measuring the changes in oxyhemoglobin and deoxyhemoglobin concentrations. It is portable and inexpensive, utilizes harmless nonionizing radiation[29] and is thus also suited for babies and infants.[30]

---

[18] Kozel et al. (2005), p. 605.

[19] Miller (2010), p. 1336.

[20] Logothetis (2008), p. 869.

[21] Walter (2009), p. 70.

[22] Chen and Chen (2011), p. 70.

[23] Lai et al. (2010), p. 292.

[24] Polikar et al. (2010), p. 6058.

[25] Wortzek et al. (2008), p. 312.

[26] See Morano and Seibyl (2003), pp. 192–193.

[27] Gründer et al. (2010), p. 97.

[28] See Wortzek et al. (2008), pp. 310–322.

[29] Hielscher et al. (2002), p. 314.

[30] Gibson and Dehghani (2009), p. 3062.

However, measurements have to be repeated frequently to minimize the high number of artifacts provoked by severe undersampling.[31] Its spatial resolution is limited in contrast to MRI, but it provides access to multiple physiological parameters. The high-speed data acquisition allows subsecond imaging of spatio-temporal changes of physiological processes which would otherwise not be accessible through other neuroimaging techniques.[32] Its clinical applications are particularly functional brain[33] and breast imaging,[34] but also arthritic finger, muscle, and small animal imaging.[35]

Diffusion Tensor Imaging (DTI) is a noninvasive MRI technique that produces pictures of biological tissues through the observation of translational molecular movement of water ("water diffusion"), so that images of cerebral white matter are generated in two and three dimensions. DTI has the potential to visualize brain structures, which are not able to be imaged noninvasively[36] or by way of MRI.[37] However, neuroimaging is not dispensable, as the technique of water diffusion has a low resolution and is highly sensitive to subject motion.[38] DTI is used to study normal and diseased brains, particularly multiple sclerosis, stroke, aging, dementia, and schizophrenia.[39]

Magnetoencephalography (MEG) is a noninvasive neuroimaging technique that is used to record the brain activity by measuring magnetic fields arising from electrical currents in the brain. The accurate localization of sources of electrical activity still represents a technical problem. However, the clinical applications of MEG are in the field of epilepsy,[40] Alzheimer's disease,[41] schizophrenia, depression, and autism.[42] A technical advance of MEG is fetal MEG (fMEG) which has the potential to record brain activity generated by the fetus in utero.[43]

Near Infrared Spectroscopic Imaging/Near Infrared Spectroscopy (NIRSI/NIRS) belongs to the group of noninvasive imaging techniques and utilizes light in the range of 700–1,000 nm to measure brain activity by monitoring blood flow changes in the front part of the brain. It is portable and allows monitoring of freely moving subjects (e.g., of athletes during exercise to assess muscle oxygenation and of

[31] Süzen et al. (2010), p. 23676.

[32] Hielscher et al. (2002), p. 314.

[33] Boas et al. (2004), p. S275.

[34] Tromberg et al. (2008), p. 2443.

[35] Gibson and Dehghani (2009), p. 3063.

[36] Bandettini (2009), p. 277.

[37] Rugg-Gunn et al. (2001), p. 531.

[38] Bandettini (2009), p. 278.

[39] See Assaf and Pasternak (2008), pp. 51–61.

[40] See Ray and Bowyer (2010), pp. 14–22.

[41] See Montez et al. (2009), pp. 1614–1619.

[42] See Williams and Sachdev (2010), pp. 273–277.

[43] See Sheridan et al. (2010), pp. 80–93.

interacting subjects or of animals to study brain activity)[44] and is thus also suitable for children.[45] Until now, it has a rather limited field of application, as NIRSI/NIRS is relatively expensive and it can only be applied to image the brain cortex.[46] Nevertheless, it is expected to shed light on several physiological issues and diseases in near future.[47]

Voxel-Based Morphometry (VBM) is a neuroimaging analysis technique and has been developed to compare changes in gray matter between different groups of subjects. For that purpose, high-resolution MRI scans are compared after segmentation and spatial normalization. However, up to date, the relationship between gray matter concentration and MRI signal intensity is not yet clearly established as well as the mechanism and timing by which gray matter concentration changes with learning, experience, or disease. As the brain is presumed to change on a temporal scale depending on a large number of influences such as aging, drug abuse, psychiatric disorders, stressful or enriching environmental factors, learning and chronic health problems,[48] VBM is primarily used for research purposes.

The research results that have been obtained by applying the technologies just mentioned indicate numerous legal implications; legislation, however, has not responded sufficiently to this challenge so far. Up until now, one can only find an adequately deep discussion in criminal law, which is not exactly surprising: First, criminal law entitles the state to the most serious interventions in personal freedom; second – and this perspective must not be forgotten when setting norms for modern technologies – the respective neuroscientific applications can be sold better in the media.

The starting point of the penologic discussion about the neurosciences is the so-called Libet-experiment,[49] which seems to unsettle the assumption that the human being in principle has a free will. Independently of the question of the methodical correctness of the Libet-experiment and irrespective of the general arguments that can be invoked against a hard or soft determinism, philosophy and, building on this, law as well have lead the "Debate about the freedom of the will" with great vigor. The reason for this is obvious: If the picture of the free will is unsettled, then the concepts of legal responsibility or diminished responsibility, too, begin to totter. In fact, also renowned colleagues have temporarily claimed to substitute the principle of legal responsibility in force by a somehow science fiction-like preventive approach. By now, however, the respective voices have lowered noticeably, and a permanent alteration of the criminal law system as such is out of discussion.

---

[44] Muehlemann et al. (2008), p. 10324.

[45] Walter (2009), p. 71.

[46] Walter (2009), p. 71.

[47] Wolf et al. (2008), pp. 062104–062109.

[48] Bandettini (2009), p. 277.

[49] Cf. Libet et al. (1983), p. 322.

More interesting are thus the discussions about potential practical applications of neuroscientific findings in the criminal trial.[50] This holds true in particular for methods of detecting lies, which are already offered especially by US-American companies.[51] Therefore, like in the case of the polygraph, not only the civil rights of the accused or the question of the voluntariness of his or her participation are at stake, but also the aspect of the reliability of the corresponding technologies.

In addition to this, clearly more subtle and already practice-relevant today is the question in how far other neuroscientific methods find entrance into penal procedures by way of expert opinions.[52] Indeed in many legal systems, medical and psychological expert witnesses seem to draw on corresponding procedures – for instance imaging techniques – without the jurists involved being really aware of it.[53] To this extent, it is therefore not the question whether the Libet-experiments unsettle our understanding of penologic guilt, but rather whether particular fMRI-data, for instance, suggest that the accused has a disposition that excludes or reduces his or her guiltiness.

Outside of criminal law, the procedural relevance of the neurosciences seems to be centered primarily upon two realms: insurance law claims on the one hand and disability issues on the other. This trend becomes understandable in view of the fear of simulators; so there is hope that by neuroscientific methods it can, for instance, be proven whether a petitioner really suffers the alleged pain or not.

However, the legal implications of neuroscientific findings reach much further than the courtroom. Fields of research such as neuroeconomics or neuromarketing that are already advertised insistently not only on the Internet raise questions about sufficient consumer protection or the limits of unfair competition. There is absolutely no doubt that applications in these fields show an enormous potential. In fact, many companies specializing in neuromarketing are consulting major companies across multiple categories – and they are determined to find out what consumers "really want". As one company takes it: "[Our] process is about digging deeper into human thoughts. We go beyond the top layer, the rational 'thinking' brain. We don't stop at the middle layer, the emotional 'feeling' brain. We dig on to the very bottom, to the primal, instinctive Reptilian brain. This is where we find the hidden treasure. [Our] research digs deeper than any other research method into a target's most primitive, feelings, instincts and habits."[54] However, the corresponding technologies could also be used by assessment centers in order to examine the "true" attitude of the applicant regarding his or her future dream-employer. And finally the employment of mind-reading machines in the war against terrorism is of course being discussed too.

---

[50] See on this for instance: Tancredi and Brodie (2007), p. 239 et seq.

[51] Regarding the situation in India: Rödiger (2011).

[52] Markowitsch (2009).

[53] See Thompson (2007), p. 341 et seq.

[54] http://www.mindcode.com/01_about_us.html (Accessed 18 April 2011).

On the other hand, the application field of so-called BCIs, which can be used in order to improve the communicative ability of disabled people, prompts without exception positive associations. By means of the construction of appropriate "dictating machines", the persons concerned might be enabled to directly express their will – be it with regard to issues of everyday life or to aspects of legal relevance such as formulating contracts, writing one's testament etc.[55] As the case may be, one day it might even be possible to communicate with people in a vegetative state.

Almost completely unrecognized by law, there is also currently a discussion going on, the juridical consequences of which can hardly be overestimated: Recent neuroscientific results suggest the suspicion that brain death is not the definite caesura that has so far been ascribed to this criterion on a regular basis – which might lead to substantial conflicts regarding, for example, the admissibility of postmortal organ transplantation.

While all of the fields mentioned concern aspects of neuroscientific research results and of corresponding applications, the assumed legal implications are already noticeable on a considerably earlier level, too, namely when it comes to generating knowledge. Thus, it is first of all the protection of the participants of scientific studies that is at stake. Regarding this, the so-called exclusion criteria are relatively unproblematic: Beyond others, people with metal implants, pacemakers, piercings, larger tattoos, or claustrophobia may not take part in fMRI-studies. However, a much more complex question is how to deal with so-called incidental findings, which primarily in the context of research with imaging techniques are the order of the day.[56] If one reflexively grants the research participant a "right not to know", then the researcher is in the danger of being faced with a dilemma not only under ethical aspects: If an incidental finding occurs, that turns out to be life threatening and the researcher lets the test person go due to his or her "right not to know" that he exerted, then in the case of a damage even claims against the researcher are possible. But if, on the other hand, the researcher undertakes a "coercive informing", then he violates the "right not to know" of the test person. A practicable solution could be to let only those test persons participate in a study who agree in advance to not exert their "right not to know". Of course, as is so often the case, here, too, the devil is in the details; for instance, the question arises who is technically qualified to examine an incidental finding with a view to its medical relevance.

The legal aspects that have just been addressed can in part be clarified individually with respect to a particular legal order. But already the broader employment of appropriate lie detection machines in a larger democratic legal order would have consequences beyond the boarders of the country concerned. The realm of territorially limited or limitable regulation is ultimately left when it comes to establishing good practice-standards or SOPs for neuroscientific research: The international

---

[55] Spranger (2009b).

[56] Cf. Illes et al. (2004).

interconnection of research projects, the financing of the corresponding projects across states and the extensive exchange of data and samples make it necessary to speak proactively about the establishment of international standards or about the adjustment of standards that already exist.

The juridical publications that exist so far refer without exception to partial aspects of a single legal order or to the synopsis of chosen aspects valid in similar legal orders. The volume at hand clearly extends this perspective: On the one hand, it presents country reports from legal orders that represent not only geographically different regions, but also different legal cultures and approaches of regulation. On the other hand, the norm-settings of relevant international organizations are analyzed with respect to the question whether fundamental principles relevant to the neurosciences can be inferred from them. This way the collected reports by thirty authors offer an excellent overview of the legal landscape in the field of neurosciences. Thus, the penetration of the material that differs from state to state shall serve as the starting point of an essential broader internationalization of the discussion.

References

Ariely D, Berns GS (2010) Neuromarketing: the hope and hype of neuroimaging in business. Nat Rev Neurosci 11:284–292

Assaf Y, Pasternak O (2008) Diffusion tensor imaging (DTI)-based white matter mapping in brain research: a review. J Mol Neurosci 34(1):51–61

Bandettini PA (2009) What's new in neuroimaging methods? Ann N Y Acad Sci 1156:260–293

Birbaumer N, Veit R, Lotze M, Herrmann C, Erb M, Grodd W, Flor H (2005) Deficient fear conditioning in psychopathy: a functional magnetic resonance imaging study. Arch Gen Psychiatry 62:799–805

Boas DA, Dale AM, Franceschini MA (2004) Diffuse optical imaging of brain activation: approaches to optimizing image sensitivity, resolution, and accuracy. NeuroImage 23: S275–S288

Chen K, Chen X (2011) Positron emission tomography imaging of cancer biology: current status and future prospects. Semin Oncol 38(1):70–86

Fattouch J, Di Bonaventura C, Strano S, Vanacore N, Manfredi M, Prencipe M, Giallonardo AT (2007) Over-interpretation of electroclinical and neuroimaging findings in syncopes misdiagnosed as epileptic seizures. Epileptic Disord 9(2):170–173

Gibson A, Dehghani H (2009) Diffuse optical imaging. Philos Trans R Soc A 367:3055–3072

Gründer G, Vernaleken I, Bartenstein P (2010) Anwendungen von PET und SPECT in der psychiatrie. Der Nervenarzt 81:97–108

Häusel HG (2007) Methoden der neuromarketing-forschung. In: Häusel HG (ed) Neuromarketing. Erkenntnisse der Hirnforschung für Markenführung, Werbung und Verkauf. Haufe, München, pp. 210–220

Hielscher AH, Bluestone AY, Abdoulaev GS, Klose AD, Lasker J, Stewart M, Netz U, Beuthan N (2002) Near-infrared diffuse optical tomography. Dis Markers 18:313–337

Illes J, Kann D, Karetsky K, Letourneau P, Raffin TA, Schraedley-Desmond P, Koenig BA, Atlas SW (2004) Advertising, patient decision making, and self-referral for computed tomographic and magnetic resonance imaging. Arch Intern Med 164(22):2415–2419

Kiehl KA, Smith A, Mendrek A, Forster B, Hare RD, Liddle P (2004) Temporal lobe abnormalities in semantic processing by criminal psychopaths as revealed by functional magnetic resonance imaging. Psychiatry Res 130:297–312

Kozel FA, Johnson KA, Mu Q, Grenesko EL, Laken SJ, George MS (2005) Detecting deception using functional magnetic resonance imaging. Biol Psychiatry 58(8):605–613

Lai V, Mak HK, Yung AW, Ho WY, Hung KN (2010) Neuroimaging techniques in epilepsy. Hong Kong Med J 16(4):292–298

Libet B, Wright EW, Gleason CA (1983) Readiness potentials preceding unrestricted spontaneous pre-planned voluntary acts. Electroencephalogr Clin Neurophysiol 54:322–325

Logothetis NK (2008) What we can do and what we cannot do with fMRI. Nature 453:869–878

Markowitsch HJ (2009) Mind Reading?: Gutachten vor Gericht. In: Stephan S, Spranger TM, Walter H (eds) Von der Neuroethik zum Neurorecht? Vandenhoeck & Ruprecht, Göttingen, pp. 132–148

Matthews PM, Honey GD, Bullmore ET (2006) Applications of fMRI in translational medicine and clinical practice. Nat Rev Neurosci 7:732–744

Miller G (2010) FMRI lie detection fails a legal test. Science 328(5984):1336–1337

Montez T, Poil SS, Jones BF, Manshanden I, Verbunt JPA, Van Dijk BW, Brussaard AB, Van Ooyen A, Stam CJ, Scheltens P, Linkenkaer-Hansen K (2009) Altered temporal correlations in parietal alpha and prefrontal theta oscillations in early-stage Alzheimer disease. Proc Natl Acad Sci USA 106(5):1614–1619

Monti MM, Vanhaudenhuyse A, Coleman MR, Boly M, Pickard JD, Tshibanda L, Owen AM, Laureys S (2010) Willful modulation of brain activity in disorders of consciousness. New Engl J Med 362:579–589

Morano GN, Seibyl JP (2003) Technical overview of brain SPECT imaging: improving acquisition and processing of data. J Nucl Med Technol 31:191–195

Muehlemann T, Haensse D, Wolf M (2008) Wireless miniaturized in-vivo near infrared imaging. Opt Express 16(14):10323–10330

Owen AM, Coleman MR, Boly M, Davis MH, Laureys S, Pickard JD (2006) Detecting awareness in the vegetative state. Science 313:1402

Owen AM, Schiff ND, Laureys S (2009) A new era of coma and consciousness science. Progr Brain Res 177:399–411

Poldrack RA (2008) The role of fMRI in cognitive neuroscience: where do we stand? Curr Opin Neurobiol 18:223–227

Polikar R, Tilley C, Hillis B, Clark CM (2010) Multimodal EEG, MRI and PET data fusion for Alzheimer's disease diagnosis. Conference Proceedings: IEEE Engineering in Medicine and Biology Society, pp. 6058–6061

Ray A, Bowyer SM (2010) Clinical applications of magnetoencephalography in epilepsy. Ann Indian Acad Neurol 13(1):14–22

Rödiger C (2011) Das Ende des BEOS-Tests? Zum Jüngsten Lügendetektor-Urteil des Supreme Court of India. Nervenheilkunde 30(1–2):74–79

Rugg-Gunn FJ, Symms MR, Barker GJ, Greenwood R, Duncan JS (2001) Diffusion imaging shows abnormalities after blunt head trauma when conventional magnetic resonance imaging is normal. J Neurol Neurosurg Psychiatry 70:530–533

Saddique SM, Siddiqui LH (2009) EEG based brain computer interface. J Softw 4(6):550–554

Sheridan CJ, Matuz T, Draganova R, Eswaran H, Preissl H (2010) Fetal magnetoencephalography – achievements and challenges in the study of prenatal and early postnatal brain responses: a review. Infant Child Dev 19(1):80–93

Sitaram R, Caria A, Veit R, Gaber T, Rota G, Kuebler A, Birbaumer N (2007) fMRI brain-computer interface: a tool for neuroscientific research and treatment. Comput Intell Neurosci: 25487

Smith SJM (2005) EEG in the diagnosis, classification, and management of patients with epilepsy. J Neurol Neurosurg Psychiatry 76:ii2–ii7

Spranger TM (2009a) Rechtliche Implikationen der Generierung und Verwendung neurowissenschaftlicher Erkenntnisse. In: Schleim S, Spranger TM, Walter H (eds) Von der Neuroethik
zum Neurorecht? Vandenhoeck & Ruprecht, Göttingen, pp. 193–213

Spranger TM (2009b) Neuroprothetik und bildgebende Hirnforschung – neue Impulse für die
Praxis des Betreuungsrechts. Betreuungsmanagement 5:206–208

Süzen M, Giannoula A, Durduran T (2010) Compressed sensing in diffuse optical tomography.
Opt Express 18(23):23676–23690

Tancredi LR, Brodie JD (2007) Is a picture worth a thousand words? Neuroimaging in the
courtroom. Am J Law Med 33(2-3):239 et seq

Tatum WO, Vale FL, Anthony KU (2008) Epilepsy surgery. In: Husain AM (ed) A practical
approach to neurophysiologic intraoperative monitoring. Demos Medical Publishing,
New York, pp. 283–302

Thompson SK (2007) A brave new world of interrogation jurisprudence? Am J Law Med
33(2-3):341 et seq

Tromberg BJ, Poque BW, Paulsen KD, Yodh AG, Boas DA, Cerussi AE (2008) Assessing the
future of diffuse optical imaging technologies for breast cancer management. Med Phys
35(6):2443–2451

Walter H (2009) Was können wir messen?. In: Schleim S, Spranger TM, Walter H (eds) Von der
Neuroethik zum Neurorecht? Vandenhoeck & Ruprecht, Göttingen, pp. 67–103

Weiskopf N, Mathiak K, Bock SW, Scharnowski F, Veit R, Grodd W, Goebel R, Birbaumer N
(2004) Principles of a brain-computer interface (bci) based on real-time functional magnetic
resonance imaging (fMRI). IEEE Trans Biomed Eng 51:966–970

Williams MA, Sachdev PS (2010) Magnetoencephalography in neuropsychiatry: ready for application? Curr Opin Psychiatry 23(3):273–277

Winterer G, McCarley RW (2011) Electrophysiology of schizophrenia. In: Weinberger DR,
Harrison PJ (eds) Schizophrenia. Wiley-Blackwell, Oxford, pp. 311–333

Wolf M, Ferrari M, Quaresima V (2008) Progress of near-infrared spectroscopy and topography
for brain and muscle clinical applications. J Biomed Opt 12(6):062104

Wortzek HS, Filley CM, Anderson CA, Oster T, Archiniegas DB (2008) Forensic applications of
cerebral single photon emission computed tomography in mild traumatic brain injury. J Am
Acad Psychiatry Law 36:310–322

Neuroscience and Law: Australia

Leanne Houston and Amy Vierboom

Abstract The Australian legal system has not been receptive to new neuroscientific technology. Current case law and legislative provisions demonstrate the hurdles imposed by the rigorous admissibility standards.

1 Introduction

Structural neuroimaging studies such as CT scans and diagnostic MRI scans are routinely admitted as evidence in civil and criminal trials in Australia. Studies are proffered as evidence in, for example, cases determining the presence of brain injury due to trauma, (R v Jeong Ming Foo [2008] NSWSC 587) declaration of brain death due to pathology or injury, (R v KT [2007] NSWSC 83) diagnosis of brain pathology, (Tabet v Gett [2010] HCA 12) testamentary capacity and dementia (Burgess v Leech [2007] NSWSC 700) and mental illness (R v Coleman [2010] 9 NSWSC 177). It is the advances in neuroscience technology that enable non-invasive detection of brain activity using, in particular, functional MRI (fMRI) that has aroused considerable debate and interest.[1] There has been tremendous growth and widespread acceptance of fMRI in peer review scientific journals. Indeed, neuroscience has become a dominant aspect of social enquiry.[2] Investigative and exploratory fMRI neurological studies have in the course of research

Leanne Houston is a Lecturer of Law at the University of Technology Sydney (UTS). Amy Vierboom, Honours student in Law (UTS).

[1] Rosen (2007), p. 39.
[2] Moriarty (2008), p. 29.

L. Houston (✉) • A. Vierboom
University of Technology Sydney, PO Box 123, Broadway, Sydney, NSW 2007 Australia
e-mail: Leanne.Houston@uts.edu.au

T.M. Spranger (ed.), *International Neurolaw*,
DOI 10.1007/978-3-642-21541-4_2, © Springer-Verlag Berlin Heidelberg 2012

demonstrated potential legal implications.[3] As anticipation grows as new and improved imaging tools allow for more detailed imaging of the brain, so too does an awareness of limitations and challenges. Lawyers are of course very keen to introduce any neurological evidence to support their cases.

This rapid expansion of knowledge is creating difficulties in the reception of expert evidence and its ultimate utility in resolving a dispute. The role of expert evidence in the interpretation of fMRI images and the display of these images in an Australian court to judge and jury raises considerable concerns.

The focus of this chapter will be centred on neuroscience and criminal law, in particular the role of fMRI as a lie-detection tool and the evidentiary rules in place.

2   The Admissibility of Neuroscientific Methods in Australian Courts

The greatest area of attention has been focused on the role that neuroscience-based lie detection may make to criminal law.[4] Various emerging neuroscientific techniques for detecting deception have been suggested as the next generation of lie-detection tools.[5] The main technologies being electroencephalography (EEG),[6] fMRI and brain fingerprinting.[7] However, despite a relatively high profile in the media[8] and scholarly attention,[9] these technologies are rarely used in criminal proceedings worldwide and their potential is speculative.

There have been no cases in Australia to date where these techniques have been successfully introduced as evidence in both criminal and civil trials.

The primary focus for Australian courts rests with admissibility in criminal hearings. In Australia, the responsibility for evaluating the validity of scientific tests falls on the judiciary via the rules of evidence, in particular the Evidence Acts.[10] Expert testimony based on functional studies is deemed to constitute a scientific technique warranting elucidation through the provision of expert evidence and therefore subject to the strict rules of evidence.

---

[3] Baker (2009).

[4] Shen and Jones (2011).

[5] Greely and Iles (2007), p. 377.

[6] Cournos and Bavaniss (2003). See also: *Maharastra v Sharma and Khandelwal*, Sessions Case No. 508/07 (June 12, 2008).

[7] Farwell (1999); *Harrington V State*, 659 N.W.2d 509 (Iowa 2003).

[8] Leenaghan and Guerrera (2005).

[9] Dickson and McMahon (2005).

[10] *Evidence Act 1995* (Cth); *Evidence Act 1971* (ACT); *Evidence Act 1975* (NSW); *Evidence Act* (NT); *Evidence Act 1977* (QLD); *Evidence Act 1929* (SA); *Evidence Act 2001* (Tas); *Evidence Act 2008* (Vic); *Evidence Act 1906* (WA). Chapter will focus on Cth and NSW jurisdictions.

Whilst overseas cases[11] may prove persuasive in Australian jurisdictions, fMRI evidence would have to meet relevancy criteria. The threshold enquiry when considering the admissibility of expert opinion evidence, as with evidence of any kind, is to identify its relevance.

Sections 55–56 of the *Evidence Act 1995* (Cth); *Evidence Act 1995* (NSW) ("the Acts") provide:

Section 55 Relevant evidence

(1) The evidence that is relevant in a proceeding is evidence that, if it were accepted, could rationally affect (directly or indirectly) the assessment of the probability of the existence of a fact in issue in the proceeding.

(2) In particular, evidence is not taken to be irrelevant only because it relates only to –

(a) the credibility of a witness; or
(b) the admissibility of other evidence; or
(c) a failure to adduce evidence.

Section 56 Relevant evidence to be admissible

(1) Except as otherwise provided by this Act, evidence that is relevant in a proceeding is admissible in the proceeding.

(2) Evidence that is not relevant in the proceeding is not admissible.

Sections 76 of the Acts provide the general rule that operates to exclude evidence of an opinion and reflects the general common law approach:

Section 76 The opinion rule

Evidence of an opinion is not admissible to prove the existence of a fact about the existence of which the opinion was expressed.

Section 79 of the Acts provides an exception to section 76 and is relevantly in the following terms:

Section 79 Exception – opinions based on specialised knowledge

(1) If a person has specialised knowledge based on the person's training, study or experience, the opinion rule does not apply to evidence of an opinion of that person that is wholly or substantially based on that knowledge.

The general discretionary exclusions that apply to all otherwise admissible evidence must also be considered when determining the admissibility of an expert opinion. Sections 135–137 provide:

Section 135 General discretion to exclude evidence

The court may refuse to admit evidence if its probative value is substantially outweighed by the danger that the evidence might –
(a) be unfairly prejudicial to a party; or
(b) be misleading or confusing; or
(c) cause or result in undue waste of time.

---

[11] *United States Of America v Lorne Allan Semrau*, (31 May 2010), No. 07-10074, Report and Recommendations and *Wilson v. Corestaff Services*, (14 May 2010), Supreme Court, Kings County, New York State Law Reporting Bureau.

Section 136 General discretion to limit use of evidence

The court may limit the use to be made of evidence if there is a danger that a particular use of the evidence might—
(a) be unfairly prejudicial to a party; or
(b) be misleading or confusing.

Section 137 Exclusion of prejudicial evidence in criminal proceedings

In a criminal proceeding, the court must refuse to admit evidence adduced by the prosecutor if its probative value is outweighed by the danger of unfair prejudice to the defendant.

All evidence must be shown to be relevant, in that it "could rationally affect (directly or indirectly) the assessment of the probability of the existence of a fact in issue in the proceeding".[12] It is highly probable that scientific evidence, purportedly showing the likelihood that a testimony is true or false, would meet this criteria.

Section 79 raises three discrete questions to be resolved when considering the admissibility of opinion evidence; does the witness have *specialised knowledge*, is that knowledge *based on the person's training, study or experience* and is the opinion of the witness *wholly or substantially based* on that knowledge. More difficult for Australian lawyers will be demonstrating that lie-detection meets the test laid out for expert opinions under section 79 (1).

The traditional "field of expertise" test applied by the common law has the consequence that a purported expert cannot give evidence in relation to areas of knowledge that do not form part of a "formal sphere of knowledge".[13] There is a line of authority which suggests that the evidence must derive from a body of knowledge or experience that is accepted as being reliable.

Dixon CJ, said in *Clark v Ryan*:[14]

On the one hand it appears to be admitted that the opinion of witnesses possessing peculiar skill is admissible whenever the subject-matter of inquiry is such that inexperienced persons are unlikely to prove capable of forming a correct judgment upon it without such assistance, in other words, when it so far partakes of the nature of a science as to require a course of previous habit, or study, in order to the attainment of a knowledge of it ... While on the other hand, it does not seem to be contended that the opinions of witnesses can be received when the inquiry is into a subject-matter the nature of which is not such as to require any peculiar habits or study in order to qualify a man to understand it.

The judgement of King CJ in *Bonython v R*[15] is often cited both in Australia and the UK when considering the "field of expertise" test. His Honour said:

Before admitting the opinion of a witness into evidence as expert testimony, the judge must consider and decide two questions. The first is whether the subject matter of the opinion

---

[12] Section 55 *Evidence Act 1995* (Cth), (NSW) and may be found so in relation to witness credibility: s55 (2) (a) of the *Evidence Act 1995* (Cth) (NSW) specifically considers the possibility of evidence being relevant in relation to the "credibility of a witness" is expressly acknowledged.
[13] Freckleton and Selby (2009).
[14] (1960) 103 CLR 486 at 491.
[15] (1984) 38 SASR 45.

falls within the class of subjects upon which expert testimony is permissible. This first question may be divided *into two parts: (a) whether the subject matter of the opinion is such that a person without instruction or experience in the area of* knowledge or human experience would be able to form a sound judgment on the matter without the assistance of witnesses possessing special knowledge or experience in the area, and (b) whether the subject matter of the opinion forms part of a *body of knowledge or experience which is sufficiently organised or recognised to be accepted as a reliable body of knowledge or experience, a special acquaintance with which by the witness would render his opinion of assistance to the court.* The second question is whether the witness has acquired by study or experience sufficient knowledge of the subject to render his opinion of value in resolving the issues before the court.

Both *Clark* and *Bonython* confirm that the "field of expertise" requirement is concerned with the need for the opinion to derive from a "body of knowledge", which is both "organised" and "accepted". The purpose of the test is to ensure the trustworthiness and reliability of the science or technique that is to be relied upon.[16] There is a line of common law authority in Australia, which imposes a threshold requirement of evidentiary reliability before a field of knowledge upon which an opinion is based can be left to a jury.[17]

This threshold question of evidentiary reliability at common law has often been determined by reference to the approach advocated by the United States Supreme Court in *Frye v United States*.[18] The *Frye* test, or a variant of that approach, which considers whether there is "general acceptance" of a particular discipline for determining the question of reliability as part of the field of expertise rule has come to form part of the common law in Australia.[19]

Furthermore, "the concept of 'specialised knowledge' imports knowledge of matters which are outside the knowledge or experience of ordinary persons and which 'is sufficiently organized or recognized to be accepted as a reliable body of knowledge or experience'".[20]

Chief Justice Spigelman of the New South Wales Supreme Court held in *R v Tang*[21] that the meaning of "knowledge" for the purpose of section 79 is the same as that attributed by the United States Supreme Court in *Daubert v Merrell Dow Pharmaceuticals*[22] at 590:

[T]he word knowledge connotes more than subjective belief or unsupported speculation. The term applies to any body of known facts or to any body of ideas inferred from such

---

[16] Justice Peter McClellan, Chief Judge at Common Law, Supreme Court of New South Wales, "Admissibility of expert evidence under the Uniform Evidence Act", Judicial College of Victoria 1 of 42 Emerging Issues in Expert Evidence Workshop Melbourne, 2 October 2009.

[17] *R v Gilmore* [1977] 2 NSWLR 935, *Lewis v R* (1987) 88 FLR 104.

[18] 293 F 2d 1013 (1923).

[19] *R v Gallagher* [2001] NSWSC 462.

[20] *Velevski v The Queen* (2002) 187 ALR 233.

[21] *R v Tang* (2006) 65 NSWLR 681 (CCA).

[22] 509 US 579 (1993).

facts or accepted as truths on 'good grounds'... Proposed testimony must be supported by
appropriate validation- i.e., "good grounds", based on what is known.

Chief Justice Spigelman's judgement has been the subject of considerable
discussion. It remains to be seen whether the approach preferred by Spigelman
CJ in *Tang* will continue to be followed. At present, it is a persuasive case for the
inadmissibility of neuroscientific lie detection evidence.

In *R v Tang*, the accused had been convicted of robbery with an offensive
weapon; much of the prosecution's case rested on Tang's identification in video
surveillance footage. The prosecution engaged Dr Sutisno as an expert in the field
of forensic anatomy, which purportedly included expertise in "facial mapping" and
"body mapping". Only the final ground of the appeal referring to the admissibility
of "body mapping" evidence was upheld on appeal.[23]

The case emphasised that Dr Sutisno had failed to illuminate the basis for the
factual science she drew her opinions from. Chief Justice Spigelman followed
Makita Pty Ltd v Sprowles[24] in which Justice Heydon identified six useful points
for assessing the admissibility of expert opinion evidence. These include: the
expert's duty to demonstrate a "specialised knowledge"; identify the specific aspect
which the witness is an expert in; the substantial basing of the opinion in that expert
knowledge; the identification and proof of "observed", "accepted" and "assumed"
facts; proof that "the facts on which the opinion is based" are a "proper foundation
for it"; and the demonstration of those bases.[25]

Although Justice Spigelman differentiated Australian law from the application
of the *Daubert v. Merrell Dow Pharmaceuticals*[26] test in other instances, in
characterising the first limb of the section79 test His Honour held that the meaning
of "knowledge" is the same as that identified in the reasons of the majority
judgment in *Daubert*. That is: "*[T]he word 'knowledge' connotes more than subjective belief or unsupported speculation. The term applies to any body of known
facts or to any body of ideas inferred from such facts or accepted as truths on good
grounds*".[27] Therefore, while American case law on fMRI technologies does not
create precedent, it is likely that the treatment of lie-detectors under the *Daubert*
test for "knowledge" would be "instructive"[28] in the Australian discussion of
admissibility.

---

[23] 509 US 579 (1993) at 156.

[24] *Makita (Australia) Pty Ltd v Sprowles* (2001) 52 NSWLR 705 (CA).

[25] *Makita (Australia) Pty Ltd v Sprowles* (2001) 52 NSWLR 705 (CA), JA Heydon, at 85.

[26] *Daubert v Merrell Dow Pharmaceuticals Inc [1993] USSC 114;* 509 US 579 (1993) (here after
known as '*Daubert*').

[27] *R v Tang* (2006) 65 NSWLR 681 (CCA) CJ Spigelman at 138.

[28] *R v Tang* (2006) 65 NSWLR 681 (CCA) CJ Spigelman at 139.

In the *United States of America v Lorne Semrau* (2010),[29] the admissibility of fMRI lie-detection was examined under the principles laid out in *Daubert*.[30]

The defendant, Semrau, was a CEO of two corporations which contracted with psychiatrists. He was prosecuted for fraud and money laundering after allegedly indicating the services provided attracted a higher rate of reimbursement than they did. Semrau pleaded not guilty and sought the services of CEPHOS for fMRI lie-detection scans to assist. The prosecution brought a motion to exclude the evidence and on 13 May 2010 a *Daubert* hearing was conducted by Judge Pham. His report examined the evidence put forward under Evidence Rule 702 (as set out by *Daubert*) providing:

> If scientific, technical, or other **specialized knowledge** will assist the trier of fact to understand the evidence or to determine a fact in issue, a witness qualified as an expert... may testify thereto in the form of an opinion or otherwise, if (1) the testimony is based upon sufficient facts or data, (2) the testimony is the product of reliable principles and methods, and (3) the witness has applied the principles and methods reliably to the facts of the case.[31]

The Supreme Court in *Daubert* established two prongs of this test: first, that the testimony is reliable and "grounded in the methods and procedures of science and *must be more than unsupported speculation or subjective belief*".[32] The second prong examined whether the application of the methodology to the facts in question was "fit".[33]

Judge Pham, in discovering whether Dr Laken held a "specialised knowledge" within the first limb, employed the four non-exclusive factors given in *Daubert*:

(1) Whether the theory or technique can be tested and has been tested;
(2) Whether the theory or technique has been subjected to peer review and publication;
(3) The known or potential rate of error of the method used and the existence and maintenance of standards controlling the technique's operation; and
(4) Whether the theory or method has been generally accepted by the scientific community.[34]

Judge Pham found that neuroscientific lie-detection passed the testing and peer review qualifications. However, on the third point the Judge highlighted that lie-detection testing had been contained to a small sample size and accordingly error rates were specific to laboratory testing; 'real-life' error rates remain unknown. This issue, amongst others, was highlighted by the literature examined by the Judge.

---

[29] *United States Of America v Lorne Allan Semrau*, (31 May 2010), No. 07-10074, Report and Recommendations, J Pham.

[30] *Daubert v Merrell Dow Pharmaceuticals Inc [1993] USSC 114;* 509 US 579 (1993), pp. 12–15. This is made under Rule 104 of *The Federal Rules of Evidence 1975* (US).

[31] Federal Evidence Rule 702, *The Federal Rules of Evidence 1975* (US).

[32] *Daubert v. Merrell Dow Pharmaceuticals, Inc.,* 509 U.S. 579, 125L. Ed. 2d 469, 113 S. Ct. 2786. In this the court held that this rule superseded the "general acceptance" test set out in *Frye v. United States*, 293 F. 1013 (D.C. Cir. 1923).

[33] *United States v. Bonds*, 12 F.3d 540, 555-56 (6th Cir. 1993).

[34] *Daubert v. Merrell Dow Pharmaceuticals, Inc.,* 509 U.S. 579, 125L. Ed. 2d 469, 113 S. Ct. 2786 at 593–594.

While Dr Laken testified to his own standards, scholarly consensus that lie-detection was not ready for real-life application meant the court could not "adequately evaluate the reliability of a particular lie-detection examination".[35] In considering the fourth point, the Court found that the majority of literature held that lie-detection is not reliable enough for use in court.[36] On these bases, the court found the evidence to be inadmissible.

Given that Australia takes *Daubert*'s formulation of "knowledge", it is likely that fMRI lie-detection would be found inadmissible in Australia on similar grounds.[37] Furthermore, *Tang* shows the caution with which Australian courts treat new technologies. Chief-Justice Spigelman quoted *R v Gray*[38] which held:

> There is no means of determining objectively whether or not such an opinion is justified. Consequently, unless and until a national database or agreed formula or some other such objective measure is established, this court doubts whether such opinions should ever be expressed by... mapping witnesses.

It is likely that fMRI lie-detection would be subject to the same standards of scrutiny. Various Australian academics have supported such caution by courts as combined with the current rationalistic mindset, hasty inclusion could easily lead to "the contamination of criminal trials with unreliable incriminating expert opinion evidence".[39]

3   The Impact of *f*MRI Lie-Detection on the Legal System

For the purpose of investigating the impact of fMRI lie-detection on the Australian legal system, one is required to look forward to a day when such evidence could meet the rigorous admissibility standards in place. Any employment of neuroscience would require legislated regulation to mitigate the complex concerns over unfair prejudice, such as the number of expert opinions able to interpret a scan, requirements that both parties participate in determining questions used in the scan, and the balance of power between scientists and lawyers in treatment of the data.

Therefore, the first frontier of impact requires consideration of the relationship between science and law, and the conceptual limitations of using fMRI data.

Even after settling these questions, and introducing legislation there would be consequences for many central tenets of justice in the legal system; especially the jury system, the right to silence and personal privacy.

---

[35] *United States Of America v Lorne Allan Semrau*, (31 May 2010), No. 07-10074, Report and Recommendations, J Pham, at p. 31.

[36] The court looked at articles including: Simpson (2008) and Chen (2009).

[37] As set out in *R v Tang* (2006) 65 NSWLR 681 (CCA) CJ Spigelman at 138.

[38] *R v Gray* [2003] EWCA Crim 1001.

[39] Edmond et al. (2008), *University of New South Wales, University of Technology, Sydney, Australian Research Council* and *University of Western Sydney*.

4   The Conceptual Boundaries of the Use of *f*MRI Lie-Detection Data

The introduction of fMRI lie-detection into criminal cases would be accompanied by increased participation by scientists in court processes. If fMRI lie-detection was treated as conclusive evidence, the role of legal practitioners would inevitably be relegated to one of simple deference to scientific expertise.

However, it is important to recognise that scientific research looks only to the empirical correlations between deceptive behaviour and brain activity. While eventually strong correlative patterns may be found, these cannot definitively state whether a person is lying or not, as deception is behaviour of the whole person, rather than simply an operation of the brain.

Therefore, the inductive evidence provided by neuroscience, though appropriately given weight, must still be grounded in sound evidentiary and general legal analysis. Even if certain mental states were shown to be necessary for deception – they could not be said to be sufficient, as deception must involve the *subjective* criteria of intention. Therefore, while fMRI may provide inductive evidence about a subject's brain activity, the limitations on the meaning of this information only allow one to evaluate whether scans are more similar or less similar to functioning patterns associated with deception at a group and individual level. Functional MRI cannot offer testimony about lies being 'produced' or 'happening' in the brain, but is limited to heightening or lessening the probability that one was engaged in deceptive behaviour at the time of the scan.

This limitation on the use of brain scans in lie-detection would protect innocent persons who happen to have incriminating states of mind, or those who conceived the intention to lie but did not carry out the act. It ensures that where there is strong circumstantial evidence to the contrary, lie-detection would not be given undue weight; lawyers would be responsible for moderating this. Furthermore, requiring behavioural evidence and an analysis of the subjective and criterial bases of deception would ensure the protection of the right to a fair trial. While the role of lawyers would continue to be a crucial one, whether jurors, would be sufficiently capable to make and hold the necessary distinctions in dealing with fMRI evidence remains to be seen.

5   The Impact of fMRI Lie-Detection for the Jury

On the premise that a highly effective lie-detector exists, which meets the standards of *R v Tang*,[40] courts would still need to consider the impact on the role and function of the jury to "hear evidence and make decisions about facts with the guidance on

---

[40] *R v Tang* (2006) 65 NSWLR 681 (CCA).

the law from a judge".[41] With an accurate and efficient lie-detector, the traditional role of juries as "fact finders" (as affirmed in Australia and elsewhere[42]) would be reduced to a mere rubber-stamping on the work of courtroom scientists. Already, American courts have proven reluctant to admit fMRI evidence that intrudes on this central function of the jury. This intrusion would, in many respects, be analogous to the impact of DNA evidence. Finally, in cases where jury's did not follow the verdicts of lie-detection science, their currently implicit 'powers of nullification' would become explicit. Should this 'power of nullification' be elevated to an explicit and central function of the jury, this would prompt questions about the legitimacy of the jury in its current form.

Historically, the role of the jury has been fiercely protected. Juries have been entrusted with the evaluation of evidence; with legitimacy being derived from public trust in the system to do this fairly and accurately. Consequently, judges have taken a cautious approach for admitting evidence based on new fields of investigation, especially those which purportedly answer questions of fact.

Unlike DNA evidence (which connects a person's DNA with the event in question), fMRI lie-detection goes directly to the question of witness credibility or the credibility of the accused; informing the jury that a testimony was given with brain activity that correlates with deception or honesty. The problem this poses for courts in allowing juries the freedom to decide the credibility of a witness was seen in the first consideration of lie-detection technologies by an American court, in *Corestaff v Wilson*.[43] In that case, the plaintiff and defendant disagreed in their testimony about what the defendant did or did not tell a witness. The Judge identified that a crucial question in admitting the evidence was whether the assessment of the witness' credibility was "within the ken of the juror"[44]; as this is always a key function of the jury, it was found to be so. The Judge declined the motion for a 'Frye-hearing',[45] on the basis that American common law excludes expert evidence, which bolsters the credibility of a witness as it improperly intrudes on the "province of the jury".[46] In a similar way evidence from polygraphs has been consistently excluded, as judges have recognised that it poses an unwarranted threat

---

[41] Jury Service, 'The Role of Juries', *Justice and Attorney-General*, NSW Government, 31 July 2007.

[42] *Brownlee v The Queen* [2001] HCA 36; (2001) 75 ALJR 1180.or in America: *United States v. Scheffer*, 523 U.S. 303, 312–13 (1998).

[43] *Wilson v. Corestaff Services*, (14 May 2010), Supreme Court, Kings County, New York State Law Reporting Bureau.

[44] *Wilson v. Corestaff Services*, (14 May 2010), Supreme Court, Kings County, New York State Law Reporting Bureau, 2.

[45] *Frye v. United States*, 293 F. 1013 [DC 1923], remains the authoritative case in NY District Court's jurisdiction.

[46] *Wilson v. Corestaff Services*, (14 May 2010), Supreme Court, Kings County, New York State Law Reporting Bureau, p. 3.

to the jury's role of determining witness credibility.[47] American law has greatly influenced Australian thinking on the jury.[48] In *The Queen v Murdoch*,[49] where expert evidence was offered for assessing witness credibility, it was noted that a witness should "possess scientific knowledge, expertise and experience outside the ordinary knowledge, expertise and experience of the jury".[50] The high barriers of admissibility for expert evidence protect the role of the jury in assessing credibility and would therefore be a significant barrier to the incorporation of fMRI lie-detection technology.

However, historically, the law's acceptance of new technology has been based on the reliability of that technology.[51] Even the way in which the jury system came about in the middle ages when trial by ordeal was rejected, reflects the fact that while change may be resisted, when presented with a more accurate fact-finder, the public legitimacy of the legal system will depend upon acceptance of it.[52] More recently, this has been evident in the treatment of DNA evidence, where following exonerations on the basis of new scientific evidence, convictions were quashed and the authority of the jury was questioned. If and when an accurate system of lie-detection became available, the current role of the jury will be called into question. A modified role has been posited of evaluating the subjective truth of lay witness testimonies or evaluating scientific evidence within the wider body of evidence.[53] Finally, an implicit operation of juries is the quasi-legislative capacity of adjudging whether the law befits the facts. While the first of these is arguably within the traditional scope of the juror's role, it may disappear as technologies are refined.

A second remaining function of the jury would be to examine the credibility of scientific evidence within the body of evidence, in a way similar to the jury's current assessment of DNA evidence. Since 1989, the use of DNA evidence in Australia has increased rapidly,[54] and defence challenges to this evidence have passed. Recent Australian research has identified[55] that juries are 23 times more likely to convict in homicide and 33 times more likely to convict in sexual assault cases when presented with DNA evidence.[56] Furthermore, jurors find statistical evidence difficult to understand; they may be 'overawed by the scientific garb in which the evidence is presented and attach greater weight to it than it is capable of

---

[47] Seaman (2009), p. 461.

[48] *Brownlee v The Queen* [2001] HCA 36; (2001) 75 ALJR 1180 at 21: see *Williams v Florida* [1970] USSC 155; 399 US 78 (1970) at 100.

[49] *The Queen v Murdoch* [2005] NTSC 78.

[50] *The Queen v Murdoch* [2005] NTSC 78 CJ Martin at 108.

[51] Caudill (2010).

[52] Shapiro (1986).

[53] Seaman (2009), p. 475.

[54] Easteal and Easteal (1990).

[55] Goodman-Delahunty and Hewson (2010), p. 1.

[56] Briody (2004).

bearing'.[57] In keeping with the high place science holds in the community, jurors place a high level of trust in expert witnesses. A recent Victorian case in which a man was erroneously convicted of sexual assault based on contaminated DNA evidence despite overwhelming evidence showing him to be innocent inspires little faith in the ability of juries to assess scientific evidence in light of other evidence proffered in a case.[58] This complete, unquestioning trust in the science of DNA evidence would likely follow scientifically approved fMRI lie-detection evidence in the courtroom.

A blind and unregulated acceptance of fMRI lie-detection would threaten jury secrecy; the legislative[59] and common law[60] principle that ensures free and honest jury room discussion. More certainty about the jury's belief of the factual guilt or innocence of the accused would unmask the basis on which a jury acquitted or convicted them.[61] At present, the secrecy of jury deliberations also affords juries the right to convict or acquit not purely on the facts of the case, but also in the belief that a law is unjust or unfitting. In this way, the jury would be in effect exercising a quasi-legislative authority of determining what should and should not be law. The existence of their right to do this was initially recognised in *R. v. Kirkman*[62] and then in *R v Abbott*[63] where it was held, "sometimes it appears to a jury that although a number of counts have been alleged against an accused person and have been technically proved, justice is sufficiently met by convicting him of less than the full number".[64] If clear evidence, in the form of fMRI lie-detection became available to courts, a jury's disagreement on the verdict would most likely point to, not a disagreement over facts, but an assessment of the appropriateness whether the law should apply at all.

This right of juries has been implicitly recognised on various occasions by both the courts and in the public arena.[65] The power of nullification provides in "an unusual arrangement of checks and balances, [a] way of building discretion, equity, and flexibility into the legal system"[66]; therefore, reinforcing the importance of democratic consensus at the level of the court system.

---

[57] *R v Duke* 1979 22 SASSR 46, King CJ at 48.

[58] Hagan (2009).

[59] *Jury Act 1977* (NSW).

[60] *Ng v R* [2003] HCA 20; 217 CLR 521; 197 ALR 10; 77 ALJR 967 (10 April 2003).

[61] Seaman (2009), p. 427.

[62] *R. v. Kirkman* (1987) 44 S.A.S.R. 591.

[63] *R v Abbott* [2006] VSCA 100 (4 May 2006).

[64] *R. v. Kirkman* (1987) 44 S.A.S.R. 591 at 593.

[65] An example of this can be seen in the recent decision by a jury, acquitting a Queensland couple who were being tried under s225 of their Criminal Code for illegally procuring an abortion, while it is not possible to know for sure that this was a case of jury nullification, it does present a persuasive example of where the jury arguably returned a merciful verdict in application of the law: see, Wainer (2010).

[66] Kalver and Zeisel (1966).

However, the public on unveiling of this right would entail various problems. It could likely increase its usage,[67] and given the variety of existing views about the appropriateness of certain laws, increase the instances of hung juries and controversial decisions by juries swayed more by emotion than fact.

Furthermore, in *R v Abbott* it was held that the trial judge was not at liberty to instruct a jury of this implicit right as it would have, inappropriately, bestowed on him the power to instruct juries to "determine which of the laws of the land are to be enforced".[68] Finally, scholars have remarked that any open understanding of jury deliberations might cause the public to reject their verdicts.[69] Legitimacy of the jury system in the eyes of the public is based on whether it is able to fulfil its assigned role; given the problematic effects associated with promoting this remaining quasi-legislative role to a central function of the jury, the constitutional right of trial by jury might come into question. For all the shortcomings or occasional failings of the current jury system, the inclusion of fMRI evidence in this way would not prove an opportunity for reform, but rather pose a threat to its continued existence by undermining public legitimacy.

Any introduction of fMRI evidence would have stark ramifications for traditional conceptions of the jury's role and function. Embracing a system in which "fact-finding" is the domain of science would likely spark debate about the jury's continued legitimacy. Given such dramatic consequences, any introduction of fMRI lie-detection would likely be met with hesitation by courts.

6   The Right to Silence and the Right to Privacy

A further remaining task of the jury, should fMRI lie-detection become a significant part of the legal system, will likely be determining guilt where accused persons choose not to undergo lie-detection. The accused is entitled, both by common law and statute to the "right to silence" at various stages of the trial process.[70] It is accordingly unlikely that an accused could be compelled to undergo a lie-detection scan in a trial.

Many scholars have commented on the erosion of the right of the accused to silence,[71] given the adverse inferences commonly drawn by juries in the absence of a testimony. Should lie-detection evidence become commonplace in criminal trials,

---

[67] Seaman (2009), p. 484.

[68] *R v Abbott* [2006] VSCA 100 (4 May 2006) per JA Buchanan at 18.

[69] Ruprecht (1997), p. 217.

[70] S17(3) *Evidence Act 1995* (NSW).

[71] Hocking and Manville (2001).

any negative consequences for an accused, would likely be exacerbated, given the significant weight of such evidence.

The way in which judges could comment on any failure to undergo fMRI lie-detection might mirror legislative boundaries given in section 20(2) of the Acts; permitting comment, but requiring the comment "must not suggest that the defendant failed to give evidence because the defendant was, or believed that he or she was, guilty of the offence concerned".[72] Given the weight lie-detection tests would inevitably hold in the eyes of the jury, even if a judge's directions were aimed at tempering that weight, it is likely the average juror would be unable to "shut their eyes to the consequences of exercising the [right to silence]".[73]

Weissensteiner v The Queen[74] dealt with the failure to explain where facts "peculiar" to the defendant.[75] In lie-detection, it is likely that facts about the credibility of an accused's testimony will belong peculiarly to the defendant. In breaking from steps aimed to further protect the accused's right to silence, accepting fMRI lie-detection evidence would inevitably further threaten the right to silence.

At stake in this are significant privacy concerns, with highly sensitive information about one's inner brain functioning in question. Some authors have rejected including lie-detectors entirely because of the coercion that could occur, and the dangers where the information gained is about one's cognitive abilities.[76] Although privacy laws have not anticipated that the inner workings of the brain might one day constitute sensitive, available information,[77] balancing this right with the interests of justice would be a significant concern for the legislature and the courts. Indeed, it is likely that although not legally compellable, the desire of an accused to be acquitted would lead to their wavering of privacy rights in many instances, where fMRI lie-detection was admitted. This would in turn lead to concerns about the resulting prejudice for those who do not waive their privacy rights. The very real concern of highly sensitive information and unavoidable pressure to testify as an accused would need to be accounted for by legislation.

---

[72] S20 (2) *Evidence Act 1995* (Cth), (NSW).

[73] *Weissensteiner v The Queen* [1993] HCA 65; (1993) 178 CLR 217, per Mason CJ, Deane And Dawson JJ at 33.

[74] *Weissensteiner v The Queen* [1993] HCA 65; (1993) 178 CLR 217, per Mason CJ, Deane And Dawson JJ at 33 and *Azzopardi v R* [2001] HCA 25; 205 CLR 50; 179 ALR 349; 75 ALJR 931 (3 May 2001).

[75] *Weissensteiner v The Queen* [1993] HCA 65; (1993) 178 CLR 217per JJ Gaudron and McHugh at 4.

[76] White (2010), p. 258.

[77] *Privacy and Personal Information Protection Act* 1998, (NSW), found at the Office of the NSW Privacy Commissioner.

7 Admissibility Under General Exclusionary Clauses and the Regulation of fMRI Lie-Detection

What becomes clear upon reflection is just how dramatic the impact of advances in neuroscience may be in criminal cases: at stake is the constitutional right to trial by jury.[78] Indeed, the difficulties for juries in dealing with the amount of scientific evidence in making verdicts may already prove grounds for reform of its current operation.[79] Jury panels might be reconstituted as an expert witness panel, including neurologists and others more equipped to assess scientific evidence. The secondary role of the jury, as an impartial arbiter of laws, moderating them in accordance with societal standards would be the primary remaining role of a layperson jury. If this quasi-legislative function was revealed, the legal system would face the problem of having two appliers of the law (judge and jury); a host of difficulties in facilitating this capacity would endanger the enduring role of the jury. A regulated approach to fMRI lie-detection is further warranted, considering the erosion of the accused's right to silence that would likely occur.

A final hurdle of admissibility for a scientifically viable fMRI lie-detector would be the general exclusionary clauses found in sections 135, 136 and 137 of the Acts[80]; which allow courts to exclude (or limit use of) evidence that may be unfairly prejudicial, misleading or confusing or exclude evidence resulting in an undue waste of time. Section 137 requires the court to refuse evidence "adduced by the prosecutor if its probative value is outweighed by the danger of unfair prejudice to the defendant".[81]

The issue of unfair prejudice in relation to fMRI lie-detection evidence was discussed in *Semrau*, where the court turned to the comparable discretionary section, section 403,[82] whereby: "if the unfair prejudice substantially outweighs the probative value of the evidence, the evidence is inadmissible".[83] Judge Pham followed the line of argument put forward in *United States v. Sherlin*[84] where the probative value of results from a polygraph test were held to be substantially lessened by the fact that the defendant risked nothing in obtaining the test unilaterally, that is, without the "knowledge or acquiesce of the government" (who had no chance to amend or submit questions). The danger of unfair prejudice outweighed any probative value, particularly as it was being used to bolster witness credibility.

---

[78] Section 80, *The Australian Constitution 1900* (Cth).

[79] Goodman-Delahunty and Hewson (2010).

[80] S135, s136, s137 in *Evidence Act 1995* (Cth) (NSW).

[81] S135, s136, s137 in *Evidence Act 1995* (Cth) (NSW) at s137, this is the main general discretionary clause used in evidence: *R v Keenan Mundine* [2008] NSWCCA 55 (18 March 2008).

[82] *United States Of America v Lorne Allan Semrau*, (31 May 2010), No. 07-10074, Report and Recommendations, at p. 33.

[83] *United States v. Thomas*, 167 F.3d 299, 308-09 (6th Cir. 1999).

[84] *United States v. Sherlin*, 67 F.3d 1208, 1217 (6th Cir. 1995).

The judge also considered *United States v. Thomas,*[85] which held, where there is no chance of the evidence holding negative consequences for the defendant, the probative value is outweighed by prejudice. The court also considered that a lapse in time between the crime and tests might lead to unfair prejudice. For all these reasons, Judge Pham found the lie-detection evidence was not admissible under Evidence Rule 403. While not precedent, it is likely that these remarks would be persuasive in Australian courts and therefore, Australian courts might choose to exclude fMRI evidence on the basis of unfair prejudice, under section135 or 137.

Similarly courts may exclude fMRI lie-detection evidence, finding the volume of complicated data may confuse or mislead the jury, and lead them to attach more weight to scientific evidence than should be. Consequently, fMRI data would unjustly encroach on the "province of the jury" as referred to in *Corestaff v Wilson*[86]; that is the finding of fact. Furthermore, without appropriate regulation, the likelihood of different scientists asserting multiple conclusions from the same evidence might preclude evidence (or limit its use) on the basis of an onerous burden on the court or a waste of court's time.

The term "probative value" in section 137[87] has been interpreted in a number of different ways. In *Papakosmas v The Queen*[88] Justice McHugh distinguished between relevance and probative value; relevance not being concerned with reliability, whereas probative value is. This was contradicted by *Adams v The Queen*[89] where Justice Gaudron defined "probative value" only in relation to the relevance of the evidence. This reading was confirmed by Justice Simpson in *R v Keenan Mundine*[90] where he noted,

> Although contrary views have been expressed... it is not open to a trial judge, in assessing... the probative value of any piece of evidence, to take into account ... its reliability or ...the credibility of the witness through whom it is tendered.[91]

This was held by Justice Simpson on the basis that doing so "would be to attempt to anticipate the weight the jury would attach to it".[92] Given the difficulty in assessing the credibility of fMRI lie-detection and considering the undue weight that juries will attach to the evidence, a return to *Papakosmas* is necessary in the case of the inclusion of fMRI evidence. *Papakosmas* held that evidence becomes

---

[85] *United States v. Thomas,* 167 F.3d 299,308-09 (6th Cir. 1999).

[86] *Wilson v. Corestaff Services,* (14 May 2010), SC, Kings County, New York State Law Reporting Bureau, pp. 1–2.

[87] S137, *Evidence Act 1995* (NSW).

[88] *Papakosmas v R* [1999] HCA 37; 196 CLR 297; 164 ALR 548; 73 ALJR 1274 (12 August 1999).

[89] *Adam v R* [2001] HCA 57; 207 CLR 96; 183 ALR 625; 75 ALJR 1537 (11 October 2001).

[90] *R v Keenan Mundine* [2008] NSWCCA 55 (18 March 2008).

[91] *R v Keenan Mundine* [2008] NSWCCA 55 (18 March 2008) per J Simpson, at 33.

[92] *R v Keenan Mundine* [2008] NSWCCA 55 (18 March 2008) per J Simpson, at 33.

"prejudicial" where it is used in an "improper... emotional or illogical way".[93] In the case of scientific evidence, as has been shown with DNA evidence, prohibiting an improper portrayal of fMRI as conclusive evidence would be difficult to maintain without Justice McHugh's interpretation in *Papakosmas*.

The assessment of probative value would need include reviewing the credibility of fMRI scans, including reliability, the possible prejudicial circumstances in which it was attained and difficulties that the jury may have in attaching a proper weight to it. While this might seem to lead to the exclusion of fMRI evidence under the general exclusionary clauses, historically, reliable lie-detectors have not been able to be excluded from the legal system. To maintain public faith in the legal system, courts must seek out the most accurate determiner of fact to ensure justice and fairness. The introduction of such evidence, however, must not be done haphazardly, without careful consideration of the significant consequences upon the function and role of the jury, and the right of the accused to silence.

8 Capacity, Responsibility and the Impact of Neuroscience

8.1 Can fMRI Provide New Insights into Criminal Responsibility?

As fMRI improves understanding about brain functioning, it is hoped that it may provide new insights into the concept of legal responsibility and doctrines such as voluntariness[94] and *mens rea*.[95] While neuroscience has already had an impact in these areas by providing structural images in cases involving the defence of substantial impairment,[96] fMRI capabilities present the possibility of showing *functional* aberrations which may mitigate criminal responsibility.

Experiments utilising fMRI for the purposes of criminal responsibility have investigated the association between neuronal activity and various functional capacities necessary for behaving responsibly. Scientists have tested a range of sensory, motor, affective, and cognitive processes within the brain[97] to explore the effects of peer pressure, the impact of stress, the process of evaluating risks and

---

[93] *Papakosmas v The Queen* [1999] HCA 37; 196 CLR 297; 164 ALR 548; 73 ALJR 1274 (12 August 1999) per j McHugh at 92.

[94] The question of volition, or voluntariness forms traditionally forms part of the *actus reus* inquiry and looks at whether the criminal action was freely done, following principles laid out in *Woolmington v DPP* [1935] AC 462.

[95] 'Mens Rea' looks at the mental state of the accused, requiring that the crime was done with a "guilty mind". "The mens rea requirement stems from the common law notion of reserving punishment for those behaving wickedly." Brown and Murphy (2010), p. 1119, 1128.

[96] The full name of this defence is "Substantial impairment by abnormality of mind", found at s23A of the *Crimes Act 1900* (NSW).

[97] Freund (2002).

rewards and the exercise of deliberation on neuronal activity.[98] In a similar way to structural imaging, analysing fMRI results may assist scientists in developing an understanding of 'normal' brain functioning, and enable the establishment of criteria, against which 'deviations' may be identified.

More specifically, investigations are being carried out to assess the impact of certain brain lesions[99] on function and corollaries with behavioural traits.[100] Likewise, scientists are beginning to determine the specific brain aberrations associated with psychopathic behaviour[101] or susceptibility to drug addiction[102]; which may, in time, pose significant and specific legal quandaries.[103]

Importantly, alongside this evidence linking functional capacities and known behavioural conditions is increasing support for the idea that the brain has a degree of 'plasticity'; that different regions can function in various ways, and even learn to perform new functions.[104]

As with lie-detection, it is still early in determining how neuroscience may impact on our understanding of criminal responsibility. Similarly, many of the studies have reported group level analysis (and therefore may have limited validity in individual cases).[105] Even despite these and other limitations in the studies, this area may eventually provide strong empirical evidence for conclusions relating to functional capacity and, consequently, personal responsibility.

9   The Legal Admissibility of fMRI Evidence

Expert evidence regarding responsibility and capacity would be met with the same strict scrutiny as fMRI lie-detection evidence. The admissibility under section 79 of the Acts[106] would require "specialised knowledge", and methods resting on "more *than unsupported speculation or subjective belief*".[107] Again, in characterising this, the four points expounded in *Daubert*[108] provide a useful measure of admissibility.

---

[98] Mayberg (2010), p. 37, 51.

[99] Such as those from invasive tumours, infection processes and neurodegenerative disorders, Mayberg (2010).

[100] Batts (2009); Müller et al. (2008).

[101] Yang and Raine (2009).

[102] Bloom (2010).

[103] Kiehl (2010).

[104] Wandell and Smirnakis (2009).

[105] Mayberg (2010), p. 37, 40.

[106] *Evidence Act 1995* (Cth),(NSW).

[107] *R v Tang* (2006) 65 NSWLR 681 (CCA) CJ Spigelman at 138.

[108] *Daubert v. Merrell Dow Pharmaceuticals, Inc.*, 509 U.S. 579, 125L. Ed. 2d 469, 113 S. Ct. 2786 at 593–594.

The first two points regarding testing and peer review might be met by a qualified individual. However, in considering the rate of error, the professional standards in this field and whether the *"theory... has been generally accepted by the scientific community"* the literature reveals that the general consensus amongst leading neurologists is that fMRI is not yet ready for court.[109] A detailed knowledge of error rates for many specific behavioural disorders is yet unknown,[110] and debate surrounding the appropriate use of fMRI data[111] suggests that the cautionary approach of the courts is appropriate.

Given that fMRI evidence about functional capacity would mirror the way psychological or structural brain image evidence operates, the admission of this evidence, if it met the criteria of *Daubert*, would take place on analogous grounds. Provided scientists and lawyers adhered to the conceptual limitations of using fMRI data, it is unlikely courts would need to exclude the evidence on any of the grounds found in the general discretionary clauses of sections 135–137 of the Acts.[112] Aside from significant alterations in views about sentencing and punishment, the impacts of fMRI evidence for criminal responsibility would be less dramatic than those of lie-detection evidence; it follows that this evidence may be sooner admitted for use in courts.

10   The Conceptual Limitations of fMRI Data

If fMRI data is able to contribute valuable empirical data, interpretation of the images would face inevitable conceptual limitations. As aforementioned, correlation is not causation.[113] The empirical data correlating BOLD responses with human functioning cannot identify that the existence of particular mental states or certain behaviour; but rather provide a measurement of activity in highlighted regions of the brain. While this brain activity may be shown to consistently accompany certain behaviour, fMRI evidence remains *inductive* proof of accompanying behaviour, rather than *deductive* proof and therefore provides strong but not definitive evidence about the accompanying behaviour. Similarly, a lack of neuronal activity in relevant regions does not conclusively determine a lack of associated abilities; although particularly alongside behavioural evidence, this may be indicative of functional impairment. In looking at subjective mental states, often involved in responsibility questions, the problem of equating empirical data with criterial data becomes particularly evident. Sensitivity to this limitation when

---

[109] Mayberg (2010), p. 37, 51.

[110] Mayberg (2010), p. 41.

[111] Vincent (2010), in discussion of Reimer (2008).

[112] Ss 135, 136, 137 of the Acts.

[113] Aldrich (1995), 1.

considering what constitutes 'normal' brain functioning is therefore crucial to any legitimate application of fMRI data in determining responsibility.

A second, perhaps more obvious limitation exists in applying this scientific data where there is inevitably a lapse in time between a trial and the relevant event. Although investigation into capacity and responsibility may appear to provide objective criteria for determining criminal states of mind, it faces the insurmountable difficulty of not being able to examine the brain in the moment and circumstances of the crime; meaning, for example, that it is unable to be used as evidence of intention within *mens rea*.

Neuroscience therefore should not affect the law in a vacuum; rather, the joint efforts of scientists, psychologists, philosophers and legal and judicial experts are necessary for any well-founded development of law in this area.

11 The Overarching Concept of Responsibility

The criminal justice system is underpinned by the idea that people can and should be held accountable for their actions; agent responsibility, to some degree, is required for the legal system to make sense at all.[114] While "responsibility" is raised at various points throughout a criminal trial, the term "responsible" has various uses.

One meaning of "responsibility" simply attaches a person to a consequence by reason of voluntary engagement in action. For example, the statement "X is responsible for a hole in the wall" attaches X to the "hole in the wall". Usage in this way does not necessarily imply moral guilt, as the hole in the wall may have been made accidentally, but does suggest that X was free in their actions and therefore may be properly attached to the consequential hole. If, on the other hand, X had been forced to put the hole in the wall, it would follow that *the force* would be "responsible" and thus attached to the hole. This notion of responsibility may be called "attachment responsibility".

A second usage of "responsibility" denotes the ability to be held accountable; such as where an adult is responsible for their actions but a child is not. Here, "responsibility" is dependent upon the capacity for reasonable judgement in their actions. Capacity means they are legitimately considered a moral agent whose actions may be attributed to them. In contrast, a person without the complete capacities required for deliberation on how to act, or without the capacity to act in certain ways at all, is unable to provide a reasoned account of these actions, and cannot therefore be held accountable. Such meaning not only applies to general capacities, but also in specific cases where one may not be specifically qualified to

---

[114] Mobbs et al. (2007).

make decisions, such as a doctor giving legal advice. This usage has been labelled by at least one theorist as "capacity responsibility".[115]

A third way of applying "responsible" is as a description of virtuous action; one is responsible when one's actions in a specific context are just or appropriate. For instance, a child may not have the *capacity* to be responsible, but may have acted 'responsibly'.[116] This meaning has been labelled "virtue responsibility".[117]

These various concepts of responsibility appear at different times in the criminal trial process. Generally, "attachment responsibility" looks at the question of voluntariness and, the latter two meanings are used in assessing the element of *mens rea* and sentencing, respectively.

12   Voluntariness

Attribution of criminal action requires an agent, who is free and therefore can be held responsible; as established in *Woolmington v DPP* in which it was stated, "the Crown must prove [the offence]... as the result of a voluntary act of the accused".[118] It is often assumed that fMRI data will remove the possibility of truly voluntary action at all. Such challenges to 'free will' stem from materialist readings of the data, and hold to deterministic views of human nature.[119]

The roots of causal determinism are found in Hume's philosophy of causality, which purports that everything is caused[120] and cannot therefore be freely willed. His treatise has been dubbed "the founding document of cognitive science"[121] by subsequent philosophers and scientists. Modern fMRI investigations aim at showing the way in which brain activity may be held responsible for all human action. Perhaps, the most famous of these is Libet's experiments into consciousness,[122] where people were asked to move their hand, while the electrical activity in their brain, known as their "readiness potential" was monitored. He found that this electrical current preceded the conscious decisions of subjects to move their hands by up to half a second.[123]

Although much has been said about the implications of these experiments regarding conscious experience, they have also been used as evidence that action

---

[115] Vincent (2010), p. 17.

[116] Vincent (2010), p. 18.

[117] Vincent (2010), p. 17.

[118] *Woolmington v DPP* [1935] AC 462.

[119] Robert M. Sapolsky (2004).

[120] Russell (2007).

[121] Fodor (2003), p. 134.

[122] Libet (1999).

[123] Marchetti (2005).

is determined rather than voluntary.[124] Functional MRI scanning has already been recognised as a way of observing such determined movement more conclusively. Such evidence might arguably be used to exculpate people, on the basis that they were not truly free and were therefore were not acting voluntarily.

The doctrine of *compatibilism*,[125] which holds that free will and determinism may be held compatibly as beliefs, has provided determinists with a way of defending legal understanding of agency and attributing responsibility.[126] However, future research of fMRI, if interpreted to show that certain brain activity preceded experience of choosing, might undermine the availability of such a premise.[127] If the *experience* of voluntariness is shown to be "a mental mechanism that gives rise to a sense of conscious will and the agent self in the person"[128] then this might provide grounds for a defence, similar to automatism, which could show that although intention and action were present, they were beyond the control of the accused.[129]

The scientific application of brain activity imaging in the area of personal responsibility in this way would exceed the conceptual limitations of fMRI data. Brain activity would be used as the measure of free will, rather than the behavioural, criterial data pertaining to free will. Monitoring brain activity during the process of decision-making provides a measurement of functioning, but does not show "choosing" as a function itself. A materialist belief that reduces subjective, conscious human experience to the physical processes that *accompany* it cannot result in a logical framework for law. In fact, the necessity of voluntary agency and free will and the associated difficulties in their abandonment may guide scientists in their application of fMRI to legal questions of voluntariness.

Functional MRI data may provide legitimate insights into the factual presumption of voluntariness,[130] where people can be seen to have conditions affecting their consciousness and awareness in action. Beyond showing actions as being performed by agents suffering from a persistent impairment in consciousness, application of this research would be limited by the fact that research into mental states of the accused would not be contemporaneous to the crime. The problems of leaving the weighing of such complex evidence in the hands of the jury are similar to those discussed previously.

---

[124] Libet (1999).

[125] McKenna (2009).

[126] Morse (2008), p. 19, and Greene and Cohen (2004), p. 1775, 1778, come to different conclusions about how this understanding may impact on legal justice.

[127] Hodgson (2000).

[128] Wegner (2005).

[129] Sapolsky (2004), p. 1794.

[130] *R v Falconer* (1990) 171 CLR 30.

13   The Element of *Mens Rea*

The *mens rea* element of criminal investigation looks to whether one has the requisite mental state of a 'guilty mind', which is established by demonstrating that the accused performed the crime with intent, knowledge, recklessness or even mere negligence. This usually involves a subjective determination of the belief or intent of the accused, as well as an objective assessment of the reasonableness of their action in light of that belief.[131]

In addition to providing one of the bases for justification of punishment, *mens rea* "establishes a legal requirement that cannot be concretely measured".[132] Since fMRI scans test the brain activity that accompanies certain behaviours, it is thought, by some, that they may also reveal patterns relating to specific 'intention' within human brains. Cognitive behaviours (given the physical restrictions of fMRI scanners), such as conjuring up criminal intentions, are dependent on the individual's own perspective. The task is objectively understood in terms of what constitutes intention, such as purpose or foreseeing an outcome; however, it exists within the subjective experience of an individual.[133] Therefore, scholars have identified that any progress made relating to *mens rea* will require the expansion of the library of understanding generated from subjective behavioural reporting and recording contingent brain activity.[134] Given that subjective feelings inevitably vary from person to person, based on the complex interaction of circumstances, memories, openness to emotion and other personalised traits, mapping these seem beyond the technical possibilities of fMRI. Indeed, even judges using universally shared language have not been able to encapsulate the essence of intention.[135] It seems even with a far advanced understanding of the contingency of brain states and mental states, the unique *experience* of neuronal activity in an individual (relating to intention) will not be overcome with fMRI technology.

Yet again, the requirement that the assessment of mental states be contemporaneous to the criminal act in *mens rea* would also limit the possibilities of neuroscience.

---

[131] ANU Law Department (2010).

[132] Brown and Murphy (2010), p. 1119, 1130.

[133] Nagel (1974).

[134] Brown and Murphy (2010), p. 1119, 1129.

[135] See: 'Intention: Multiple Meanings', in ANU Law Department (2010).

14   Criminal Defences

Neuroscientific understandings, gained through fMRI scanning, may make inroads into legal concepts of criminal responsibility in the area of functional capacity and relevant defences. The "veil"[136] separating defences such as insanity from the positive test of *mens rea* is based on how responsibility is afforded, once *actus reus* requirements are satisfied. Capacity responsibility, which determines how one is to be held accountable, implies a similar subjective experience of intention, as it requires the capacity to deliberate and choose a certain action. Insanity or substantial impairment denies the proper capacity for the deliberation required; it is a "defence to criminal responsibility by reason of absence of *mens rea*".[137]

The case law looking at the defences of insanity and substantial impairment by abnormality of mind[138] has confirmed that "as an excuse, [insanity] reflects the fundamental moral principle *'that a person is not to be blamed for what he has done if he could not help doing it'*.[139] In her capacitarian model of responsibility, Vincent sets out the factors required; culpability depends upon one's duties, these duties arise out of what one can or cannot do, and what one can and cannot do relies in part (at least) on one's mental capacities.[140] If brain scans provided compelling evidence of mental capacities, their integration into this capacitarian understanding of responsibility would appear straightforward.

Mental capacities range from those which carry a sense of ambiguity (particularly regarding behavioural manifestation), such as reasoning, to those which can more clearly be identified with behavioural indicators, such as one's ability to foresee likely outcomes of certain behaviour (as is assumed in recklessness). Functional MRI research has been used to identify aberrations appearing consistently alongside certain well-known dysfunctional behavioural syndromes, such as antisocial personality disorders or more serious forms of depression.[141] Neuroscience, in the same way, may, by determining areas of the brain consistently active when making moral judgements, elucidate the areas required for this functioning.

The doctrine of insanity, was set down in *M'Naghtens's case*[142]; requiring,

---

[136] Brown and Murphy 2010, p. 1119, 1129.

[137] *R v S* [1979] 2 NSWLR 1 as discussed *Hunter Area Health Service & v Presland* [2005] NSWCA 33 (21 April 2005) per JA Santow at 312, accessed at http://www.austlii.edu.au/cgi-bin/sinodisp/au/cases/nsw/NSWCA/2005/33.html?stem=0&synonyms=0&query=insanity.

[138] s23, Crimes *Act 1900* (NSW).

[139] H.L.A. Hart, *Punishment and the Elimination of Responsibility*, (1962), Athlone Press at 20, as discussed *Hunter Area Health Service & v Presland* [2005] NSWCA 33 (21 April 2005) per JA Santow at 312.

[140] This is made clear in any discussion of omission – we cannot be held accountable for those things we could not have done: Vincent (2010).

[141] Mayberg (2010), p. 37, 38.

[142] M'Naghten's case [1843] UKHL J16 (19 June 1843).

to establish a defence of insanity, it must be clearly proved that, at the time of committing the act, the party accused was labouring under such a defect of reason, from disease of the mind, as not to know the nature and quality of the act he was doing; or, if he did know it, that he did not know he was doing what was wrong.

The first premise, the time of the act, presents another problem for neuroscientists. Introducing fMRI into insanity defences requires specificity about the time of "disease of mind" being examined, as brain functioning capacity may be affected by static (i.e. stable, long-term, fixed aberrations), episodic (e.g. epilepsy or bipolar manic depression) or progressive (such as dementia) aberrations.[143] While in the first of these instances, scans may hold significant weight, and be able to show a persistent abnormality; in the second two instances, fMRI would be persuasive in assessing certain functional impairment, only alongside other contemporaneous evidence.

The second premise of "defect of reason" from "disease of mind" refers to concepts upon which brain imaging shines limited light. Even with a wealth of understanding of the brain activity accompanying the behavioural conditions of reason, reason is not a static capacity, but rather subjective in the way that people exercise it with different capability and qualification. Even beyond reaching the 'age of reason', education in 'thinking' greatly affects the exercise of one's reason, or ability to understand. These functions of reasoning and understanding cannot be fully grasped by fMRI, as the brain processes associated with these functions cannot elucidate the subjective aspect present in attaching meaning to this data. However, if someone was severely lacking brain function in areas commonly associated with the behaviour of 'reasoning', they might argue that this amounted to a "disease of the mind", as brain capacities are necessary for exercising powers of the mind. Therefore, in some limited cases, where it is likely that mere examination of behaviour would reveal insanity anyway, fMRI would provide compelling evidence to support this severely reduced functioning.

The statutory defence of substantial impairment by abnormality of mind, which lessens a sentence of murder to manslaughter, requires the accused to show that:

at the time of the acts or omissions causing the death concerned, the person's capacity to understand events, or to judge whether the person's actions were right or wrong, or to control himself or herself, was substantially impaired by an abnormality of mind arising from an underlying condition...[144]

In *R v Dusan Maric,*[145] in an attempt to define the meaning of "substantial" it was held that, "although the impairment may be less than total it must be more than trivial".[146] The difficulty in defining "substantial" is further exacerbated by section 23 (2)'s prohibition of opinion evidence. Functional MRI may develop an objective

---

[143] Mayberg (2010), p. 37, 40.

[144] S23A (1) (a) of *Crimes Act 1900* (NSW).

[145] *R v Dusan Maric* [2009] NSWSC 346 (1 May 2009).

[146] *R v Dusan Maric* [2009] NSWSC 346 (1 May 2009), as per Harrison J at 35.

measurement of impairment, by testing the actual capacity of those parts of the brain, which would usually be active during the behaviour of making moral judgements. While psychological analysis is commonly accepted, accused persons face difficulties where they have not previously sought psychological help and cannot provide weighty evidence on the continuity of any condition. Functional MRI is on the path to being able to give significantly accurate answers to the status of mental conditions, that is, whether something is "underlying" as opposed to "transitory",[147] and thus may assist in these cases.

The introduction of neuroscientific evidence regarding mental capacities may, however, invite an alternative approach to responsibility than held by Australian courts. In cases where conditions of insanity, such as hallucinations or delusions are not present, and fMRI evidence instead reveals a deficiency in functioning associated with moral understanding, such as in the case of psychopaths, it has been argued that these people are fundamentally bad, rather than mad and therefore should not be excused on this basis. Maibom, has argued for this approach, saying "to excuse psychopaths from moral blame is tantamount to excusing them for being bad... presumably, we do not intend with our system of law to exculpate those whose disorder primarily consists in being bad".[148] While opening the courts to fMRI would open this use of scans to a prosecution, responses to this approach point out that this confuses "capacity responsibility" with "virtue responsibility" (a description of how one has acted based on fMRI evidence) and thus deals with elements of the sentencing process, rather than the guilt assessment process.[149]

It is important that the scientific evidence is not regarded as an objective measure of a capacity, but rather as an objective measurement of brain functioning associated with certain cognitive moral capacities. Given that generally, the capacitarian view of assigning responsibility is facilitated by allowing for mitigation of responsibility where impairment exists, the increased clarity offered by fMRI may see statutory or common law defences for other criminal actions be established. For example, if strong empirical evidence suggested an association between lesions in the frontal cortex with violent behaviour, this might form a basis for a defence against assault as an "underlying condition" and decreases in one's ability to "control" physical responses might be identified. Furthermore, research into more subtle reactions of persons may eventually lead to empirical evidence suggestive of one's propensity to be "provoked" to the point of losing self-control. Such application of fMRI would require a sophisticated level of understanding by juries, to ensure scientific evidence was duly scrutinised.

If these developments had the effect of somehow reducing all criminal activity to brain defects, this would set the legal system on a path to a watered-down determinism of sorts. Such impacts would undoubtedly demand a review of the

---

[147] s23 (8) of *Crimes Act 1900* (NSW).

[148] Maibom (2008).

[149] Vincent (2010), p. 17.

sentencing of criminals, which would need to balance these concerns in light of the interests of justice.

15 Non Determinism, Sentencing and the Importance of Moral Education

If abnormalities in functioning were able to be effectively equated with any criminal peculiarity, moral responsibility might be framed in terms of mental defects rather than moral culpability.[150] However, while fMRI may impact on sentencing, neuroscientific evidence of non-determinism ensures that the fundamental concept of responsibility could not be laid aside entirely. In the face of criminal behaviour, the need for reparation for both victims and society hails from the innate desire for justice, upon which the legal system gains its force. While courts need to examine a criminal's "virtue responsibility" and retain punitive measures, a deeper understanding of the brain may eventuate in opportunities to implement philosophies of restorative or transformative justice.

The call to embrace treatment-based programs of justice, rather than traditional punitive ones, is a strong one. Jessel and Moir write,

> ...with the growing knowledge that crime is ... a function of biology... Evil may be...no more sinister than a matter of loose connections... Is it practically possible to discard the traditional concept of justice based on guilt and punishment and replace it with a 'medical model' based on prevention, diagnosis and treatment?[151]

This view is held by many who conclude that if brain lesions can be associated with criminal behaviour, then knowledge of capacity and responsibility should lead to the treatment of crime as an illness, and therefore medicate, rather than punish offenders. In fact, if a sense of "biological determinism"[152] could be shown to exist, retributive punishment for offences committed outside of one's control would make little sense.[153]

While fMRI revealing impaired capacities may affect the assessment of one's "capacity responsibility", this model of medical treatment reflects a view of determinism, which requires the legal system to abdicate the concept of free will, and thus abandon agency with responsibility. Applying this in the extreme, mandatory fMRI scanning, and locking up those with brain abnormalities, in a pre-emptive movement against crime, would seem reasonable.

However, what the current, though incomplete, understanding of brain capacities and their impact on behaviour, from fMRI, assures one of is that abnormalities are

---

[150] Calls for a therapeutic approach are discussed in: Hodgson (2000).

[151] Moir and Jessel (1995).

[152] Sapolsky (2004), p. 1795.

[153] Norrie (1991).

not determinative. In a review of findings from various studies looking at common behavioural brain lesions, Mayberg writes, "Interestingly, most patients with these types of lesions do not display antisocial or criminal behaviour and not all criminals show such brain abnormalities..."[154] Similarly, neuroscientific research looking into drug addiction has found that neurotransmitter systems play a significant part in specific addictive patterns, however, "biological vulnerabilities do not exonerate the person for responsibility for their addictive state since it is their choice to use the drugs, once or multiple times".[155]

That brain abnormalities may or may not be a characteristic of criminals convicted of the same crime or, likewise, that certain brain lesions may be present in both the offender and the 'just citizen' proves that at a level human agency plays a crucial part in the determination of human behaviour. This is not to disregard the recognition of disadvantage or the reduced responsibility that may eventuate from understanding of impairments of the mind. Indeed, there are many positive impacts such knowledge may have on the legal system.[156]

A legitimate reading of Jessel, Moir and others, who would adopt a therapeutic system of justice, might see their concerns as stemming from the notion that justice should be about responding to the needs of offenders and transforming society into a better place in the process. Such aims might have to reject the common appeal to view the system *consequentially*,[157] and enforce practices based merely on the merit of their outcomes: deterrence, prevention and rehabilitation. Rather, responding in this way has often been labelled "restorative justice" or "transformative justice".[158] Neuroscience, in revealing that abnormalities are not determinative of criminal behaviour, has shown that there is a level at which moral education about decision-making is increasingly important; a strengthening of the just character in citizens. As Moir and Jessel write, "identifying the cause...loads us with the responsibility of doing something about it – treating the offender".[159] To the extent that wider society may be blamed for 'producing' criminals, a response is demanded in the form of wider moral education, particularly for those marginalised and disadvantaged, which might include programs focusing on the value of thinking reasonably, opportunities to be inspired to foster ambitions, and general encouragement to develop a sense of justice.

Furthermore, recent developments in neuroscience reveal the potential of the brain for 'plasticity'; the ability to change structurally and functionally.[160] While there are limitations, and prerequisites – such as healthy brain tissue and the

---

[154] Mayberg (2010), p. 37, 39.

[155] Bloom (2010), p. 44.

[156] Doidge (2009).

[157] Sapolsky (2004), p. 1795

[158] Ashworth (2002), also see: Mertus (1999).

[159] Moir and Jessel (1995).

[160] Schaechter et al. (2006).

capacity for motivation and focus, this means that some functions lost due to abnormalities in certain regions of the brain may be regained.

Where neuroscience further develops and gives scientists, working with psychologists, a deeper understanding of brain plasticity, those who are viable candidates but who lack certain helpful capacities required for moral decision-making might further benefit from neurological exercises aimed at restoring these.

There would of course be opposition to such change, and significant difficulties in deciding upon *which* moral code should be adopted. However, any therapeutic system would bring similar questions, given the diversity of brain structures and the diversity of functioning patterns. Moreover, there are inherent dangers posed in preferring certain brain functional structures, if science was given domain to 'fix' abnormalities. The diversity and creativity that may be intrinsically reliant on what may appear as 'abnormalities' lead one to reject therapeutic modification of the human brain.

Finally, not all sense of "punishment" could be done away with, even in a system of restorative justice, which aims at rehabilitating offenders. The non-determinism of brain structure and function requires that notions of free will, intention and fundamental choice in action remain. Not only does some form of punishment respond to the needs of victims, being a form of recompense, but it is required to affirm the fundamental tenets of the legal system and enforce the sense of 'justice' or giving of someone's 'due'. This would further deter a number of malicious abuses that might come out of a purely therapeutic system. Ultimately, the legal system recognises the inherent value of self-determinism, the great dignity attached with being "master of one's destiny", by punishing where one removes that ability from another, and upholding the fact that there are boundaries to the exercise of freedom.

In this way, the advances of neuroscience would act to further confirm principles held dear to the functioning of the legal system, while enhancing the benefits for society that may be achieved through the justice system. A deeper understanding of human agency will better enable legislators and the juridical system to respond to the needs of their community, and encourage a sense of justice that fosters law-abiding citizens.

16 Conclusion

In Australia, the legal system is currently reluctant to accept neuroscientific techniques such as fMRI, EEG and brainfingerprinting evidence. This may change in the future.

Professor Henry Greely's words support the Australian perspective too[161]:

---

[161] Professor Henry Greely (Stanford Law School), "Can MRIs Help Solve Crimes" National Public Radio interview, May 14, 2010 with Paul Raeburn.

Well, I think, certainly our ability to read minds, and it's a strong term, but I think it's an accurate one here, through fMRI and other neuroimaging, is going to get better and better, both from technical advances and through statistical advances. Whether it will ever be good enough to be used in a courtroom remains to be seen.

References

Aldrich J (1995) Correlations genuine and spurious in Pearson and Yule. Stat Sci 10(4):364–376, 1

ANU Law Department (2010) Components of criminal offences: mens rea. Aust Natl Univ. http://law.anu.edu.au/criminet/notes.html. Accessed September 2010

Ashworth A (2002) Responsibilities, rights and restorative justice. Br J Criminol 42(3):578–595

Baker G (2009) Neuroscience and the law: real potential with a healthy dose of caution. West Virginia Law Rev:3(2)

Batts S (2009) Brain lesions and their implications in criminal responsibility. Behav Sci Law 27:261–272

FE Bloom (2010) Does neuroscience give us new insights into drug addiction? A judge's guide to neuroscience: a concise introduction, vol 1. University of California, Berkeley, pp 42–44

Briody M (2004) The effects of DNA evidence on homicide cases in court. Aust N Z J Criminol 37:231–252

Brown T, Murphy E (2010) Through a scanner darkly: functional neuroimaging as evidence of a criminal defendant's past mental states. Stanford Law Rev 62(4):1119–1130

Caudill DS (2010) Expert scientific testimony in courts: the ideal and illusion of value-free science. The Pantaneto Forum:39. Accessed at: http://www.pantaneto.co.uk/issue39/caudill.htm

Chen I (2009) The court will now call its expert witness: the brain. Stanford Lawyer 81:1

Cournos F, Bavaniss DL (2003) Clinical education and treatment planning: a multimodal approach. In: (Wiley) (June 9, 2003). Tasman A, Kay J, Lieberman JA (eds) Psychiatry, vol 478, 2nd edn

Dickson K, McMahon M (2005) Will the law come running? The potential role of 'brain fingerprinting' in crime investigation and adjudication in Australia. JLM 13:204

Doidge N (2009) Science writer Norman Doidge's right brain quizzes his left brain about their newly discovered elasticity. The Australian. http://www.fastforword.com.au/Content_Common/ns-Dr-Norman-Doidge-Speaks-to-The-Australian-about-Brain-Plasticity.seo. Accessed 16 May 2009

Easteal PW, Easteal S (1990) The forensic use of DNA profiling. Trends and Issues in Crime and Criminal Justice, vol. 26. Australian Institute of Criminology, Canberra. Accessed at: http://www.aic.gov.au/publications/current series/tandi/21-40/tandi26.aspx

Edmond G, Biber K, Kemp R, Porter G (2008) Law's looking glass: expert identification evidence derived from photographic and video images. Curr Issues Crim Justice 20:37–77

Farwell L (1999) Farwell brain fingerprinting: a new paradigm in criminal investigations. Brain fingerprinting laboratory Inc: see http://www.brainwavescience.com/. Accessed 17 February 2011

Fodor J (2003) Hume variations. Oxford University Press, New York, p 134

Freckleton I, Selby H (2009) Expert evidence: law, practice, procedure and advocacy, 4th edn. Lawbook Co, Sydney, p 52

Freund HJ (2002) fMRI studies of the sensory and motor areas involved in movement. Adv Exp Med Biol 508:389–395

Goodman-Delahunty J, Hewson L (2010) Enhancing fairness in DNA jury trials. Trends and issues in crime and criminal justice, vol 392. Australia Institute of Criminology, Canberra, pp 1–6

Greely HT, Iles J (2007) Neuroscience-based lie detection: the urgent need for regulation. Am J Law Med 33:377

Greene J, Cohen J (2004) For the law, neuroscience changes nothing and everything. Philos Trans R Soc B Biol Sci 359:1775–1778

Hagan K (2009) DNA fiasco: rape conviction quashed. The Sydney Morning Herald. http://www. theage.com.au/national/dna-fiasco-rape-conviction-quashed-20091207-kfc3.html, October 2010. Accessed 8 December 2009

Hocking BA, Manville LL (2001) What of the right to silence: still supporting the presumption of innocence, or a growing legal fiction? [2001] Mq LawJl 3; (2001) 1 Macquarie Law Journal 63

Hodgson D (2000) Guilty mind or guilty brain? Criminal responsibility in the age of neuroscience. Aust Law J 74:661–680

Kalver H, Zeisel H (1966) The American Jury. New Society, 25 August 1966, p 290. Accessed via: Law Reform Commission, The Jury's Verdict, Discussion Paper 12 (1985) - Criminal procedure: the jury in a criminal trial. http://www.lawlink.nsw.gov.au/lrc.nsf/pages/DP12CHP9

Kiehl K (2010) Can Neuroscience Identify Psychopaths? in A Judge's Guide to Neuroscience: a Concise Introduction, 47 (sage center for the study of the mind, 2010)

Leenaghan N, Guerrera O (2005) Call for brainfingerprinting. The Age (30 September 2005)

Libet B (1999) Do we have free will? J Conscious Stud 6(8–9):47–57

Maibom HL (2008) The mad, the bad, and the psychopath. Neuroethics 1:167–184

Marchetti G (2005) Commentary on Benjamin Libet's Mind Time. The temporal factor in consciousness. Accessed at: http://www.mind-consciousness-language.com/Commentary%20Libet%20Mind%20Time.pdf

Mayberg H (2010) Does neuroscience give us new insights into criminal responsibility? A judge's guide to neuroscience: a concise introduction. University of California, Berkeley, pp 37–51

McKenna M 2009 Compatibilism. Stanford encyclopaedia of philosophy. http://plato.stanford. edu/entries/compatibilism/. Accessed 5 October 2009

Mertus J (1999) From leal transplants to transformative justice: human rights and the promise of transnational civil society. Accessed at: http://www.auilr.org/pdf/14/14-5-2.pdf

Mobbs D, Lau HC, Jones OD, Frith CD (2007) Law, Responsibility, and the Brain. PLoS Biology 5(4):e103

Moir A, Jessel D (1995) A mind to crime. Michael Joseph, London

Moriarty JC (2008) Flickering admissibility: neuroimaging evidence in the U.S courts. Behav Sci Law 26:29–49

Morse S (2008) Determinism and the death of folk psychology: two challenges to responsibility from neuroscience. Paper given at Deinard Memorial Lecture in Law and Medicine, pp 1–36

Müller JL, Sommer M, Döhnel K, Weber T, Schmidt-Wilcke T, Hajak G (2008) Disturbed prefrontal and temporal brain function during emotion and cognition interaction in criminal psychopathy. Behav Sci Law 26:131–150

Nagel T (1974) What is it like to be a bat? The Philos Rev 83:435–450

Norrie A (1991) Law, ideology and punishment. Kluwer, Dordrecht, pp 148–149

Reimer M (2008) Psychopathy without (the language of) disorder. Neuroethics 1:185

Rosen J (2007) Grey matters. The Sydney Morning Herald, April 6–8 2007, p 39

Ruprecht CH (1997) Are verdicts, too, like sausages?: lifting the cloak of jury secrecy. University of Pennsylvania Law Rev 146:217

Russell P (2007) Hume on free will. Stanford encyclopaedia of philosophy. http://plato.stanford. edu/entries/hume-freewill/. Accessed 14 December 2007

Sapolsky RM (2004) The frontal cortex and the criminal justice system. Philos Trans R Soc Lond B Biol Sci 359(1451):1787–1796

Schaechter JD, Moore CI, Connell BD, Rosen BR, Dijkhuizen RM (2006) Structural and functional plasticity in the somatosensory cortex of chronic stroke patients. Brain 129: 2722–2733

Seaman J (2009) Black Boxes. Emory Law J 58:427–484. Accessed at: http://www.law.emory. edu/fileadmin/journals/elj/58/58.2/Seaman.pdf

Shapiro BJ (1986) "To A Moral Certainty": theories of knowledge and Anglo-American Juries 1600–1850. Hastings Law J 38:153

Shen F, Jones O (2011) Brain scans as Evidence: Truths, Proofs, Lies and Lessons. (February 23, 2011) Mercer Law Rev, Vol 62, 2011

Simpson J (2008) Functional MRI lie-detection: too good to be true? J Am Acad Psychiatry Law 36(4):491–498

Vincent N (2010) Madness, badness and neuroimaging-based responsibility assessments. Law Neurosci Curr Legal Issues 13(1):15–17

Wainer J (2010) Abortion case proves need for law change. Sydney Morning Herald. http://www.smh.com.au/opinion/society-and-culture/abortion-case-proves-need-for-law-change-20101017-16p0u.html. Accessed 18 October 2010

Wandell BA, Smirnakis SM (2009) Plasticity and stability of visual fieldmaps in adult primary visual cortex. Nat Rev Neurosci 10:873–884

Wegner DM (2005) Who is the controller of controlled processes?. In: Hassin RR, Uleman JS, Bargh JA (eds) The new unconscious. Oxford University Press, New York, pp 19–20

White AE (2010) The lie of fMRI: an examination of the ethics of a market in lie-detection using functional magnetic resonance imaging. HEC Forum 22(3):253–266

Yang Y, Raine A (2009) Prefrontal structural and functional brain imaging findings in antisocial, violent, and psychopathic individuals: A meta-analysis. Psychiatry Res 174:81–88

Country Report: Austria

Marianne Johanna Hilf and Karl Stöger

Abstract In Austria, there has so far not been much research on the impact of neuroscience on the law. As a consequence, this contribution will try to discuss how some of the major neurolegal questions identified in other states might be dealt with under Austrian law. We will focus on the following areas: First, the legal framework for neuroscientific research. In this respect, we will pay most attention to the involvement of ethics committees on the one hand, and to the legal consequences of "incidental findings" on the other hand. Second, we will try to tackle some of the questions arising from the use of neuroscientific assistive technologies and "neuro-enhancement". Third, we will visit the discussion on whether recent neuroscientific findings on determinism put the concept of prosecution based on individual guilt into question. Finally, the legal framework on the use of neuroscientific techniques in criminal and civil procedure law will be drafted.

We are very grateful to our colleague Christian Kopetzki (Vienna) for many useful comments. The usual disclaimer applies.

M.J. Hilf (✉)
Institut für Strafrecht, Strafprozessrecht und Kriminologie, Universität St. Gallen,
Tigerbergstrasse 21, CH-9000 St. Gallen, Switzerland
e-mail: marianne.hilf@unisg.ch

K. Stöger (✉)
Institut für Österreichisches, Europäisches und Vergleichendes Öffentliches Recht,
Politikwissenschaft und Verwaltungslehre, Universität Graz, Universitätsstrasse 15/C3, 8010 Graz, Austria
e-mail: karl.stoeger@uni-graz.at

T.M. Spranger (ed.), *International Neurolaw*,
DOI 10.1007/978-3-642-21541-4_3, © Springer-Verlag Berlin Heidelberg 2012

1   Introduction

Neuroscientific questions are the object of regular research in Austria, including interdisciplinary approaches. This has resulted in the establishment of the Austrian Neuroscience Association, which holds annual conferences[1]; several regional research groups have also been formed.[2] However, the law faculties have so far not joined the club: "Neurolaw" defined in this contribution as a set of legal questions raised by the ongoing developments in neuroscientific research and treatment[3] is a (research) topic that Austrian jurists still have to discover.[4]

For this, three reasons can be given: First, the number of Austrian jurists working on questions of medical law is – with the exception of the slightly more popular field of medical malpractice – rather small compared to bigger European countries, especially Germany. Second, the Austrian legal tradition is still dominated by a positivist approach and, as a consequence, predominantly focusing on the interpretation of existing legislation. In contrast to this, neurolegal research will often have to deal with questions about how the law should in the future adapt to neuroscientific developments. Third, it should be taken into account that some questions raised by neuroscience have already been discussed in respect to other technologies – for example, the question of incidental findings is addressed by the Austrian Gene Technology Act (see below). Some questions might therefore have been judged by legal researchers as being a bit "more of the same".

As a consequence, this contribution is not a report on ongoing legal research or existing case-law. Instead, we will discuss how some of the major neurolegal problems identified by the German doctrine might be dealt with under Austrian law. The recourse to German doctrine and/or jurisprudence is not uncommon in Austria as some main aspects of the Austrian and the German legal order are to a certain extent similar – which also holds true for medical law. At the same time, the methods used to address legal problems differ remarkably – as has been mentioned before, the Austrian legal tradition, at least in public and criminal law, is still a rather positivist one. Accordingly, while it makes sense to refer to the German legal order when identifying some major neurolegal questions, the answers to these questions will have to be found along different lines.

The questions which we will discuss are the following: First, the legal framework for neuroscientific research. It seems that in Germany, most attention in this context is paid to the legal consequences of "incidental findings"; however, we will try

---

[1] Cf. the homepage at http://www.univie.ac.at/ANA/php/index3.php. Accessed 17 February 2011.

[2] For Styria, cf. the "Initiative Hirnforschung Steiermark"; http://www.gehirnforschung.at/. Accessed 17 February 2011.

[3] This definition corresponds to the one proposed by Schleim et al. (2009), p. 8.

[4] So far, only a few articles in Austrian legal journals have been published in which questions of neuroscience were explicitly addressed: Two of them, however, were written from a political and/ or philosophical perspective (Gehring 2007; Kampits 2008) while the other ones only shortly mention the potential consequences of new neuroscientific findings for the concept of a free will (Pronay 2008; Ganner 2010; see also – as early as 1998 – Jescheck 1998).

to reach out a bit further than this. Second, questions arising from the use of neuroscientific assistive technologies and "neuro-enhancement". Third, we will visit the discussion on whether recent neuroscientific findings on determinism put the concept of "Schuldstrafrecht" (prosecution based on individual guilt) into question. Finally, the legal framework on the use of neuroscientific techniques in criminal and civil procedure law (including its constitutional foundations) will be drafted.

This contribution will – in conformity with the traditional positivist approach in Austrian jurisprudence – focus on the existing legal framework and will not cover "soft law" (e.g. recommendations by International Organizations) or scientific recommendations/guidelines/directives (e.g. by medical associations).[5] The authors do not underestimate the influence of these sources on the neuroscientific community. However, while soft law and recommendations/guidelines/directives might be of relevance when certain actions or omissions are assessed as to their conformity with legal terms such as "proper medical care" or "good clinical practice", they are – in contrast to legal norms – not per se binding. Whether such standards have been met or not, is therefore not a question of law, but one of fact.[6]

2   The Framework for Neuroscientific Research, Consequences of Incidental Findings

2.1   General Remarks

To date, there are no specific legal rules concerning neuroscientific research on human beings. Consequently, the general legal framework for research on human beings applies. This framework cannot be found in a specific statute, but consists of "pieces" which are scattered all over the Austrian legal order (e.g. provisions in the Civil Code, the Hospital Acts, the University Act or in the Data Protection Act)[7] – with several fundamental rights standing in the background. Different parts of this framework apply according to the type of research performed (especially medical research vs other research) and to the institution where it takes place (e.g. universities, university hospitals, other hospitals). As to the fundamental rights which are of relevance for (neuroscientific) research, we may only point out that the Austrian Constitution, namely Art. 17 of the Basic Law on the General Rights of Nationals,[8] guarantees the freedom of scientific research and teaching to any

---

[5] For an introduction (based on the example of standards within the medical profession), see Damm (2009).

[6] Kopetzki (2010a), p. 9.

[7] For an overview cf. Kopetzki (2010b).

[8] Staatsgrundgesetz über die Allgemeinen Rechte der Staatsbürger, Imperial Law Gazette No. 142/1867 as amended.

person, whether professional scientist or not. This freedom makes rules which specifically constrain research and teaching unconstitutional; however, research and teaching must (still) be exercised in conformity with the (general) laws of the land, especially those protecting the rights of other persons.[9] This is of some importance if research requires the participation of other human beings. Among the most important rights in this context one must name the right to private life under Art. 8 ECHR (the ECHR has been integrated into the Austrian constitution), the right of freedom from torture and inhuman or degrading treatment under Art. 3 ECHR and the right to protection of data. In contrast to other states, the Austrian Constitution does not contain a fundamental right protecting human dignity as such. However, case-law[10] and doctrine largely agree that some, if not most elements that are understood to form the right to human dignity in other states, can be found in other fundamental rights of the Austrian Constitution – with Art. 3 ECHR somehow acting as a fall-back clause.[11] Neuroscientific research, especially imaging techniques, does or might in the future allow some more or less precise insight into the mind of a human being which is normally hidden to the outside which, in our opinion, means that the possible intrusion into the private sphere under Art. 8 ECHR is particularly intense. As the fundamental rights of the Austrian Constitution not only enshrine rights of individuals against the state, but also impose on the state a duty to protect these fundamental rights of individuals, these rights are of importance not only for research in institutions controlled by the state, but also in the sphere of "private research". It should also be noted that Austria is not a signatory state to the Convention on Human Rights and Biomedicine, CETS No. 164 and its additional protocols.

Generally speaking, research on human beings or the use of data for research purposes requires (in the vast majority of cases) the informed consent of the person in question[12] and/or compliance with the provisions of the Data Protection Act.[13] It should be borne in mind that the agreement to participate in a research project will create some kind of contract between the person or the institution performing the research and the participant (we will come back to this point later). Consequently,

---

[9] For an overview cf. Berka (1999), para. 587.

[10] In this context, it should also be noted that the Constitutional Court has stated that experiments with human beings, even if they have declared their consent, interfere with their "human dignity" and therefore must aim for results which are beneficial for humanity. Cf. Decisions by the Constitutional Court No. 13.635/1993.

[11] Cf. Berka (1999), para. 378.

[12] In general terms, such consent is indispensable if the physical or mental integrity of a person or his or her personality rights – the existence of which is deduced from the notion of "inborn rights" in Sect. 16 of the Austrian Civil Code (Allgemeines Bürgerliches Gesetzbuch, "Justizgesetzsammlung" No. 946/1811 as amended) could be affected by the planned research. However, in certain situations it is difficult to answer whether this is the case or not: For a recent example cf. Kopetzki (2011), pp. 21–22. There are also some specific provisions which dispense from the necessity of an informed consent: e.g. Sect. 43a Pharmaceuticals Act. For an overview concerning medical research, cf. Burgstaller and Schütz (2003), paras. 100–124.

[13] Datenschutzgesetz 2000, Federal Legal Gazette No. 165/1999 as amended.

specific legal questions arise if participants cannot give a valid consent themselves (e.g. due to their age or the state of their mental health).[14]

In the following, we will focus on three aspects of the legal framework which are, in our opinion, of specific relevance for neuroscientific research: First, the need of certain research projects to be assessed by an ethics committee; second, the relevance of incidental findings for the researchers and third, whether probands are obliged to disclose the results of neuroscientific examinations (whether incidental or not) to third persons.

2.2 Assessment By an Ethics Committee?

In principle, any person wishing to perform a research project is free to do so as long as he or she acts in conformity with the law. However, in certain cases, a research project must not be started unless it has been assessed by an ethics committee. This can also apply to neuroscientific research. The most relevant provisions in this respect are on the one hand Sect. 30 of the Universities Act[15] and on the other hand Sect. 8c of the Federal Hospital Act[16] (a provision more or less repeated by corresponding provisions in the provincial Hospital Acts, which are applicable to individual cases while the Federal Hospital Act is only binding for provincial legislators). According to these provisions, any Medical University – which in Austria are by law separated from other universities – has to appoint an ethics committee which has to assess clinical trials of pharmaceuticals and medical products, the application of new medical methods and applied medical research involving human subjects. Hospitals which are not a part of a medical university also have to appoint an ethics committee, either exclusively or jointly with other hospitals. These ethics committees have – as those of medical universities to assess clinical trials of pharmaceuticals and medical products, the application of new medical methods and applied medical research involving human subjects; addition, the Federal Hospital Act assigns them the task to assess research on nursing projects as well as new nursing concepts or methods.[17] Finally, any clinical trials of pharmaceuticals or medical products taking place outside of a Medical University

---

[14] For the general limits of a procurator's power to give his or her consent to research on a (mentally) disabled person cf. Sect. 284 of the Civil Code. This provision has more or less enshrined pre-existing case-law: cf. Bernat (2001). For specific clinical research, cf. Sects. 42–43 Pharmaceuticals Act and Sects. 51–52 Medical Devices Act.

[15] Universitätsgesetz, Federal Legal Gazette No. 120/2002 as amended.

[16] Bundesgesetz über Kranken- und Kuranstalten; Hospitals and Health Resorts Act, Federal Legal Gazette No. 1/1957 as amended.

[17] If the hospital in question is part of a Medical University, the ethics committee of this University can assume the responsibilities (including the ones concerning nursing projects) of the ethics committee under the Hospital Acts.

or a hospital have to be assessed by an ethics committee which is responsible for the respective federal province (Sect. 41 Pharmaceuticals Act,[18] Sect. 58 Medical Devices Act[19]). These provisions can also apply to neuroscientific research in which pharmaceuticals ("neuro-enhancement", see chapter 3 below) or neurological implants (which are medical devices; see below as well) are tested.

The ethics committees under the above-mentioned Acts are acting independently from instructions and are of multidisciplinary character. In the Pharmaceuticals Act and the Medical Devices Act, there are some very clear instructions (especially concerning persons with limited capacity to act and minors) as to when the ethics committee has to give a negative assessment.[20] As to other assessments by ethics committees, the legal background remains disappointingly vague: Neither is there full agreement on the legal quality and the legal effects of a (negative) assessment of an ethics committee,[21] nor is it entirely clear which (legal) standards have to be applied by an ethics committee when assessing the admissibility of a research project.[22] The legal discussions on these questions have recently intensified, but they are far from being solved.[23] Another very important question to be answered in this context is which type of research has to be qualified as "medical" under the Universities Act or the Hospitals Acts (and, as a consequence, requiring an assessment by the respective ethics committee). Medical research is certainly given if the research activities in question legally require the participation of doctors. This is the case if the research in question comprises activities which must only be performed by doctors under Sect. 2 para. 2 of the Doctors Act.[24] One of these activities is the examinations of human beings (including an indirect analysis of bodily fluids and tissues) to confirm or rule out the existence of bodily or mental illness or of dysfunctions, disabilities, deformities or aberrations which are of pathological character. Should any of these activities – or any other activity[25] mentioned in Sect. 2 of the Doctors Act – be part of a neuroscientific research project, this project has to be qualified as medical research and requires the participation of a doctor who has to carry out the above-mentioned activities. Whether this is true or not, can only be decided on a case-to-case basis.

---

[18] Arzneimittelgesetz, Federal Legal Gazette No. 185/1983 as amended.

[19] Medizinproduktegesetz, Federal Legal Gazette No. 657/1996 as amended.

[20] Sections 42–43 Pharmaceuticals Act; Sects. 51–52 Medical Devices Act.

[21] For a recent overview, cf. Eberhard (2010).

[22] See, e.g. Pöschl (2010), p. 98.

[23] For a very recent overview over the ongoing discussion cf. the contributions in Körtner et al. (2010), especially those by Eberhard, Kopetzki and Pöschl. For a field manual on the functioning of ethics committees (which, however, does not really contain information on the theoretical legal discussions) cf. Druml (2010).

[24] Ärztegesetz, Federal Law Gazette No.169/1998 as amended.

[25] We may only point out that some of these tasks can also be delegated to other medical staff, e.g. nurses, as long as they act under the supervision of a doctor: The extraction of blood would be a practically relevant example.

While the law of the state limits the participation of ethics committees more or less to "medical" research as described above, the (non-medical) universities which enjoy some freedom as to their internal organization, have increasingly begun to enshrine in their charters provisions about their own ethics committees which shall assess whether research projects are "ethically maintainable". To give an example, the Faculty of Psychology of the University of Vienna had established, as early as 2006, an ethics committee for psychological research projects. In January 2011, the tasks of this committee were transferred upon a university-wide ethics committee.[26] The University of Graz established an ethics committee in 2008: This committee has the task to assess, amongst others, any research project which includes (medical or non-medical) examinations "which may affect the physical or mental integrity, the private sphere or other important rights or interests of the probands or their family members".[27] If the committee comes to the conclusion that the project is not ethically maintainable (a very unclear criterion[28]), it has to issue a negative statement and inform the rectorate which has to render a final decision on the admissibility of the proposed research.[29] It might be questioned to which extent such university rules are in conformity with the freedom of research and teaching. It goes without saying that most neuroscientific research projects will probably meet the requirements for an assessment by the committee, especially in respect of the protection of the private sphere of the proband.

To sum up, neuroscientific research projects will increasingly require an assessment by an ethics committee of the relevant institution. This assessment will regularly bring "soft law" (recommendations etc.) into play and raise difficult questions if the committee's decision is a negative one. However, as mentioned above, the legal discussion of the role of ethics committees in science is only about to take off.

2.3 Incidental Findings

After having discussed some legal requirements for neuroscientific research, we will now turn to possible (legal) consequences of such research: The advance of neuroscientific imaging techniques has resulted in a remarkable rise of "incidental findings", defined as unanticipated results of (neuroscientific) research which make it advisable to seek medical assistance to rule out (or to confirm) potential negative

---

[26] Cf. http://www.univie.ac.at/ethikkommission/index.php. Accessed 17 February 2011.

[27] Cf. Sect. 3 para. 2 of the sub-section concerning the ethics committee of the Charter of the University of Graz: http://www.uni-graz.at/zvwww/gesetze/satzung-ug02-17.html. Accessed 17 February 2011.

[28] See the critical assessment by Pöschl (2010), pp. 98–99.

[29] Cf. Sect. 6 of the sub-section concerning the ethics committee of the Charter of the University of Graz.

consequences of these findings to the mental or physical health of the proband.[30] In the German doctrine, there has already been some discussion about how to legally handle such incidental findings.[31] In Austria, however, the subject has not yet emerged at full scale.

One reason for this is probably found in the Gene Technology Act.[32] Genetic research can also yield incidental findings, and this case has been explicitly addressed by this Act, in a way that does not leave too much space for discussion. The Gene Technology Act draws a sharp line between testing for medical purposes and testing for scientific purposes.[33] Testing for medical purposes takes place to rule out or confirm a predisposition for a disease or to confirm a manifest disease. In these cases, the patient has the right to refuse that he or she be informed of the results (Sect. 69 para. 5 of the Act). However, as will be shown, this provision only refers to results concerning the specific (potential) disease which has been the cause for testing. This conclusion can be drawn from Sect. 71 of the Gene Technology Act which refers to all types of genetic testing (including testing for research).[34] This provision clearly states (in para. 1 sub-section. 2) that "the proband has to be informed of unexpected findings of direct clinical relevance or of results the communication of which he or she has explicitly demanded. The communication, especially in cases where the proband has not asked for the results, has to be arranged in such a manner that it will not upset the proband. In borderline cases, the information may be completely withheld". Apart from the very vague last sentence, this provision is quite clear: While the proband has a right to refuse information about the results of a planned search, he or she has – with one exception – to be informed of unexpected results if these are found during the test. Section 70 of the Gene Technology Act is also of some relevance for the question of incidental findings as it obliges a doctor to recommend to a person to suggest genetic testing to his or her relatives if the test results (planned or incidental) of this person show a "serious risk" of a disease for the relatives.

However, two aspects must not be overlooked: First, genetic testing always legally requires the involvement of doctors and second, it is not clear whether the cited provisions of the Gene Technology Act should be regarded as specific solutions for a specific problem or rather as merely specifying a general solution for the problem of incidental findings which can generally be deduced from the Austrian legal order.

---

[30] For Germany, see e.g. Schleim et al. (2007) who also cite other publications. It is nevertheless expected that improved diagnostic methods might again reduce the number of incidental findings requiring further medical examinations.

[31] Cf. once more Schleim et al. (2007); Spranger (2009a), pp. 194–197

[32] Gentechnikgesetz, Federal Law Gazette No. 510/1994 as amended.

[33] Cf. Stelzer and Havranek (2007), p. 681.

[34] On this provision and its history in general see Stelzer and Havranek (2007), p. 684.

In our opinion, there are some valid arguments pointing toward the second solution.[35] The starting point for this is Sect. 95 of the Criminal Code[36] which states that a person (including members of a research team) is criminally liable if he or she deliberately does not "in case of a casualty [ ] provide the aid required to save a person from deadly peril, considerable bodily injury or damage to health". This obligation must not be overstretched; [37] however, if a neuroscientific test shows the risk of an acute (!) danger to one's health, this has to be qualified as a "casualty"[38] and creates an obligation to inform the proband. In contrast to this, a mere predisposition for a certain disease or health risk as such is not sufficient to constitute a "casualty" and therefore does not lead to an obligation to act under criminal law.[39] Furthermore, Sect. 95 does not, in our opinion, oblige researchers to regularly check test results for incidental findings.

Second, if a doctor is part of the research team, he or she has to act according to professional standards. Section 49 of the Doctors Act obliges doctors to diligently examine or treat any healthy or ill person without difference if they have accepted to give medical advice to or to treat this person. While, again, this obligation must not be overstretched, a doctor who participates in neuroscientific research has, in our opinion, to some extent accepted to supervise the condition of the proband during the research and to interpret the findings of the research – however, this acceptance as such is limited to the purpose of the research and does not oblige the doctor to actively search for potential risks to the proband's health. If, however, he or she incidentally detects some findings which in his or her expert opinion require further clarification, he or she will have to inform the proband.

Third, if incidental findings are withheld, this could lead to a damage arising in the sphere of the proband or with third persons. This brings the rules of civil law, especially of contractual and non-contractual liability, into play. While it is commonly recognized that there is no general duty under current legislation to prevent damages to others by one's own action, such a duty is to some extent accepted if the law provides for specific fiduciary duties or for duties which intent to protect the interest of third persons, or if such a duty can be deduced from a contract.[40] Such duties might, for example, be deduced from the above-mentioned Sect. 49 of the Doctor's Act. Furthermore, one's consent to participate in research as a proband

---

[35] In respect to a right not to be informed of test results cf. Bernat (1995), pp. 42–43. See also Bernat (2002), pp. 191–192 who claims that the rules of the Gene Technology Act do not really go beyond the results that an interpretation of the general medical law would yield anyway.

[36] Strafgesetzbuch, Federal Law Gazette No. 60/1974 as amended. In Germany, a similar approach has been proposed by Schleim et al. (2007), p. 1042 with Sect. 323c of the German Criminal Code as a starting point.

[37] For an analysis of the specific relevance of this provisions for doctors see Kienapfel (1979), p. 597.

[38] Jerabek (2010), para. 4.

[39] Jerabek (2010), para. 4.

[40] Koziol (1997), para. 4/60; Kodek (2010), paras. 4–5.

will create some kind of contract between the proband and the researcher or the research institution.[41] According to the specific contents of this (oral) contract, it could be argued on a case-by-case basis that the researchers have undertaken an obligation to report incidental findings to the proband.

In addition, some civil law authors come to the conclusion that a person is also under an obligation to act (in our case: to inform the proband) if the risk of a severe damage to a third person's (in or case: the proband's) goods (including health or life) can be averted easily and without risk to oneself.[42] While the existence of such a far-reaching obligation is disputed,[43] it is clear that if such a duty existed, critical incidental findings during neuroscientific research would perfectly match this formula.

This brings us to the last question which is whether the proband has a right to refuse to be informed of such findings. Taking the arguments mentioned above into account, this will not be the case if the findings are sufficiently "severe", in other words, if they may point toward a serious health risk – in these cases, the research team is under an obligation to inform. However, findings of minor significance can probably be withheld in conformity with the above-mentioned arguments if the proband so wishes. Should, however, the research team erroneously classify findings as being of minor importance while there has been an obligation to inform, the negligent breach of this obligation may well result in a duty to pay damages.

To sum up, we can say that there is some kind of obligation to inform the proband of incidental findings at least in acute or very severe cases; however, the legal bases for this obligation are a bit sketchy which might – to strengthen legal certainty – make it advisable to create a provision similar to that in Sect. 71 of the Gene Technology Act which states comparatively clearly under which circumstances a person has to be informed of incidental findings. However, as far as an obligation to inform exists, their violation has to be judged according to the standard rules of liability law or, at the most, of criminal law.

Finally, it should not be overlooked that in those cases where an ethics committee has to give its approval to a certain research project, the ethics committee could – with good reasons due to the legal uncertainty which we have just shown – argue that the research plan has to contain clear guidelines how to handle the problems of incidental findings.

---

[41] In the German doctrine, Schleim et al. (2007), p. 1042 are not very clear on this point. While it is true that there will probably be no contract on medical treatment in a research project, some kind of contract will exist by which the proband agrees to co-operate with the researchers. As every contract under civil law, such a contract will create mutual obligations. It will be necessary to designate their contents by way of interpretation of the contract according to the general rules of contract law.

[42] Cf. Koziol (1997), para. 4/61 (at the end of the para.); see also Kodek (2010), para. 4.

[43] Reischauer (2007), para. 3.

2.4 Obligation to Disclose Neuroscientific Test Results?

Finally, incidental findings also raise the question whether a person is obliged to inform a potential employer or insurer about a potential health risk which has been discovered in the course of a neuroscientific examination or – on a more general basis – whether an employer or an insurer has the right to demand that a person accepts such an examination. In this context, special attention must be paid once again to the Gene Technology Act, in particular its Sect. 67, which bans employers, insurers and their agents from collecting, demanding, accepting and making use of the results of genetic analyses of employees, job-seekers, insured persons or persons wishing to take out insurance. In the context of occupational safety, the Workers' Protection Act[44] regulates – in Sects. 49–59 – the obligation of employers either to have employees or jobseekers examined by doctors if their (planned) occupation might pose a risk to their health or (for occupations with a comparatively lower, but still significant risk to the employer's health) to offer them the possibility to undergo such examinations. Apart from these cases, it is not entirely clear – neither in practice nor in theory – to what extent employees are obliged to disclose information about their health (including disclosure by participation in examinations organized by the employer) to potential or actual employers[45] – the answer also depends on the occupation in question and the data the disclosure of which has been demanded (e.g. a simple medical check vs. a complex blood test). The disclosure of health-related data to insurers is regulated by Sect. 11a of the Insurance Contract Act[46] which only – in accordance with the Gene Technology Act – explicitly excludes the use of data from genetic testing.

The constitutional background to this problem is primarily the right to private life (Art. 8 ECHR) and the right to data protection (Sect. 1 Data Protection Act). As the law currently stands, the results of neuroscientific examinations are – due to the absence of specific provisions – legally treated no different than other health data. However, as with health data in general, it will have to be taken into account that certain results of such examinations are of much more personal character than "standard" data such as weight, height, or blood pressure. Still, it is a fact that data from neuroscientific examinations do not enjoy the explicit legal protection which the Austrian legislator has given to genetic data.

Hence, the question which remains to be answered – and this also requires a medical perspective which cannot be given here – is whether the results of (certain) neuroscientific examinations (especially imaging techniques) are comparable to that of genetic testing: This would have to be answered in the positive if (or from a present perspective: as soon as) such techniques yielded results with an informative

---

[44] ArbeitnehmerInnenschutzgesetz, Federal Law Gazette No. 450/1995 as amended.

[45] See e.g. Marhold and Friedrich (2006), pp. 12–14; Löschnigg and Schwarz (2003), pp. 185–188; 706 (especially p. 186).

[46] Versicherungsvertragsgesetz, Federal Law Gazette No. 2/1959 as amended.

quality similar to that of genetic testing; e.g. if they allowed some kind of forecast about the potential (mental) health status of a person in the future. If this were the case, the absence of a specific provision for neuroscientific examination results as compared to the above-mentioned provision of Sect. 67 Gene Technology Act could well be regarded as unconstitutional under the Austrian non-discrimination clause (Art. 7 of the Austrian Federal Constitution[47]) which obliges the Austrian legislator to enact comparable legislation for comparable factual situations.[48]

3 Assistive Technologies and Neuro-Enhancement

Neuroscience is also contributing to the development of certain assistive technologies, which can support disabled, chronically ill or elderly people and which are linked with the brain or the nervous system (e.g. stimulators implanted in the brain [with "deep brain stimulation" being a practical example] or brain/neuronal-computer interaction devices [BNCI][49]). In contrast to this, "neuro-enhancement" is in this report understood as the improvement of "cognitive, emotional and motivational" functions of healthy persons.[50] Neuro-enhancement at present often relies on the use of pharmaceuticals,[51] but could – at least in the years to come – also be reached by "technical" means such as surgery. Admittedly, the line between the two areas can be difficult to draw in certain cases. From a legal point of view, the main question is whether such neuroscientific interventions can still be judged as therapeutical or as going beyond this. In this respect, the legal situation in Austria is quite comparable to the German one. In both cases, interventions will fall under the legal framework for medical treatment. However, broadly speaking, any intervention which is not a therapeutical one, requires a higher level of informed consent of the patient and must be judged as not being "contra bonos mores" (for more details see below).

If the intervention is to be judged as therapeutical ("Heilbehandlung" as defined by Sect. 110 of the Criminal Code), it will require the informed consent of the patient on the one hand and it must be in conformity with the medical standards

---

[47] Bundes-Verfassungsgesetz, Federal Law Gazette No. 1/1930 as amended.

[48] For a comprehensive overview cf. Pöschl (2008).

[49] For some examples of present and possible future assistive technologies (not limited to neuroscientific ones), cf. the report by the Austrian Bioethics Commission on "Assistive Technologies: Ethical Aspects of the Development and Use of Assistive Technologies", dated 13 July 2009. A bilingual (German-English) version can be downloaded from http://www.bka.gv.at/DocView.axd?CobId=39411. Accessed 17 February 2011.

[50] This definition has already been used by Heuser (2009), p. 49.

[51] Cf. Heuser (2009), pp. 49–50; Rosenau (2009), p. 69.

which – to cite the two most important examples[52] – the Doctors' Act or the Hospital Acts of the Austrian provinces[53] require for medical treatment. This means that new techniques will have to be scrutinized very closely as to their conformity with medical science before being administered on a patient.[54] It goes without saying that interventions which completely change the personality of a patient or lead to a complete loss of his or her free will will regularly have to be judged as not being in conformity with medical standards as they can no longer be qualified as therapeutical treatment.[55] Furthermore, the administration of "new medical methods" in hospitals and university hospitals will – as already discussed above – require the consent of the ethics committee of this institution[56]; the same applies to trials of new medical devices, irrespective of where these trials takes place (on this point, see below). However, from a legal perspective, there is no specific difference between neuroscientific techniques and other forms of medical intervention.

The same holds true for non-therapeutic interventions ("neuro-enhancement"): These are also governed by the general framework for this type of intervention. Generally speaking, there are three main differences between therapeutic and non-therapeutic interventions: First, the extent of information about the planned intervention which has to be given to the patient to make his or her consent valid is higher for non-therapeutic ones. Second, while a therapeutic intervention is admissible as long as it is in conformity with medical standards, a non-therapeutic intervention is inadmissible if it is "contra bonos mores" (irrespective of the consent of the patient). Third, while a therapeutic intervention without consent is – as long as it is carried out lege artis – only regarded by the Austrian Supreme Court as an "unlawful therapy" (Sect. 110 Criminal Code), a non-therapeutic intervention without consent is punishable as bodily harm (Sects. 83–88 of the Criminal Code). The latter provisions, however, only apply if the intervention can be

---

[52] For specific medical treatments, specific statutes can apply, e.g. the Artificial Procreation Act, Federal Law Gazette No. 275/1992 as amended. Neuroscientific treatment will, though, in most if not all cases only fall under the "standard" rules of the Hospital Acts and the Doctors' Act.

[53] As has been mentioned above, there is one Federal Hospital Act which regulates the basic questions of Austrian Hospital law while the provincial acts go into detail and are directly applicable. In general, they demand that a patient only be treated according to the "principles and recognized methods of medical science". On the meaning of this standard, cf. Stöger (2008), pp. 641–648.

[54] It should be noted that new medical methods which are nevertheless tested with the intention to help the proband in question ("Heilversuch") are classified as therapeutical interventions irrespective of the fact that there might also be a strong scientific interest in their performance: Burgstaller and Schütz (2003), paras. 105–110.

[55] In this case, it is not possible to give one's valid consent to this type of „treatment". As a consequence, such interventions legally count as bodily harm – except in cases where there is neither a bodily injury nor any impairment of a person's health (Burgstaller and Fabrizy 2002, para. 5).

[56] See again Sect. 30 Universities Act and Sect. 8c of the Federal Hospital Act.

classified as an injury to the body or an impairment to a person's health which raises questions as to the categorization of certain non-invasive neuroscientific techniques without effects on a person's health as such – in the German doctrine, there has already been some discussion concerning this point[57]; and insofar the Austrian legal situation seems similar.[58] In our opinion, it is the second point – the ban on interventions which are contra bonos mores – which might prove to be most controversial in respect to neuro-enhancement technologies. One should only think of technologies which either lead to changes in the personality of a person or which give it a cognitive advantage – e.g. in exams – over others. How intensely must such a change or advantage be to be judged as contra bonos mores, making it a crime to perform it on a person or to demand that it be performed onto oneself (at least if the technique includes an injury to the body or an impairment of health)? There can be no general answer to this question at this place; however, this question can probably be regarded as the starting point to intense legal discussions once neuro-enhancement techniques become more effective and more easily available.

Assistive technologies and neuro-enhancement do also raise some other questions from a legal point of view. Some of these questions have been shortly addressed in a report on the use of assistive technologies which has been produced by the Austrian Bioethics Commission, an expert advisory board of the Federal government.[59] Though the report only focuses on assistive technologies (whether neuroscientific or not), some of the problems identified might as well, at least in the future, be of relevance for certain forms of technically advanced neuro-enhancement. The first problem is obviously that of responsibility for malfunction or wrong interpretation of the results of such equipment and its consequences.[60] Second, assistive technologies will regularly be used to help people who are not in a position to declare an informed consent. Though the Austrian legal order contains provisions in this respect, it will have to be discussed whether the potentially grave consequences of the use of such technologies require specific provisions on consent by third persons or whether the legislation in force can sufficiently deal with this problem.[61] Third, implants necessary for assistive-technologies and neuro-enhancement will regularly be medical devices under the (Austrian) Medical Devices Act.[62] However, the law of medical devices is to a large extent dominated

---

[57] Cf. Rosenau (2009), pp. 80–81 who refers to a presentation by R. Merkel at the Meeting of German Criminal Law Teachers 2009.

[58] Cf. Burgstaller and Fabrizy (2002), paras. 5–11.

[59] Report by the Bioethics Commission on "Assistive Technologies: Ethical Aspects of the Development and Use of Assistive Technologies", dated 13 July 2009. For details, see footnote 50 above.

[60] Report by the Bioethics Commission, p. 33.

[61] Report by the Bioethics Commission, p. 35.

[62] Report by the Bioethics Commission, p. 38.

by EU legislation.[63] As a consequence, it will sooner or later become necessary to (comprehensively) address the use of such technologies at the level of European legislation. In this respect, it should be borne in mind that clinical trials of new medical devices regularly require the consent of an ethics Committee irrespective of where this research takes places (in other words, not limited to hospitals or universities – see Sect. 58 of the Medical Devices Act). Fourth, if technologies of this kind lead to the collection of patients' data, data protection law comes into play. [64] Taking these considerations into account, it is not surprising that the Bioethics Committee has recommended in its report that a study regarding the necessary legal arrangements be conducted.[65]

Such a study will have to take into account that the use of neuroscientific assistive technologies can also have an impact on a person's legal capacity to act. The absence of this capacity is – broadly speaking – a prerequisite for, amongst others, the appointment of a procurator,[66] the activation of a health care proxy[67] or the right of the next of kin to act for a person in everyday business.[68] The capacity to make a will is also closely connected to the "general" capacity to act.[69] Neuroscientific assistive technologies – e.g. computers transforming thoughts into speech – could in the future enable persons who currently have to be regarded as incapable to act to regain at least some minimal (legal) capacity to act.[70] However, this does in our opinion not mean that the notion of capacity to act as such will have to be changed. What will change, though, are probably the standards for judging whether this capacity is still given or not, and this will probably be a task for the courts. The legislator could facilitate this task by explicitly defining situations in which persons using assistive technologies must be regarded as possessing the capacity to act. Of course, the choice between leaving the solution to the courts or to the legislature instead will be a political one as there are certainly valid legal arguments for both solutions.

Finally, a specific problem of "neuro-enhancement" has to be addressed: It has already been mentioned that such techniques could be regarded contra bonos mores if they give a person an unfair advantage as compared to his or her peers. The question remains how to deal with the outcomes of such techniques if they are nevertheless performed, like the use of effective pharmaceuticals or implants in exams. In our opinion, the current legislation does already offer some solutions:

---

[63] See above all Council Directive 93/42/EEC of 14 June 1993 concerning medical devices, OJ 1993 L 169/1 as amended.

[64] Report by the Bioethics Commission, pp. 35–36.

[65] Report by the Bioethics Commission, p. 37.

[66] Section 268 of the Civil Code.

[67] Sections 284f–284h of the Civil Code.

[68] Sections 284b–284e of the Civil Code.

[69] Sections 566–568 of the Civil Code.

[70] In Germany, this aspect of neuroscientific assistive technologies has already been discussed by Spranger (2009c).

For example, Sect. 74 para. 2 of the Universities Act provides for an exam to be declared null and void if the candidate has obtained the result by fraud, e.g. by the use of illegitimate aid.[71] This doubtlessly covers the use of secret notes as well as the use of (effective) neuro-enhancing technologies.[72]

To sum up, it seems quite probable to us that the use of assistive technologies and neuro-enhancement does or at least will raise some interesting legal questions; however, the current legal framework probably already contains solutions for the majority of these questions so that the further development of these techniques will not lead to a "legal vacuum".

4   Determinism and Criminal Law

One of the major legal discussions in connection with new neuroscientific findings focuses on the question whether the concept of a "free will" is still valid.[73] The Austrian Criminal Code, as most comparable codifications in other states, states that only who acts culpably must be punished (Sect. 4 Criminal Code). As a consequence, the concept of guilt is of eminent importance in Austrian criminal law. However, this concept requires that the perpetrator can be accused of not having acted otherwise than he or she did. A strongly determinist approach, however, will doubt that a person had this chance as his or her actions or omissions were predetermined by (neuronal) factors beyond one's control.

The crucial question is therefore which concept of guilt has been laid down in Austrian criminal law. Guilt can either be defined in a way that a (mentally) "healthy" (in our present understanding) person is blamed for having acted in a different way than expected from the "average" person ("Charakterologischer Schuldbegriff").[74] Whether he or she can be blamed for being different than average, is not of relevance for this concept of guilt. Another definition of guilt focuses on blaming a person for having chosen to do wrong even though he or she could have chosen not to do so ("Sozial-ethischer Vorwurf").[75] It has been rightly stated that elements of both approaches can be found in the Austrian Criminal Code

---

[71] For a commentary on this provision, cf. Perthold-Stoitzner (2010).

[72] For a German perspective on this problem, cf. Rosenau (2009), p. 76.

[73] The German literature on the subject is – in contrast to the Austrian – quite abundant. For two contributions which give a very instructive overview, cf. Streng (2007) and Paeffgen (2010) (see also the literature list in this contribution after para. 230j which clearly shows that the discussion on the importance of the concept of a free will in criminal law dates further back than "modern" neuroscientific findings on the question). For a Swiss perspective, cf. the contributions in Senn and Puskas (2006).

[74] Tipold (2005), paras. 10–11.

[75] Tipold (2005), para. 13.

as the legislator seems to have tried to avoid a clear decision on this point.[76] In other words, it is possible to read the law both ways. And that is what the doctrine does, even though the majority has opted for the first reading which focuses on the deviation from the "average" person.[77] This concept would not be put into question even neuroscience could prove that our reactions in certain situations really are beyond our control. In other words: The concept of guilt that can (even though not exclusively) be found in the Austrian Criminal Code does not require the idea of a free will to remain (fully) valid.

This result is not changed by constitutional considerations: On the one hand, it is not entirely clear whether the Austrian constitution would hinder the legislator from creating a criminal law that is not based on the concept of guilt at all.[78] On the other hand, the idea of defining guilt as deviation from average is probably even enshrined in the Constitution as Art. 142 of the Federal Constitution which provides for a number of supreme state organs to be tried before the Constitutional Court for violation of the Constitution or other statutes.[79]

Taking all this into account, it seems very improbable that new neuroscientific findings concerning the non existence of a free will could "undermine the concept of guilt"[80] in Austrian criminal law.

5   "Mind-Reading Technology" in Criminal and Civil Procedures

The advancement of neuroscientific examination techniques, especially imaging techniques, has led to a growing interest in their possible use in criminal or civil procedures.[81] There is a certain expectation that new technologies could in the future be used as a viable alternative to the polygraph the use of which is either illegal in some countries – as this is currently the case in Austria (see below) – or met with some suspicion as to its reliability in others.

---

[76] Tipold (2005), paras. 15–16.

[77] Tipold (2005), para. 10.

[78] For further references on this discussion, see Tipold (2005), paras. 44–48.

[79] On this discussion, Lewisch (1993), pp. 267–268.

[80] Spranger (2009b), pp. 42–43 and Spranger (2009d), p. 1035 comes to a similar conclusion for German criminal law. See also Paeffgen (2010), paras. 230i–230j; Streng (2007), pp. 690–691; further Günther (2009) with a similar result on pp. 236–239. Hochhuth (2005), p. 753 comes to the general conclusion (not limited to criminal law) that the legal concept of a free will can be upheld.

[81] This interest has certainly been intensified by the appeal judgment of the Indian Supreme Court No. 1267/2004 (5 May 2010) in the case Selvi and Ors. vs State of Karnataka in which the Court declared that the compulsory administration of "certain scientific techniques, namely narcoanalysis, polygraph examination and the Brain Electrical Activation profile (BEAP) test" violates the "right against self-incrimination". Lower courts had before accepted some of these techniques as admissible.

The use of neuroscientific techniques in the context of establishing the facts in a criminal or civil case is partially a constitutional question as it touches on several fundamental rights.

In respect to the forced examination of the accused in criminal procedures, the Austrian Constitutional Court has deduced from Art. 90 of the Federal Constitution (which states that the opening of any criminal procedure requires an indictment) that no person must be forced to incriminate him- or herself in criminal procedures "by making the own body available as evidence".[82] Some Austrian commentators have interpreted this broadly by concluding that no person has to accept medical examination during which samples of bodily fluids or tissues are taken against his or her will. Others have argued that there is a difference between the duty of the accused to actively contribute to the obtainment of evidence – which falls under the right not to incriminate oneself – and the mere obligation to tolerate the obtainment of such material by medical examination.[83] These authors regard the latter obligation as not falling under the right not to incriminate oneself, but under Arts. 8 and 3 of the ECHR (right to private life; prohibition of inhuman and degrading treatment) which means that the admissibility of such examinations is subjected to a proportionality test.[84]

The latter opinion is closer to the jurisprudence of the ECtHR which is of specific importance to Austria as the ECHR has – as already mentioned above – been incorporated into the constitutional order as one of the two main catalogues of fundamental rights. In the *Saunders* case, the ECtHR has – by reference to the right of fair trial under Art. 6 of the Convention – stated that

> The right not to incriminate oneself is primarily concerned, however, with respecting the will of an accused person to remain silent. As commonly understood in the legal systems of the Contracting Parties to the Convention and elsewhere, it does not extend to the use in criminal proceedings of material which may be obtained from the accused through the use of compulsory powers but which has an existence independent of the will of the suspect such as, inter alia, documents acquired pursuant to a warrant, breath, blood and urine samples and bodily tissue for the purpose of DNA testing.[85]

Taking this ECtHR judgment[86] and the above-mentioned Austrian doctrine into account, there seems to be a clear answer – already on the constitutional level – as to the legal admissibility of forced mind-reading techniques in Austrian criminal procedure law: The forced administration of such techniques on the accused would touch the core of the right not to incriminate oneself as described by the ECtHR as the accused could not remain "silent" but would give away some of his

---

[82] See VfSlg (Decisions by the Constitutional Court) 10.976/1986.

[83] For a comprehensive overview with further references, see Birklbauer (2010), paras. 8–10; further Müller (2001).

[84] See Birklbauer (2010), para. 15.

[85] ECtHR 17.12.1996, 19187/91 Saunders v. United Kingdom, para. 69

[86] For a comprehensive overview over the ECtHR case-law on the right not to incriminate oneself under Art. 6 ECHR cf. Harris et al. (2009), pp. 260–264.

and her thoughts.[87] The gravity of such an examination would consequently go beyond that of the forced obtainment of tissues and bodily fluids and cannot be justified by a proportionality test as it would effectively eliminate a person's right to remain silent and force it to "actively" contribute to the obtainment of evidence. Accordingly, Art. 90 of the Federal Constitution as interpreted by the Constitutional Court and Art. 6 ECHR as understood by the ECtHR do not permit the forced use of such techniques on the accused.

This result on the constitutional level is echoed by the Criminal Procedure Act:[88] According to its Sect. 164 para. 4 (in conjunction with Sect. 245 para. 2), the "freedom of the accused person to make up his mind and to manifest his will [ ] must not be impaired by any measures taken, least of all by intrusions into his physical integrity". Legal practice and doctrine therefore qualify the forced use of polygraphs, narcoanalysis, hypnosis or veritaserum as illegal under these provisions.[89] In our opinion, this applies to the use of neuroscientific "mind-reading" techniques as well.

However, there is no complete agreement as to the voluntary use of such techniques. A small minority of authors regard them as admissible if the accused demands their application while legal practice and the majority of authors reject this claim.[90] There are two lines of argument for this rejection: The first – and stricter – one states that even voluntary submission to such techniques deprives the accused of the possibility to freely "make up his mind and to manifest his will" at any time.[91] This is certainly true for any technique which does not allow for the accused to stop an interrogation at any time without already having given away some information. Neuroscientific imaging techniques can be considered as problematic in this respect as the accused possibly cannot hinder a measurable brain reaction when confronted with a certain question – even if he or she demands that the interrogation be stopped. Another line of argument rejects the voluntary use of such techniques, especially the polygraph, because it is simply not regarded as reliable enough and consequently cannot contribute to the establishment of the facts of a case.[92] While this argument would become invalid once neuroscientific "mind-reading" techniques become sufficiently precise and interpretable, the argument that such techniques could impair the freedom to make up one's will would remain valid. It is therefore probable that they will be met with skepticism by legal practice and doctrine even if they reach sufficient reliability to deliver precise data on the thoughts of an accused person.

---

[87] Berka (1999), para. 379 states that the use of "veritaserum" (which has an effect comparable to mind-reading by exposing one's thoughts to a third person) would constitute a violation of Art. 3 ECHR.

[88] Strafprozeßordnung, Federal Law Gazette No. 631/1975 as amended.

[89] See Achammer (2009), para. 23; Kirchbacher (2009b), para. 72 with further references.

[90] For further references, cf. Kirchbacher (2009b), para. 72; Achammer (2009), para. 23.

[91] Kirchbacher (2009a), para. 38.

[92] Cf. the arguments of Pilnacek (2002), pp. 480–481.

As to the administration of "mind-reading" examinations on witnesses in criminal procedures, it should be borne in mind that as long as they do not deliver reliable and interpretable results, any use of these results against the accused is in our opinion incompatible with his or her right to a fair criminal trial under Art. 6 ECHR. From the perspective of the witness, the forced use of such techniques touches on his or her right to private life under Art. 8 ECHR[93] (and might also be classified as degrading treatment under Art. 3 ECHR) as one's thoughts are forcibly exposed to the court. Furthermore, a witness has the right to remain silent if the giving of evidence might expose him- or herself to a risk of criminal prosecution[94] – a right which is derived from the above-mentioned fundamental right of freedom from self-incrimination. Neuroscientific "mind-reading" might – depending on the quality of its results – also prove to be problematic in this respect. This could even cover the voluntary submission to such techniques.

When discussing the use of neuroscientific "mind-reading" techniques in civil procedure law, two of the above-mentioned fundamental rights are again of specific relevance: Arts. 6 and 8 of the ECHR. Unless such techniques become reliable and interpretable, their acceptance by a court in a civil procedure is not compatible with the right to a fair trial. This can also be deduced from Sect. 272 of the Civil Procedure Act[95] which states that the judge has to decide the case according to his and her own and diligent consideration of evidence, which in our opinion precludes a court from using evidence that has been gained by a technique which does not deliver reliable results.[96] Accordingly, this even applies if the parties agree on the use of this technique. However, the latter case will not lead to practical problems: the parties to a civil procedure are free to undergo such examinations in private and to agree, as a result, on certain facts. If, as a consequence, they declare before the court that they have agreed on these facts (Sect. 266 Civil Procedure Act), the court is – generally speaking – obliged to base its judgment of these agreed facts. Should neuroscientific techniques become sufficiently reliable and interpretable one day, their use in civil procedures will have to be judged in the light of Art. 8 ECHR as such techniques – as mentioned above – constitute an intrusion into the very private sphere of a person. In this context, it should be mentioned that the Austrian legislation in force in general does not approve of the use of force for medical examinations in civil procedure

[93] A witness must under Art. 8 ECHR not be forced to testify: See already Birklbauer (2010), para. 13.

[94] Section 157 para. 1 Criminal Procedure Act.

[95] Zivilprozessordnung, Imperial Law Gazette No. 113/1895 as amended.

[96] See also the decision by the German Federal Court of Justice, VI ZR 327/02 (23 June 2003) in which it regarded the result of a polygraph test as an unsuitable piece of evidence in a civil procedure.

law,[97] the sole exception being procedures concerning ancestry where moderate force is permissible.[98]

6 Concluding Remarks

As we stated at the beginning of this contribution, it is not really a "country report" in the sense that it reports on ongoing legal research or existing case-law. However, we hope to have made clear that there are a number of open legal questions in respect to neuroscientific research in Austria. Technological advances will probably make it necessary to give more comprising and detailed answers to these questions than we could give in the context of this text. In this respect, we hope that this report might on the one hand be useful for researchers wishing to get a first insight into the legal problems of neuroscience in Austria, but that on the other hand it also increases the interest of Austrian jurists in these problems.

References

Achammer C (2009) § 7 StPO. In: Fuchs H, Ratz E (eds) Wiener Kommentar zur StPO (loose-leaf and online). Manz, Wien

Berka W (1999) Die Grundrechte. Springer, Wien-New York

Bernat E (1995) Recht und Humangenetik – ein österreichischer Diskussionsbeitrag. In: Deutsch E et al (eds) Festschrift für Erich Steffen. De Gruyter, Berlin, pp 33–56

Bernat E (2001) Die Forschung an Einwilligungsunfähigen. Recht der Medizin:99–105

Bernat E (2002) Schutz vor genetischer Diskriminierung und Schutzlosigkeit wegen genetischer Defekte: die Genanalyse am Menschen und das österreichische Recht. In: Byrd et al (eds) Jahrbuch für Recht und Ethik X. Duncker & Humblot, Berlin, pp 183–216

Birklbauer A (2010) Vorbemerkung zu §§ 118, 123, 124 StPO (neu). In: Fuchs H, Ratz E (eds) Wiener Kommentar zur StPO (loose-leaf and online). Manz, Wien

Burgstaller M, Fabrizy E (2002) § 83, StGB. In: Höpfel F, Ratz E (eds) Kommentar zum StGB (loose-leaf and online), 2nd edn. Manz, Wien

Burgstaller M, Schütz H (2003) § 90 StGB. In: Höpfel F, Ratz E (eds) Kommentar zum StGB (loose-leaf and online), 2nd edn. Manz, Wien

Damm R (2009) Wie wirkt Nichtrecht? Genesis und Geltung privater Regeln am Beispiel medizinischer Professionsnormen. Z Rechtssoziologie 30:3–22

Druml C (2010) Ethikkommissionen und medizinische Forschung. Ein Leitfaden für alle an medizinischer Forschung Interessierte. Facultas WUV, Wien

Eberhard, H (2010) Forschungskontrolle durch Ethikkommissionen – von der kollegialen Beratung zur staatlichen Behörde? In: Körtner et al. (eds) Ethik und Recht in der Humanforschung. Springer, Wien-New York, pp 152–176

---

[97] Kopetzki (2007), p. 43.

[98] Section 85 para. 3 of the Act on non-contentious legal proceedings (Außerstreitgesetz), Federal Law Gazette I No 111/2003 as amended.

Ganner M (2010) Die Freiheit is a Vogerl. Psychische Krankheit - eine Herausforderung für die Gesellschaft im Allgemeinen und für Angehörige, Richter, Ärzte und Pflegepersonal im Besonderen. Österreichische Richterzeitung:2–5

Gehring P (2007) Ein Organ wie jedes andere? Zur Rechtspolitik der Hirnbildverwendung und der Hirnmanipulation. Juridikum:200–204

Günther K (2009) Die naturalistische Herausforderung des Schuldstrafrechts. In: Schleim et al (eds) Von der Neuroethik zum Neurorecht? Vandenhoeck&Ruprecht, Göttingen, pp 214–242

Harris D et al (2009) Law of the European convention on human rights, 2nd edn. Oxford University Press, Oxford

Heuser I (2009) Psychopharmaka zur Leistungsverbesserung. In: Ethikrat D (ed) Der steuerbare Mensch. Über Einblicke und Eingriffe in unser Gehirn. Deutscher Ethikrat, Berlin, pp 49–55

Hochhuth M (2005) Die Bedeutung der neuen Willensfreiheitdebatte für das Recht. (Deutsche) JuristenZeitung 15/16:745–753

Jerabek R (2010) § 95 StGB. In: Höpfel F, Ratz E (eds) Kommentar zum StGB (loose-leaf and online), 2nd edn. Manz, Wien

Jescheck H (1998) Wandlungen des strafrechtlichen Schuldbegriffs in Deutschland und Österreich. Juristische Blätter:609–619

Kampits (2008) Menschenwürde und Menschenrechte. 100 Jahre Richtervereinigung. Österreichische Richterzeitung:61–76 (with pictures on pp 67–74).

Kienapfel D (1979) Die Hilfeleistungspflicht des Arztes nach deutschem und österreichischem Strafrecht. In: Kaufmann A et al (eds) Festschrift für Paul Bockelmann zum 70. Geburtstag. CH Beck, München, pp 591–601

Kirchbacher K (2009a) § 164 StPO. In: Fuchs H, Ratz E (eds) Wiener Kommentar zur StPO (loose-leaf and online). Manz, Wien

Kirchbacher K (2009b) § 245 StPO. In: Fuchs H, Ratz E (eds) Wiener Kommentar zur StPO (loose-leaf and online). Manz, Wien

Kodek G (2010) § 1294 ABGB. In: Kletecka A, Schauer M (eds) ABGB-ON. Online-Kommentar zum ABGB. Manz, Wien

Kopetzki C (2007) Invasive Verfahren aus rechtlicher Sicht, Teil I. In: Gesellschaft der Gutachterärzte Österreichs (ed) Zumutbare Maßnahmen in Diagnostik und Therapie bei der Begutachtung. Forschung und Praxis der Begutachtung 68/2006. Gesellschaft der Gutachterärzte Österreichs, Wien, pp 33–58

Kopetzki C (2010a) Behandlungen auf dem "Stand der Wissenschaft". In: Pfeil (ed) Finanzielle Grenzen des Behandlungsanspruchs. Manz, Wien, pp 9–46

Kopetzki C (2010b) Braucht Österreich eine Kodifikation des biomedizinischen Forschungsrechts? In: Körtner U et al (eds) Ethik und Recht in der Humanforschung. Springer, Wien-New York, pp 56–89

Kopetzki C (2011) Epidemiologische Studien in der Intensivmedizin - Neuland und Graubereich für Ethikkommissionen? Juristische Stellungnahme. Recht der Medizin:19–22

Körtner U et al (eds) (2010) Ethik und Recht in der Humanforschung. Springer, Wien-New York

Koziol H (1997) Österreichisches Haftpflichtrecht, vol I. Manz, Wien

Lewisch P (1993) Verfassung und Strafrecht: verfassungsrechtliche Schranken der Strafgesetzgebung. WUV-Verlag, Wien

Löschnigg G, Schwarz W (2003) Arbeitsrecht, 10th edn. ÖGB, Wien, New edition (2011) in print

Marhold F, Friedrich M (2006) Österreichisches Arbeitsrecht. Springer, Wien-New York

Müller R (2001) Neue Ermittlungsmethoden und das Verbot des Zwangs zur Selbstbezichtigung. Europäische Grundrechte-Zeitung 2001:546–559

Paeffgen U (2010) Vorbemerkungen zu §§ 32ff StGB. In: Kindhäuser U et al (eds) Strafgesetzbuch, vol 1, 3rd edn. Nomos, Baden-Baden

Perthold-Stoitzner B (2010) § 74 Universitätsgesetz. In: Mayer H (ed) Universitätsgesetz 2002, 2nd edn. Manz, Wien

Pilnacek C (2002) Die Zulässigkeit des freiwilligen Lügendetektor- und Polygraphen-Tests im Strafverfahren - eine Erwiderung. Österreichisches Anwaltsblatt:479–481

Pöschl M (2008) Gleichheit vor dem Gesetz. Springer, Wien-New York

Pöschl M (2010) Von der Forschungsethik zum Forschungsrecht: Wie viel Regulierung verträgt die Forschungsfreiheit? In: Körtner U et al (eds) Ethik und Recht in der Humanforschung. Springer, Wien-New York, pp 90–135

Pronay EC (2008) Richterliche Ethik zum Ausprobieren – 100 Jahre Richtervereinigung. Österreichische Richterzeitung:82–84

Reischauer R (2007) § 1294 ABGB. In: Rummel (ed.) Kommentar zum ABGB, vol IIa, 3 rd edn. Manz, Wien

Rosenau H (2009) Steuerung des zentralen Steuerungsorgans - Rechtsfragen bei Eingriffen in das Gehirn. In: Ethikrat D (ed) Der steuerbare Mensch. Über Einblicke und Eingriffe in unser Gehirn, Deutscher Ethikrat, Berlin, pp 69–82

Schleim S et al (2007) Zufallsfunde in der bildgebenden Hirnforschung. Nervenheilkunde 26(11):1041–1045

Schleim S et al (ed) (2009) Von der Neuroethik zum Neurorecht? Vandenhoeck&Ruprecht, Göttingen, pp 7–21

Senn M, Puskas D (eds) (2006) Gehirnforschung und rechtliche Verantwortung. Fachtagung der Schweizerischen Vereinigung für Rechts- und Sozialphilosophie, 19. Und 20. März 2006, Universität Bern. Franz Steiner Verlag, Stuttgart

Spranger T (2009a) Rechtliche Implikationen der Generierung und Verwendung neurowissenschaftlicher Erkenntnisse. In: Schleim et al (eds) Von der Neuroethik zum Neurorecht? Vandenhoeck & Ruprecht, Göttingen, pp 193–213

Spranger T (2009b) Das gläserne Gehirn? Rechtliche Probleme bildgebender Verfahren. In: Ethikrat D (ed) Der steuerbare Mensch. Über Einblicke und Eingriffe in unser Gehirn. Deutscher Ethikrat, Berlin, pp 35–47

Spranger T (2009c) Neuroprothetik und bildgebende Hirnforschung. Neue Impulse für die Praxis des Betreuungsrechts. Betreuungsmanagement 4:206–208

Spranger T (2009d) Der Einsatz neurowissenschaftlicher Instrumente im Lichte der Grundrechtsordnung. Juristenzeitung 21:1033–1040

Stelzer M, Havranek B (2007) Gentechnikrecht. In: Holoubek M, Potacs M (eds) Handbuch des österreichischen Wirtschaftsrecht, vol II. Springer, Wien-New York, pp 631–689

Stöger K (2008) Ausgewählte öffentlich-rechtliche Fragestellungen des österreichischen Krankenanstaltenrechts. Manz, Wien

Streng F (2007) Schuldbegriff und Hirnforschung. In: Pawlik M et al (eds) Festschrift für Günther Jakobs. Carl Heymanns Verlag, Köln

Tipold A (2005) § 4 StGB. In: Höpfel F, Ratz E (eds) Kommentar zum StGB (loose-leaf and online), 2nd edn. Manz, Wien

*Brain*zil Imaging: Challenges for the Largest Latin American Country

Henrique Moraes Prata and Márcia Araújo Sabino de Freitas

Abstract This article investigates the neurolaw in Brazil, the largest country in Latin America. It concludes that though studies in neuroscience have great exponents in the country, there is not, still, a Brazilian neurolaw – probably owing to the traditional distance the Brazilian law has in relation to sciences and the social reality. But, already counting with several mentions of aspects of neuroscience in court decisions, albeit low qualified, and in face of some recent studies on neurolaw that are being produced, there are signs of a promising future development of the area. This article also brings an overview of the country to foreigners: its organizational structure, the way some of its important institutions acts, recent data about scientific research, and the main rules that must be followed by those who wish to do research in the country.

Henrique Moraes Prata is PhD Student at University of São Paulo Law School, Brazil; Master in Comparative Law at University of Bonn, Germany (2005); Graduated in Law (2000) and Philosophy (2011) at University of São Paulo; Researcher at Getulio Vargas Foundation Law School of São Paulo and Member of the Court of Ethics and Deontology of the Bar Association of the State of São Paulo. Márcia Araújo Sabino de Freitas is Master Student at University of São Paulo Law School, Brazil; Graduated in Law at the Federal University of Minas Gerais (2009), Brazil; Researcher at Getulio Vargas Foundation Law School of São Paulo.

H.M. Prata (✉)
Alameda Joaquim Eugênio de Lima, 881, cj. 909, CEP 01403-001 São Paulo, Brazil
e-mail: hmp@hmp.adv.br; henrique.prata@usp.br

M.A.S. de Freitas (✉)
University of São Paulo Law School, São Paulo, Brazil
e-mail: asf.marcia@usp.br

T.M. Spranger (ed.), *International Neurolaw*,
DOI 10.1007/978-3-642-21541-4_4, © Springer-Verlag Berlin Heidelberg 2012

1   Introduction

The discussion about neurolaw in Brazil is incipient and neuroscience still running out of most juridical circles, though there are great researches carried out by Brazilian neuroscientists. There are almost no literature and specific studies about neurolaw in the country yet.

On the other side, it is remarkable that judges and lawyers are using in the courts "evidences" from magnetic resonance imaging (fMRI), electroencephalography (EEG) and magnetoencephalography (MEG), as well as theories and findings of neuroscience, But, in most cases at random and without any clinical criteria, link with the particular case,[1] or methodical and statistical analysis.

In the first part of this article, there is an overview of the Brazilian institutional setting, in which arises the conclusion that, bigger than the extension of the country are its complexities. In the second part, several judicial decisions mentioning neuroscience will be reported and the most important legal document for doing research in Brazil will be presented.

It is hard task try to capture a picture of neurolaw in Brazil, the reason why the word "challenge" was used already in the title of the article. Even harder are the institutional knots that keep the country stuck. Despite these obstacles, the overview of science in the country shows that Brazil has much to be proud of and that there is enormous potential for the development of national science.

2   Brazilian Institutional Setting

2.1   Political Structure and Regional Inequalities

The Federal Republic of Brazil has territory of 8.5 million km^2,[2] divided into 26 States and a Federal District. The States are grouped into five Regions: South (*Rio Grande do Sul, Santa Catarina,* and *Paraná* States), Southeast (*São Paulo, Rio de Janeiro, Espírito Santo,* and *Minas Gerais* States), Middle-West (*Mato Grosso, Mato Grosso do Sul,* and *Goiás* States, plus the Federal District), North (*Tocantins, Pará, Amapá,* Roraima, *Amazonas, Acre,* and *Rondônia* States) and Northeast (*Bahia, Sergipe, Alagoas, Pernambuco, Paraíba, Rio Grande do Norte, Ceará, Piauí,* and *Maranhão* States) (Fig. 1).

---

[1] The same happened with the use of DNA evidences in the early 1990s. Read the following text of the Superior Court of Justice called "STJ's decisions legitimate DNA tests as a tool for the search of justice" may be interesting to get a feeling of the use valuation of evidences (and presumptions) of new technologies by the courts: Superior Tribunal de Justiça (2010).

[2] Atlas National Geografic: Brasil (2008), p. 6.

Fig. 1 Brazil's political division map – states and federal district

There are serious inequalities between these regions. Historically, Brazil grew from the coast to the interior, so the majority of the population and the big cities are near the cost. While the North has more than 45% of the country total area and the Southeast about 10%, the Southeast account for about 55% of Brazilian PIB[3] and more than 42% of Brazilian population, in contrast to the North rates: about 5% of PIB and less than 8% of Brazilian inhabitants. Usually, the South and the Southeast have better performance in social and economic development indexes. The Northeast has the lowest Human Development Index (HDI) rates of the country.[4]

Most of the Amazon rain forest is on the North Region, plus a part in *Mato Grosso* State and a part in *Maranhão* State.

Regarding the form of government, Brazil is a presidential republic with a bicameral parliament (Chamber of Deputies and Senate).

The capital of Brazil is *Brasília*, which is located in the Federal District.

---

[3] The Brazilian PIB (*Produto Interno Bruto*) is a national measurement similar to the GDP (Gross Domestic Product).

[4] Atlas National Geografic: Brasil. (2008), p. 37.

2.2   The Legislature

Brazil has civil law tradition with a very complex legal system: there are great many
laws, a lot of kinds of norms, many sources of regulation, and some doubts about
which rule should be used in a specific case.

Among these norms, remarks are: the Federal Constitution of 1988, known as the
"citizen Constitution," owing to the many rights it has fixed; the Civil Code of 2002
that substituted the old Code of 1916; the Child and Adolescent Statute of 1990; the
Consumer's Code of 1990 and the Penal Code of 1940. Brazil is now facing reforms
on its Civil Process Code and on its Penal Process Code.

Although other institutional bodies can make some regulation (usually about
specific topics), and the Executive and the Judiciary also do some kind of regulation, in general, the Legislature is the competent to make laws. Actually, the name
"law" (*lei*) is technically used to designate the rules that are made by the Legislature
through the legislative process.

There are three levels of the Legislature: municipal, state, and federal. In the
municipal and the state levels, the parliament is made of only one house, but in the
federal level, there is a bicameral parliament divided into Chamber of Deputies and
Senate.

The deputies in the Chamber represent the population (in theory, that means that,
the more population one State has, the more deputies it will have in the Chamber)
and the senators represent the States (a State will always have three senators in the
Senate, no matter how many inhabitants it has). This bicameral system was designed
to correct regional inequalities, but a political possibility to do a bad use of it was
created, and, owing to a rule that states a minimal (8) and a maximum (70) number
of deputies in the Chamber for each State, a different practical inequality appears. In
the scramble for power among local elites, a bad use of the fixed number of senators
was already made in Brazil's history: the creation or the merger of States in the aim
to get more chairs in the Senate or to take out chairs of the opposition. The practical
inequality is that the less populated States, together, can be stronger in the parliament than the most populated ones, what means that a President usually gets elected
by the most populated States, but has to govern for the less populated ones.

Analyzing how much the population trusts in Brazilian institutions, the last
report of the Brazilian Index of Confidence in Justice (*ICJBrasil*), founded out
that only 8% of the respondents trusts in the political parties and just 20% thinks the
Parliament (*Congresso Nacional*) is trustworthy.[5] Probably, this is a reflection of
many corruption acts and abuses that unfortunately have been committed by some
congressmen in the conduction of public affairs.

Generally speaking, for a bill to become law, it has to be formally proposed,
approved by the Chamber of Deputies, by the Senate and then pass through an
Executive deliberation, which means that the President has veto power.

---

[5] Relatório ICJBrasil (2010) 3° trimestre 2010.

However, a lot of laws in Brazil are made by the Executive, through the fast track of "provisional acts" (*medidas provisórias*) – Executive emanations that can come into force immediately and demand a quick response of the Legislature to become laws. Theoretically, according to article 62 of the Constitution,[6] the provisional acts should only be used in cases of relevance and urgency and have some material limits. But, in practice, they give great power to the Executive, due to some abuse of the legal possibilities to use them.

The legislation of Brazil can be found at the official website of the Presidency, at the following link: http://www4.planalto.gov.br/legislacao.

2.3 The Judiciary

Brazil has a large and complex judicial structure, which includes: a Federal Supreme Court that acts as a Constitutional Court; four Superior Courts that are the last word on the issues of their competence; three specialized Regional Courts and specialized Judges, to decide about specific subjects; five Regional Federal Courts acting in "federal issues"; one Justice Court per State and one to the Federal District – which totals 27 Justice Courts; a lot of Law and Federal Judges and a system of Special Conciliation Courts for the small claims (Fig. 2).

The Federal Supreme Court (*Supremo Tribunal Federal – STF*) is the final appellate court, which, in principle, decides only constitutional issues. It is composed of 11 ministers that do not need to have career in the Judiciary and are appointed by the President after being approved by the Senate.

The Superior Courts, Regional Courts, and its Judges that are specialized in electoral, labor and military issues belong to the Specialized Justice, while the Superior Justice Court, the States Courts of Justice, the Regional Federal Courts and its Judges belong to the Ordinary Justice. When a conflict is not about military, electoral or labor issues, the residual competence, what means the majority of the processes, is of the Ordinary Justice.

While the Superior Electoral, Labor and Military Courts determine the last decision on those specific subjects, in the same hierarchy, the Superior Justice Court gives the final interpretation when the question is about federal law. It comprises 33 ministers, which need to be approved by the Senate and appointed by the President. Some of them have to come from the Judiciary, some from the Brazilian Bar Association and some from the "Public Ministry" (*Ministério Público*).[7]

---

[6] Constitution of the Federative Republic of Brazil of 1988.

[7] The Public Ministry, which houses the prosecutors, is, according to the article 127 of the Constitution, an institution that shall defend the legal order, the democratic regime and the inalienable social and individual interests.

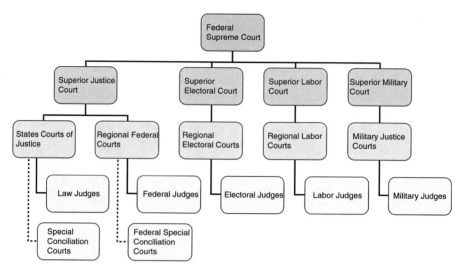

Fig. 2 Brazil's judiciary structure

General speaking, the States and Regional Courts act as appellate courts of judges' decisions, both in Specialized Justice and in Ordinary Justice.

The Federal Judges and the Regional Federal Courts are, overall, the competent to decide conflicts in which the Federal Government, its independent agencies and federal public companies are parts. The residual competence is of the Judges of Law and the States Courts of Justice.

Although the Special Conciliations Courts (*Juizados Especiais*) are attached to the Regional Federal Courts and to the States Courts of Justice, the second ones are not appellate courts to the first ones; the Special Conciliation Courts have their own appellation system. The small claims of their competence can be about civil, consumer, social security, and criminal matters, among others. While in the other bodies of justice is necessary to use a lawyer to join with action, in most of Special Conciliation Courts' cases, the use of a lawyer is an option of the litigant part.

Brazil faces a serious problem of slowness of justice, excessive litigation and, besides that, most population does not trust in the Judiciary. In the year of 2009, about 25 million new processes got in the Judiciary.[8] The report of the Brazilian Index of Confidence in Justice (*ICJBrasil*) says the main reason of using the Judiciary is related: first, to labor issues; second, to consumer's questions; and third, to family subjects.[9] According to data from the same report, for 89% of respondents, the Judiciary solves conflicts in slowly or very slowly way and for 69%, it is difficult or very difficult to use. Besides that, 63% of them think the Judiciary is not honest or has few honesty; 61% think it is not independent or has

---

[8] Conselho Nacional de Justiça (2010).

[9] Relatório ICJBrasil (2010). 3° trimestre 2010.

few independence (that means they think it is not impartial in the judgments) and 55% says the Judiciary is not competent to solve conflicts or has few competence. Only 33% of the respondents said the Judiciary is trustworthy or is very trustworthy.[10]

In this context, it is common that the Judiciary decides the processes in "block-decisions": the same judgment model applied to many cases. If properly used, this mechanism could have good effects, but, in many situations, judges and Courts use it without caution and just in order to get rid of the numerous processes that fill the Judiciary. So, as many of these cases do not get to real solution with that kind of block-decision, the conflict does not finish when the process ends in the Judiciary, turning the decision without practical effect, and making appear some big injustices.

The Constitutional Amendment nº 45, in the year of 2004, trying to face some of the Judiciary problems, stated as constitutional right the speedy trial; created the National Council of Justice (*Conselho Nacional de Justiça – CNJ*), with the aim to control the administrative and financial activities of the Judiciary and the fulfillment of the functional duties of judges; changed the Judiciary structure, reducing the number of Courts; changed the competence of Courts to relieve the overloaded ones; and created the possibility of giving binding effect to precedents of the Federal Supreme Court.

It is important to point that, although Brazil, traditionally, has a civil law system, over time, several of the many of changes that have been made in the law, created instruments that are typical of the common law system, such as the precedents with binding effect. In the currently Civil Process Code reform, a lot of proposals have to do with the common law system, in an attempt to give speed to the Judiciary. So, Brazil has, increasingly, a mixed system between the civil and the common law traditions.

Despite some reforms in the law being welcome, in Brazil is common to try to reply to complex problems just making changes in the law, what does not solve the majority of them. It seems the solution to the problems of the Judiciary undergoes more through a change of mentality and improvement of education – both to the population and to the Judges – than a change in the law.

3 Neuroscience Studies in Brazil

3.1 Overview of Science in Brazil

The concern about research in Brazil is growing, especially since the opening of the country's economy to importations in the 1990s, which forced companies, due to

---

[10] Relatório ICJBrasil (2010) 3° trimestre 2010.

competition, to invest in technological development. Currently, owing to the development of the economy, the trend is that these investments grow further. The fact that in 10 years (between 1997 and 2007), the number of Brazilian papers in indexed journals more than doubled to 19,000 per year confirms this hypothesis. Brazil now is the 13th country in the world in publishing, overtaking the Netherlands, Israel, and Switzerland.[11] Also, Brazil is in evidence among the neighboring countries on this matter: it represents more than 60% of investment in research in Latin America, and Brazilians write half of the papers from all over Latin America.[12]

Other good news come from the last evaluation of postgraduate programs in the country, made by the Coordination of Improvement of Higher Education Personnel (*Coordenação de Aperfeiçoamento de Pessoal de Nível Superior – CAPES*) in the year of 2010. It showed not only a quantitative increase in master's and doctorate's courses in the country, but also a significant qualitative increase: over the past 3 years, there was an increase of 20% in the number of new courses, and the number of courses with the maximum score increased 26% in the same period.[13]

However, the regional inequalities, already mentioned, also echo here: most science continues to occur in some southern states, and the University of *São Paulo*, alone, produces about a quarter of all scientific publications of the country.[14] Another evidence of that is about Amazon rain forest studies: besides its importance to the country and its huge size, it is still little studied by Brazilians. The majority of publications about the Amazon do not have a Brazilian as author, it has only one resident Ph.D. archaeologist and, notwithstanding its many rivers, it has no naval engineers.[15]

Another problem that scientists and researchers of Brazil still face is about obsolete regulation and practices that difficulties the importation of materials and investments in research. Absurd cases are reported by scientists: reagents that can take so many months to arrive that get to the lab already expired, and a lab instrument getting trapped in customs for months, delaying the research time.[16]

Nevertheless, the increasing concerns about research in Brazil make believe that these serious problems tend to, over time, improve for a better scenario. Besides that, as the neuroscientist *Sidarta Ribeiro* said in an interview, even in today's developed countries, science did not begin in the most favorable environment.[17]

---

[11] Regalado (2010).
[12] Regalado (2010).
[13] Antônio (2010).
[14] Regalado (2010).
[15] Regalado (2010).
[16] Regalado (2010).
[17] Regalado (2010).

At last, it is important to mention that most of research in Brazil is focused in agriculture, ecology, and infectious diseases.[18]

3.2 The Edmond and Lily Safra International Institute of Neuroscience of Natal

A group of Brazilian neuroscientists – *Miguel Nicolelis*, *Sidarta Ribeiro* and *Claudio Mello* – that were doing research at the Duke University, in the United States, idealized the Edmond and Lily Safra International Institute of Neuroscience of Natal (ELS-IINN), which aims to be a global reference for biomedical research and scientific education. A modest city in Brazil Northeast, in the State of *Rio Grande do Norte*, far from the traditional scientific production centers, was chosen to house the Institute with the purpose to make science a social and educational transformation instrument, reducing Brazilian's regional inequalities. The ELS-IINN received the largest donation in the history of Brazilian science of Ms. *Lily Safra*, which explains the homage to her and Mr. *Edmond Safra* in the name of the institute.[19]

Headed by *Miguel Nicolelis*, the Alberto Santos Dumont Association for Research Development (AASDAP) administrates the ELS-IINN. It is financial supported by the Brazilian Ministries of Science and Technology, through the Financier of Studies and Projects (*Financiadora de Estudos e Projetos – FINEP*) and the Coordination of Improvement of Higher Education Personnel (*Coordenação de Aperfeiçoamento de Pessoal de Nível Superior – CAPES*), and receive donations from the Duke University and the *Sirio-Libanês* Hospital. Moreover, due to partnership with this hospital, the ELS-IINN has a neuroscience lab inside of one of its institutes, in the city of *São Paulo*.[20]

Some research lines worked at the ELS-IINN are: brain–machine interface, neuronal basis of sensorimotor learning and tactile perception, sleep and memory, animal communication, functional neuroimaging, adult neurogenesis, cell-based therapies, epilepsy, brain oscillations, neuropathology, education and cognition, neural plasticity, and Parkinson's disease.[21]

---

[18] Regalado (2010).

[19] The Edmond and Lily Safra International Institute of Neuroscience of Natal (2010).

[20] The Edmond and Lily Safra International Institute of Neuroscience of Natal (2010).

[21] The Edmond and Lily Safra International Institute of Neuroscience of Natal (2010).

3.3 Education in Neuroscience

Since the 1990s, Brazil has postgraduate programs in neurosciences,[22] most of them focused only in life and health sciences.

Although the importance of neurolaw studies in other countries is recognized, in Brazilian Law Schools there are no disciplines dealing with it, both in postgraduate and in graduate programs. Worse, usually it is a matter that is not even mentioned in the Law Schools' classrooms.

Despite the neurolaw studies' setting, there are signs that this may change. Some proposals of post-graduate programs in neurosciences enable the admission of students of other areas than life and health sciences, and there are already bachelor Law students in some of them. That shows the awareness of the importance of neurolaw studies in Brazil is increasing, albeit slowly.

It is important to mention that the Federal University of Rio Grande do Norte, which has a postgraduate program in neurosciences, is partner of the ELS-INN. Also relevant to refer that the Federal University of Minas Gerais will soon confer Master degree in Neurosciences to two bachelor Law students, both studying some aspect of the relationship between neuroscience and law.

3.4 Brazilian Neuroscientists and the Importance of Their Researches to Law

Brazil has great neuroscientists and it is an injustice not to be able to quote them all, but, owing to the matter of this article, will be mentioned the ones that studies subjects related to Law's questions, such as *Miguel Nicolelis, Ivan Izquierdo,* and *Jorge Moll.*

Miguel Nicolelis[23] made some scientific discovers that changed the course of neuroscience studies in the world. After studying the electrical activity of the brain of mice, he began to study the brain–machine interface, making some experiences in which a rat could be able to drink water moving, with its thoughts, a little robot arm. Then, *Nicolelis* made almost the same with a monkey, which moved a robot arm merely using the thought. He likes to say it was the first time a primate brain freed from its body.[24] The next step was to make a monkey play videogames moving a joystick only with its mind, for a whole hour. Finally, a monkey was capable to command with its mind the walk of a robot that was in Japan and without anything holding it. These researchers about brain–machine interactions, in maybe

---

[22] Links to all Brazilian post-graduate programs can be found at: O Cérebro Nosso de Cada Dia (2010).

[23] For further information, access Nicolelis' Lab (2010).

[24] The Edmond and Lily Safra International Institute of Neuroscience of Natal (2010).

a very close future, may allow treatments for Parkinson disease and help paralytic people to walk.[25]

These scientific discovers are of interest to Law: looking at the progresses of brain–machine interactions, it is possible to rethink the idea the Law has of the human person. If the person is characterized by having a human body by substrate[26] to do all civil relationships and criminal acts, once the substrate can be part or even entirely a robot, we have a big modification made by science in Law's assumptions. To realize the scope of this potential concept's changing, one must consider that the legal system is based on, made by and created for the human person, which made all acts relevant to Law through the body.

Ivan Izquierdo, an Argentine-born Brazilian, studies the mechanisms of memory formation. In the last years, he demonstrated that "both the formation and the evocation of memories of short and long duration are strongly modulated by means related to emotional life (...)"[27] and that the hyperactivity of the amygdala caused by stress can produce the familiar "blanks."[28]

Studies about the set and the formation of memories are very important in Law cases' solutions. In many Brazilian judgments, as the next chapter will discuss, the "blanks" or the modification in memories caused by stressing or very emotional situations are the key for the decision to not consider a witness testimony.

Jorge Moll[29] studies the processing of moral judgments and emotions in the brain. He made research about the neural basis of human altruism, neurogenetics and cognitive neuroscience of psychopathy and psychological and neural bases of belongingness. His very interesting Ph.D. thesis is about the neural substrates of basic and moral emotions, and, in it, he expressly mentions the importance of people's moral sense for the operation of legal systems.[30]

The researches that discuss the physical substrates of the moral and find out changes in the physical substrates can modify somebody's moral sense, calls into question if the free will really exists:

> (...) is it possible to argue, today, there is free will? The implications for the law are obvious: if there is no free will, all the civil law, based on the idea of autonomy, and the entire criminal law, founded on the notion of guilt (personal responsibility), must be rethought.[31]

---

[25] The Edmond and Lily Safra International Institute of Neuroscience of Natal (2010).

[26] A Brazilian study of the person's idea and its relation with the body can be found at: Stancioli (2010).

[27] Free translation of: "tanto a formação como a evocação das memórias de curta e de longa duração são fortemente moduladas por vias relacionadas com a vida emocional (...)" (Izquierdo 2004).

[28] Izquierdo (2004).

[29] For further information, access the Labs-D'Or Center for Neuroscience (2010).

[30] Moll Neto (2003).

[31] Free translation of: "(...) é possível defender, hoje, que existe livre-abítrio? As implicações para o Direito são óbvias: se não há livre-arbítrio, todo o Direito Civil, calcado na ideia de autonomia da vontade, e todo o Direito Penal, fundado na noção de culpabilidade (responsabilidade pessoal), terão de ser repensados" (Horta 2010).

The neuroscience studies about neural bases of psychopathy and brain imaging are used in the judgments of criminals, to analyze the psychopath's mind. In Brazil, although the brain imaging is not so used in the Courts as it happens in other countries, there are academic studies being made about it.

These Brazilian studies, among others held in the world, question the basis of Law, and states as necessary to rethink it through the light of new scientific discoveries. After all, the Law should follow the development of society, so as to always be reformulated to avoid losing real effectiveness in face of the advancements of science.

4 Neuroscience in the Courtroom

A search in the Federal Supreme Court, the Superior Court of Justice, 26 State Justice Courts, and also in the Justice Court of the Federal District,[32] showed that only in three State Courts[33] reference to the term "neuroscience" have been made.

The use of the neuroscience by the Courts is aleatory, as it will be shown below. In fact, there is not a discussion regarding the use of neuroscientific evidences or even highlighting that something new and different is related to neuroscience, even in those decisions that mention it (11 in both the States of *Rio Grande do Sul* and *Santa Catarina,* and 27 in *São Paulo*).

4.1 Decisions of the Court of Justice of Rio Grande do Sul

The 11 decisions of the Superior Court of the State of *Rio Grande do Sul*[34] mentioned neuroscience within criminal processes related to drug addiction. In (just) one of its paragraph it is affirmed that:

> The WHO (World Health Organization), in a recent report ('Neuroscience of Psychoactive Substance Use and Dependency'), has made a review about contemporary studies related to the role of the brain in chemical dependency, classifying the problem as a brain dysfunction as any other neurological disturbance[35]

---

[32] The searches made with the term *"neurociência,"* and updated December 27, 2010, included only civil and criminal decisions in the largest period available for online search in each Court's website, which means after the year 2000, most of the time.

[33] Namely, *Rio Grande do Sul, Santa Catarina* and *São Paulo*. Not even the State of *Rio Grande do Norte*, where the IINN-ELS is, has dealt with the subject in the courts.

[34] Decisions from 2005 to 2008.

[35] Free translation of the text that can be accessed in: Tribunal de Justiça do Rio Grande do Sul (2010).

4.2 Decisions of the Court of Justice of Santa Catarina

The 10 out of 11 court decisions of the Court of Justice of *Santa Catarina*[36] have to do with administrative law within civil lawsuits, namely the regulation of the profession of optometry. At a paragraph that is the same in all of them, neuroscience is used as an argument to the existence of new professions:

> I believe, however, that it would be an exercise of anachronism to judge this case based on decrees of 1932 and 1934. I don't have any doubt, except for the non expression of the National Health Counsel, that what is being analyzed is the recognition of a running course by the Ministry of Education, absolutely regular for many years. And that would be really kafkian at this point, based on a decree clearly old in terms of health technology, that we simply eliminate these professions and together with them many others with total international recognition, such as, for instance, neuroscience[37]

4.3 Decisions of the Court of Justice of São Paulo

Memory and its related issues brought "neuroscience" to 27 court decisions in the State of *São Paulo*.[38] However, did not come into question the need for regulation of specific neuroscientific evidences. That is because, as it was said in the introduction of this article, its use is made at random.

Therefore, as it was said that is common that the Judiciary decides in "block-decisions," it will be transcribed only one of these decisions, since all are equal. Follows part of a vote of the reporting judge of one of the criminal appeals[39] in which neuroscience was used, together with other arguments, to impugn the value of an eye witness and dismissal of the defendant. This vote shows how that the uniqueness of neuroscience is not distinguished by the judiciary:

> The victim, however, is always taken by great emotion at the moment of the offense and, accordingly, does not have the opportunity to visualize, in details, the agents of the conduct. Under these circumstances will seek, after remembering facts, dialog with them and occasional blanks of the memory will be suppressed, been substituted by suggestions about the event. The judiciary psychology warns that the witness, after gathering the primitive sensation, suffers series of actions tending to alter the image of the occurred.
>
> In this sense, it cannot be affirmed that the victim testimony, in any situation, is sufficient prove to condemn. The offended, even not wanting to lie, may be induced to accept as real some events from which it had no perfect memory. So, if the crime involves unknown characters, only identified by generic physical traces, may, when introduced to someone some time after recognize the person because the suspect is involved in a similar crime. This memory could be fruit of what is called 'conditioned effect' in psychology.

---

[36] Decisions from 2006 to 2010.

[37] Free translation of the text that can be accessed in: Tribunal de Justiça de Santa Catarina (2010).

[38] Decisions from 2003 to 2009.

[39] Free translation of the text that can be accessed in: Tribunal de Justiça de São Paulo (2010).

Professor *Hélio Gomes* explain it this way: 'It may also occur the change of the fact even by a witness who paid great attention; due to its confidence in prior observations, assuming seeing and hearing what really didn't happened. The observer did not hear or see what really happened, but what, according to its experience and habit, should have occurred, because it's what always occurs. This could be called psychological 'conditioned effects'.

Recent studies of neuroscience have revealed the little precision of the memory, founded in facts occurred some time ago. The Brazilian neuroscientist *Suzana Herculano-Houzel*, master at the Case Western Reserve University, in the United States, PhD at *Université Paris VI* and researcher for four years at the German *Max-Planck Institute*, is currently working at the Life Museum at the *Oswaldo Cruz* Foundation, in *Rio de Janeiro*, and affirms: 'That it is possible to remember what has not happened is no new' and continues: 'How is it possible to have a register in the brain of something that has not occurred? The first reason is that the brain is not a video camera – remembering is not replaying the film of senses. On the contrary: it is believed, nowadays, that the memory is reconstituted each time it is 'called', putting together memories distributed across the brain.' And at the end states: 'It happens also that imagination, or more exactly mental visualization, the 'seeing with the eyes of the mind', use exactly the same circuits as the 'outside vision' and those circles are also part of the memory. That means that mental visualizations particularly lived may let traces undistinguished of those let by a real event (*O cérebro nosso de cada dia – Descobertas da neurociência sobre a vida cotidiana*. 4ª ed. Rio de Janeiro: Vieira & Lent, 2002, p. 134-135).[40]

In the same direction, explains Dr. John J. Ratey, associate at the Psychiatric clinic of the Harvard Medical School in the United States: 'The formation and remembering of each memory are influenced by the state of mind, environment and '*gestalt*' at the moment in which memory is formed or recovered. That is why the same event may be remembered in different ways by different persons. One person is not necessarily 'right' and the other 'wrong'. Memories also change as we change, as time goes by. New experiences change our attitudes and, therefore, what and how we remember'. And continues: 'Memory – from two minutes, two years, two decades – come and go at each vigil hour. The parts of it are linguistic unities, emotions, beliefs and actions and here we have, immediately, our first surprising conclusion: because our daily experiences alter constantly these connections, a memory is minimally each time we remember it'. And concludes: 'The operational memory allows us to work from moment to moment. But its nature may change with time, even within short intervals. New researches about how we transfer the information from the short duration to the long duration are presenting an important revelation: we can never be sure about anything we remember' (*O cérebro – um guia para o usuário*. Translation of *Alvaro Cabral*. Rio de Janeiro: Objetiva, 2002, p. 209-222).

As a matter of fact, the victim may be induced to recognize why the face of someone looks familiar to him or she: have seen it before, without keeping conscience of the fact, or have seen someone like it, who could be taken for him in a different situation. However, for a context transfer, tie it in to the committed crime.

This possibility makes what psychology calls wrong attribution, which may occur in many ways, but in this situation happens by means of the unconscient transfer phenomenon. Daniel L. Schater, chief of the Department of Psychology of the Harvard University in his book 'The seven sins of the memory – how the mind forgets and remembers', explains that a witness, in this case, 'wrongly attribute a familiarity to a face to the wrong source because, unconsciously, transfer the memory from the person from one context to another'. The author cites an experience made in which participants watched a movie about a robbery that showed an innocent pedestrian in another separated scene. Latter, they sometime

---

[40] This citation of the book "Our everyday brain – Founds of the neuroscience in the everyday life" (free translation of the title) is in all of the 26 decisions analyzed.

identified the pedestrian as the bandit because many of the participants believed, incorrectly, that both were the same person (*Os sete pecados da memória – como a mente esquece e lembra*. Translation of *Sueli Anciães Gunn*. Rio de Janeiro: Rocco, 2003, p. 118-119).

It all reveals what is no new: the fallibility of the memory when reconstructing facts even when recently perceived. Indeed, the more vague and generic are the traces observed in a person, the bigger the possibility to identify it through another, which would look alike and, for the observer, looks the same.

In another decision, the same State Court of Justice mentioned the Laboratory for Neuroscience of the Psychiatric Institute of the University of *São Paulo*,[41] particularly a study from Dr. *Alexandrina Maria Augusto da Silva Meleiro*, in a case of an appeal in a civil liability suit of the family of a patient who committed suicide toward a hospital, and, in the end, the appeal was not granted:

Evaluate the grade of seriousness of a suicide try may be a hard task because it comprehends intentional acts of self-aggression that do not result in death, from discrete and hidden threatening acts to one´s own life, some of them with the objective of gaining attention, to non specific ideas such as 'life is not worth it' or 'I'd rather be dead', to specific ideas that follow the intention to die and/or a suicide plan. The genuine suicide intention is frequently ambivalent in relation to death and the seriousness of the purpose may vary. Those patients have changing humor, with a feeling of inutility, lack of hope, lost of self-esteem and desire to die. There is the feeling of the three 'is'[42]: intolerable (cannot stand anymore), inescapable (there is no way out) and unfinishable (without an end). Potential suicide may regret and seek help after the act is committed. The relief, after a try, makes the person reflect about its act [43]

There is also a case of condemnation of a radiologist doctor[44] that had a clinic for "psychosomatic neuroscience" (*sic*) and orthomolecular medicine in which treatments for neuropsychiatry, neurophysiology, orthomolecular medicine, and brain mapping were made. The condemnation was due to the fact that the patient tried to commit suicide during a treatment for depression considering that the doctor was not a psychiatrist and orthomolecular treatments had been prohibited by the Federal Council of Medicine[45] and, therefore, his acting was judged unskillfulness, derelict, and imprudent.

---

[41] For further information: Neurosciences' Lab (2010).

[42] As from the Portuguese writing of the words.

[43] Free translation of the text that can be accessed in: Tribunal de Justiça de São Paulo (2010).

[44] Tribunal de Justiça de São Paulo (2010).

[45] Resolution n. 1938 of the Federal Council of Medicine, dated February 5, 2010, substituted other legal documents related to the subject. In few words, the resolution says that supplements are only indicated for people with deficiency of substances. A Portuguese version of the norm can be found at http://www.portalmedico.org.br/resolucoes/cfm/2010/1938_2010.htm (accessed 27 December 2010).

5 Doing Research in Brazil

There are no specific norms for neuroscientific research in Brazil and, therefore, the legal framework for research in general involving human subjects applies.[46] Besides being signatory of the most important international documents that regulate all kinds of research, Brazil has its own rules regarding the issue, as listed below. Note that there is not a law made by the Legislature dealing with general research in the country, only rules of specific administrative bodies of each area, what means there is not a universal rule.

- Resolution n. 196 of the National Health Council (*Conselho Nacional de Saúde*), dated October 10, 1996[47] – the main legal National Health Council instrument regarding research involving human subjects in the country;
- Resolution n. 41 of the National Agency of Sanitary Surveillance (*ANVISA*), dated April 28, 2000, regarding data registration for the accreditation of entities and companies on pharmaceutical, bioassessibility and/or bioequivalence trials;
- Resolution n. 240 of the National Health Council, dated June 5, 1997, regarding the representation of users in the committees for ethics in research;
- Resolution n. 251 of the National Health Council, dated August 7, 1997, regarding the approval of research norms involving human subjects for the area of new medicines, vaccines and diagnostic tests;
- Resolution n. 292 of the National Health Council, dated July 8, 1999, about researches coordinated abroad or with foreign participation involving the sent of biological material to another country;
- Regulation of the Resolution n. 292/99 (approved by the National Health Council on August 8, 2002);
- Resolution n. 301 of the National Health Council, dated March 16, 2000, contrary to modifications in the Convention of Helsinki on item II.3 and the use of placebo;
- Resolution n. 303 of the National Health Council, dated July 6, 2000, with complementary norm to the area of assisted reproduction, establishing sub-areas that must be analyzed committees on ethics and the national committee on ethics;
- Resolution n. 304 of the National Health Council, dated August 9, 2000, about researches involving indigenous people;

---

[46] Although a high number of critics, even from the side of well-known neuroscientists, as Sidarta Ribeiro [Folha de São Paulo (2010)], it is forbidden to remunerate research participants in the country.

[47] The original version of the Resolution 196/96 of the National Council of Health (1996) and an English translation made by the municipality of the city of *São Paulo* is available at Resolution n. 196 of the National Council of Health (1996). This translation is used in this article every time parts of the Resolution are transcribed.

- Resolution n. 391 of the National Agency of Sanitary Surveillance, dated August 9, 1999, approves the technical regulation for generic medicines; and
- Law n. 8.501, dated November 30, 1992, about the utilization of cadaver (non recalled) in studies or researches; among others.

5.1 Resolution n. 196 of the National Health Council

To do research on human beings in Brazil, is necessary to observe the requirements of the Resolution n. 196 of the National Health Council.

According to it, there must be at least one Committee for Ethics in Research – collegiate bodies of consultative, deliberative or educational nature – in each institution where can take place researches on human beings. Any research involving human subjects must be submitted to the appreciation of one of these Committees and the researcher must receive a positive decision of the competent Committee before beginning the research. After that, and always following the approved project, the researcher must draw up and submit partial and final reports, keep certain documents for a while and provide data that can be requested at any time by the Committee.[48]

There is also a National Committee for Ethics in Research (*CONEP*) that "is responsible for reviewing all ethical aspects of research involving human subjects, as well as adapting and updating pertinent guidelines and norms".[49] In some specific themes, all researches must pass also through the analysis of *CONEP* after the positive opinion of the competent Committee.

The preamble of the Resolution n. 196 helps to know what rules and principles have to be followed when conducting a research:

This Resolution is based on the main international documents that gave rise to declarations and guidelines on research involving human subjects: the Nuremberg Code (1947); Declaration of Human Rights (1948); Declaration of Helsinki (1964, and its later versions dated 1975, 1983 and 1989); International Agreement on Civil and Political Rights (UN, 1966, approved by the Brazilian National Congress in 1992); Proposed International Guidelines for Biomedical Research Involving Human subjects (Council for International Organizations of Medical Science/World Health Organization 1982 and 1993); and International Guidelines for Ethical Review of Epidemiological Studies (CIOMS, 1991). It also meets the provisions of the Constitution of the Federative Republic of Brazil of 1988 and related Brazilian legislation: Consumer Rights Code; Civil Code and Penal Codes; Statute of Children and Adolescents; Basic Health Law n° 8.080 of 19 September 1990 (establishing the terms for health care and the organization and operation of corresponding services); Law n° 8.142 of 28 December 1990 (community participation in the management of the Unified Health System); Decree n° 99.438 of 7 August 1990 (organization and competence of the National Health Council); Decree n° 98.830 of 15 January 1990 (collection of scientific material and data by foreigners in Brazil); Law n° 8.489 of 18 November 1992 and Decree n° 879 of 22 July 1993 (removal of tissues, organs and other

---

[48] Resolution n. 196 of the National Council of Health (1996).
[49] Resolution n. 196 of the National Council of Health (1996).

parts of the human body for humanitarian and scientific purposes); Law n° 8.501 of 30 November 1992 (utilization of cadavers); Law n° 8.974 of 5 January 1995 (use of genetic engineering techniques and release of genetically modified organisms into the environment); Law n° 9.279 of 14 May 1996 (regulates the rights and duties pertaining to industrial property); and pertinent statutes.

This Resolution includes, from the point of view of the individual and communities, the four basic principles of bioethics: autonomy, non-maleficence, beneficence, and justice, among others, and aims at ensuring the rights and duties of the scientific community, the research subjects and the State.

The contextual nature of these consideration requires that periodical reviews of this Resolution be made, according to the needs of the technical-scientific and ethics areas.

It is further emphasized that each thematic area and each modality of research must both respect the principles set forth in this text and meet all specific regulations and sectorial requirements.[50]

Keeping in mind that Brazil is very bureaucratic and that the Resolution n. 196/96, as other legal documents above mentioned, goes into details of documents and processes for the execution of a research in the country, it is strongly recommended to take a closer look at the text of these norms in order to comprehend the details.

Although it would be interesting to transcript the whole Resolution n. 196/96 in this text, it will only be highlighted its prescriptions about the mandatory freely given and informed consent for those participating in a research.

5.2   Informed Consent According to Resolution n. 196 of the National Health Council

One of the most important requirements for doing research with human beings in the country is to get from the research subjects their informed consent, in which they freely agree to participate in the research. According to the Resolution, prerequisites for the valid informed consent are:

V.1 Accessible language must be used in providing the prospective subjects information about the research, always including the following points:

a) rationale, aims and methods to be used in the research; b) any foreseeable risks or discomfort to the subject, as well as benefits that might reasonably be expected, associated with participation in the research;

c) existing alternative methods;

d) medical follow up and care to be provided to the subjects of research, as well as the identity of those responsible for such actions;

e) assurance of information about the methodology, before and during the research, including the possibility of inclusion in a control or placebo group;

f) freedom of the individual to refuse participation or withdraw his/her consent, at any time during the research, without any penalty or loss of benefits to which he/she would otherwise be entitled;

g) extend to which confidentiality of records will be maintained, so as to safeguard the privacy of the research subjects;

---

[50] Resolution n. 196 of the National Council of Health (1996).

h) forms of reimbursement of current expenditures resulting from participation in the research; and

i) types of indemnity to cover possible injury resulting from the research.

IV.2 The terms of freely given and informed consent must meet the following requirements:

a) they must be drawn up by the main researcher and express compliance with each of the above mentioned requirements;

b) they must be approved by the Committee for Ethics in Research that evaluates the research;

c) they must be signed by or identified with the fingerprint of each and every research subject or their legal guardians; and

d) an original and a copy must be signed by the research subject, the latter to be kept by the research subject or his/her legal guardian and the former to be filed.

IV.3 In the event there is any hindrance to the freedom of or access to the information required by the research subject for giving adequate consent, the following requirements must be fulfilled:

a) in research involving children and adolescents, individuals who are mentally ill or disturbed, and persons with substantially impaired or diminished autonomy, the choice of said research subjects must be clearly justified and specified in the research protocol, which must be approved by the Committee for Ethics in Research and meet all the requirements of freely given and informed consent, through the legal guardian of the prospective research subject, without detriment to the right of information of the individual, within the limits of his/her capacity of understanding;

b) freedom of consent must be particularly guaranteed to those individuals who, although adults and capable, are exposed to specific conditioning or to the influence of authority, specially students, military personnel, employees, prison inmates, inmates of rehabilitation centres, shelters, homes, religious or other institutions, ensuring them complete freedom to participate, or not, in the research, without any retaliation;

c) in the event it is impossible to record the freely given and informed consent of the research subject, such fact must be duly documented, with an explanation of the causes and the technical opinion of the Committee for Ethics in Research;

d) research on individuals diagnosed as brain dead can only be carried out after meeting the following conditions:

- document proving brain death (death certificate);

- explicit consent of the relatives and/or legal guardian, or prior statement by the individual;

- total respect for the dignity of the human subject, and not mutilation or violation of the body;

- no additional financial burden for the family;

- no deleterious effect to other patients awaiting admission or treatment;

- possibility of obtaining scientific knowledge which is relevant, new, or unobtainable through other means;

e) in communities with a different culture, including Indigenous communities, prior consent must be obtained from the community, through its leaders, without foregoing, however, efforts to obtain individual consent;

f) when the merit of the research depends on some restriction of information to the subjects, such fact must be duly stated and justified by the researchers, and submitted to the Committee for Ethics in Research. The data obtained from such research subjects cannot be used for purposes other than those contemplated in the protocol and/or terms of consent.[51]

---

[51] Resolution n. 196 of the National Council of Health (1996).

The Resolution also states that it is prohibited to require for the research subject waiver of the right to compensation for injury. The form of consent must not contain any clause that removes this responsibility, or which involve the research subjects give up their legal rights.[52]

It is noteworthy that the model of term of informed consent that is going to be used in the research will be rigorously reviewed by the Committee for Ethics in Research, so it is important to carefully make the term in straight obedience to the Resolution's requisites.

6 Conclusion

Neuroscience will change the Law dramatically. Concepts such as competence, responsibility, free-will, mental illness and lie, among thousands of others, will never be the same again. Many legal presumptions to which jurists sticks are going to be proven as inappropriate and may be left behind.

But, the continental European tradition, which influenced most of Brazilian law schools, favors expositive classes and avoids subjects not treated by the law. Rather Rome with its *codex* and cases than discussions about endeavors and incertitudes brought by new technologies, as if keeping it out of the classroom would be a solution, and not a denial, of problems. It reflects itself in the mentality of the majority of jurists and the small amount of empirical legal studies in Brazil. In fact, most jurists are very conservative, closed, and reluctant to refresh their idea of law science as part of a living in a fast changing world. That is, despite all neuroscience findings that question legal dogmas, even those made by brilliant Brazilian researchers, the Law, mostly, seems to continue ignoring the reality changes and remains locked on its formalities and abstract classifications.

Meanwhile, new neuroscientific techniques enter the courtroom without being really understood, and are spread not owing to the existence of real discussions about neurolaw in the courts, but due to the block-decisions.

However, the mere verification that neuroscience is mentioned in some decisions, albeit not in a very consistent way, in addition to the fact that studies are beginning to emerge in the country about neurolaw, signalizes a promising future development of the area in Brazil.

After all that was said, it is a strong wish to see Brazilian science fulfilling its role to be, as *Nicolelis* states in his Manifest,[53] an effective agent of social and economic transformation in Brazil. In this scenario, it is the law to provide conditions for this to occur and work to foster science development in the country.

---

[52] Resolution n. 196 of the National Council of Health (1996).

[53] The Edmond and Lily Safra International Institute of Neuroscience of Natal (2010).

Disclosure Statement

The authors are not aware of any affiliations, memberships, funding, or financial holdings that might be perceived as affecting the objectivity of this review.

Acknowledgments The authors are grateful for the encouragement of *Lie Uema do Carmo* and the kind support of *Getulio Vargas Foundation Law School of São Paulo*.

References

Antônio D (2010) Mais e melhor. Minas faz ciência 42:6–11

Atlas National Geografic: Brasil (2008) vol 2. National Geografic Society, São Paulo

Conselho Nacional de Justiça (2010) Justiça em números 2009: indicadores do Poder Judiciário – Conselho Nacional de Justiça. Panorama do Judiciário Brasileiro. Brasília. http://www.cnj.jus. br/images/conteudo2008/pesquisas_judiciarias/jn2009/rel_sintetico_jn2009.pdf. Accessed 15 Dec 2010

Constitution of the Federative Republic of Brazil of 1988. http://www.planalto.gov.br/ccivil_03/ Constituicao/Constituicao.htm. Accessed 20 Nov 2010

Folha de São Paulo (2010) Cientistas do país querem remunerar voluntários. http://www1.folha. uol.com.br/ciencia/800873-cientistas-do-pais-querem-remunerar-voluntarios.shtml. Acessed 27 Dec 2010

Horta RL (2010) Direito e Neurociências, Neurodireito: o que é isso? Blog da Sociedade Brasileira de Neurociências e Comportamento. http://blog.sbnec.org.br/2010/07/direito-e-neurociencias-neurodireito-o-que-e-isso/. Accessed 19 Sept 2010

Izquierdo I (2004) Os mecanismos das diversas formas de memória. In: Lent R (ed) Cem bilhões de neurônios: conceitos fundamentais de neurociência. Atheneu, São Paulo

Labs-D'Or Center for Neuroscience (2010). http://sites.google.com/a/neuroscience-rio.org/www/ home. Accessed 19 Sept 2010

Moll Neto JN (2003) Substratos neurais das emoções básicas e morais: investigação de indivíduos normais com ressonância magnética funcional. PhD thesis – University of São Paulo Medicine School, São Paulo

Neurosciences' Lab (2010) University of São Paulo. www.neurociencias.org.br. Accessed 27 Dec 2010

Nicolelis' Lab (2010) http://www.nicolelislab.net. Accessed 19 Sept 2010

O Cérebro Nosso de Cada Dia (2010) http://www.cerebronosso.bio.br/posgraduacao-em-neuro/. Accessed 19 Sept 2010

Regalado A (2010) Brazilian science: riding a gusher. Science 330:1306–1312. 3 December. http://www.natalneuro.org.br/noticias_brasil/pdf/2010_12_science.pdf. Accessed 19 Dec 2010

Relatório ICJBrasil (2010) 3º trimestre 2010. Escola de Direito de São Paulo da Fundação Getulio Vargas. http://www.direitogv.com.br/subportais/RelICJBrasil3tri2010.pdf. Accessed 20 Dec 2010

Resolution n. 1938 of the Federal Council of Medicine, of February 5, 2010. http://www. portalmedico.org.br/resolucoes/cfm/2010/1938_2010.htm. Accessed 27 Dec 2010

Resolution n. 196 of the National Council of Health, of October, 10, 1996. Original Portuguese version. http://conselho.saude.gov.br/resolucoes/reso_96.htm. Accessed 27 Dec 2010

Resolution n. 196 of the National Council of Health, of October, 10, 1996. English translation made by the municipality of the city of *São Paulo*. http://www.prefeitura.sp.gov.br/cidade/ secretarias/upload/saude/arquivos/comiteetica/Reso196_English.pdf. Accessed 27 Dec 2010

Stancioli B (2010) Renúncia ao exercício de direitos da personalidade ou como alguém se torna o que quiser. Del Rey, Belo Horizonte

Superior Tribunal de Justiça (2010) Decisões do STJ legitimam exame de DNA como ferramenta em busca da Justiça. http://www.stj.jus.br/portal_stj/publicacao/engine.wsp?tmp.area=398&tmp.texto=97374&tmp.area_anterior=44&tmp.argumento_pesquisa=dna. Accessed 27 Dec 2010

The Edmond and Lily Safra International Institute of Neuroscience of Natal (2010) http://www.natalneuro.org.br. Accessed 19 Sept 2010

The Edmond and Lily Safra International Institute of Neuroscience of Natal (2010) Manifesto da Ciência Tropical: um novo paradigma para o uso democrático da ciência como agente efetivo de transformação social e econômica no Brasil. http://www.natalneuro.org.br/noticias_brasil/2010-11novembro02.asp. Accessed 6 Dec 2010

The Edmond and Lily Safra International Institute of Neuroscience of Natal (2010) Temos que construir uma ciência tropical. http://www.natalneuro.org.br/imprensa/pdf/2010-03-almanaque.pdf. Accessed 19 Sept 2010

Tribunal de Justiça de Santa Catarina (2010) Appeal n. 2008.070584-3, dated October 14, 2010. http://app.tjsc.jus.br/jurisprudencia/acnaintegra!html.action?parametros.frase=&parametros.todas=neuroci%EAncia&parametros.pageCount=10&parametros.dataFim=&parametros.dataIni=&parametros.uma=&parametros.ementa=&parametros.juiz1GrauKey=&parametros.cor=FF0000&parametros.tipoOrdem=data&parametros.juiz1Grau=&parametros.foro=&parametros.relator=&parametros.processo=&parametros.nao=&parametros.classe=&parametros.rowid=AAARykAAvAAAFMZAAc. Accessed 27 Dec 2010

Tribunal de Justiça de São Paulo (2010) Appeal n. 505.967-4/9-00, vote n. 18419, dated July 2, 2009. https://esaj.tjsp.jus.br/cjsg/getArquivo.do?cdAcordao=3971909&vlCaptcha=ymtut. Accessed 27 Dec 2010

Tribunal de Justiça de São Paulo (2010) Appeal n. 992599-0/3, vote n. 9603, dated September 25, 2007. https://esaj.tjsp.jus.br/cjsg/getArquivo.do?cdAcordao=3527735. Accessed 27 December 2010

Tribunal de Justiça de São Paulo (2010) Appeal n. 993.06.018537-3, vote n. 1529, dated December 12, 2009. https://esaj.tjsp.jus.br/cjsg/getArquivo.do?cdAcordao=4194901&vlCaptcha=uHEaa. Accessed 27 Dec 2010

Tribunal de Justiça do Rio Grande do Sul (2010) Appeal n. 70023478787, dated July 17, 2008. http://www1.tjrs.jus.br/site_php/consulta/consulta_processo.php?nome_comarca=Tribunal+de+Justi%E7a&versao=&versao_fonetica=1&tipo=1&id_comarca=700&num_processo_mask=70023478787&num_processo=70023478787&codEmenta=2448045. Accessed 27 Dec 2010

Research Ethics Challenges in Neuroimaging Research: A Canadian Perspective

Ciara Toole, Amy Zarzeczny, and Timothy Caulfield

Abstract Neuroimaging research continues to engage the imaginations of scientists, members of the media, and the general public. As an area of human subject research, it also raises a number of research ethics issues that, while not necessarily unique to neuroimaging, offer particular challenges in this growing domain. Here, we consider a number of the key research ethics issues that are emerging as being of central importance to the continued development of this field. We will situate our discussion within the Canadian framework, but many of the issues raised will have broad jurisdictional relevance. While providing a comprehensive examination of all of the research ethics issues implicated by neuroimaging research is beyond the scope of this review, it is hoped that this paper will serve as a useful overview and guide to researchers, research ethics boards, and others interested in neuroimaging research.

1 Introduction

Neuroimaging research, and the belief that it "offers a direct picture of the human brain at work" (Illes et al. 2006b, p. 149), continues to engage the imaginations of scientists, members of the media, and the general public. Research in this field covers a broad spectrum of issues, from more traditional clinical applications such as investigating brain structure, pathology or injury, to probing matters of higher-order

Ciara Toole, BHSc, LLB, Faculty of Law, University of Alberta. Amy Zarzeczny, LLM, Assistant Professor, Johnson-Shoyama Graduate School of Public Policy, University of Regina. Timothy Caulfield, LLM, FRSC, Canada Research Chair in Health Law and Policy, Professor, Faculty of Law and School of Public Health, Senior Health Scholar with the Alberta Heritage Foundation for Medical Research and Research Director, Health Law Institute, University of Alberta.

C. Toole • A. Zarzeczny • T. Caulfield (✉)
Faculty of Law, Law Centre, University of Alberta, Edmonton, AB T6G 2H5 Canada
e-mail: tcaulfld@law.ualberta.ca

cognition (e.g., emotion and personality – Canli and Amin 2002; lying – Langleben et al. 2002, etc.). Over time, we have witnessed a continual growth of interest in this exciting and, at times, controversial field. This mounting interest is reflected in a number of domains including the numbers of research papers published in this area,[1] a steady growth in related articles appearing in the popular press (Racine et al. 2006), and the emergence of what appears to be a continually expanding market for commercial neuroimaging services (Racine et al. 2007).

The broad and diverse uses for neuroimaging technologies give rise to a host of ethical challenges, which have generated an increasingly well-developed body of literature, and have formed the basis for the expanding field of neuroethics (Illes 2006; Illes and Bird 2006). Furthermore, as an area of human subject research, neuroimaging also raises a number of research ethics issues that, while not necessarily unique to neuroimaging, offer particular challenges in this growing domain.

2 Neuroimaging Technologies: A Brief Overview

There are a large number of different neuroimaging modalities, including (but not necessarily limited to): electrophysiology (electroencephalogram, EEG); evoked response potential (ERP) technology; positron emission tomography (PET)/single-photon emission-computed tomography (SPECT); functional magnetic resonance imaging (fMRI), and EEG-based brain fingerprinting. Each of the foregoing has its own strengths and weaknesses, which we will not review here. However, fMRI has emerged as the tool of choice for examining the neural correlates of behavior and disease (Haynes and Rees 2006; Bandettini 2009).

As a fairly noninvasive technique which measures relative increases in oxygenated blood flow to the brain, fMRI research has grown exponentially to include studies of cognition and human brain organization (Logothetis 2008). It has also provided the basis for many more novel and potentially controversial studies in recent years, including functional neuroimaging studies of affection and emotion (Canli et al. 2000; Canli and Amin 2002), neural processing of race information (Hart et al. 2000; Lieberman et al. 2005), lie detection (Langleben et al. 2002; Davatzikos et al. 2005; Hakun et al. 2009), substance abuse (Kaufman 2001), love (Bartels and Zeki 2000), impulsivity (Congdon and Canli 2008), consumer behavior (McClure et al. 2004), grief (O'Connor et al. 2008), vicarious reward through game shows (Mobbs et al. 2009), and antisocial or violent behavior (Kiehl et al. 2001; Lee et al. 2009), among many others. While most of the research ethics issues we will discuss will be broadly applicable to different types of imaging technologies, because of its comparably high profile and attendant challenges, where appropriate our discussion will highlight fMRI to provide necessary context.

---

[1] Poldrack has estimated an average of 30–40 papers on fMRI neuroimaging are published every week (2008, p. 223).

Apart from the various ethical, legal, and social implications of many of the foregoing fields of inquiry, it is important to note that the science of fMRI (as is true for other neuoimaging modalities) faces its own challenges. Despite its increasingly broad applications, at the most basic level fMRI remains an indirect measure of brain activity (Poldrack 2008), and there has been considerable debate about the legitimacy of fMRI studies that show strong correlations between neural activation and complex behavior (e.g., Vul et al. 2009). fMRI's reliance on the blood oxygen level dependent (BOLD) signal to measure increased blood flow to particular regions of the brain during certain mental activities (Illes et al. 2006b) and corresponding inferences of linear activation of particular mental processes and suppression of others, raises significant questions (Poldrack 2008; Jackson 2006). Results can be impacted by any number of confounding factors including age or brain pathology, subject behavior (e.g., movements of the head, swallowing), and others (Desmond and Chen 2002). Furthermore, common use of group data, the choice of control conditions, the possibility of artifacts, selective statistical analyses, and the inability to distinguish between neuronal activation and inhibition, are all limitations on the generalizability and broad applicability of many fMRI studies and conclusions (Logothetis 2008; Rosen and Gur 2002; Desmond and Chen 2002).

3 Emerging Ethical Issues

Despite the foregoing scientific limitations to fMRI research, it (and other forms of neuroimaging research) continues to receive considerable traction in both published research and the popular media. This attention and the increasingly diverse applications – both existing and prospective – associated with this burgeoning field are giving rise to a host of new and familiar ethical issues. While in-depth consideration of these topics is not strictly within the scope of this piece, it is worthwhile to highlight a number of the key issues that are likely to help shape the overall development of the field, including the operation of research ethics in this domain.

As has been noted in other realms of biomedical research, such as stem cell research and genetics, it has been suggested that there is a certain degree of hype ("neurohype") developing around neuroimaging research, stemming from various sources (Caulfield et al. 2010). This hype is potentially concerning for a number of reasons, including the worry that it may contribute to deterministic and/or fatalistic beliefs on the part of members of the public who, because of the kinds of information neuroimaging research purports to tell us about ourselves, and about the differences between us, may be swayed by the "allure of determinism" (Green 2006, p. 111). This concern is heightened by the apparent power that neuro-language and neuroimages have over members of the public. Research indicates that neuroscientific explanations and neuroimages have a remarkable influence on nonexperts' understandings of psychological phenomena (Weisberg et al. 2008). Thus, despite the clear limits of fMRI research (noted above), one worry is that research participants and members of the public may overinterpret research results and their implications.

In addition, as is true in other domains of biomedical research, it appears that at least some neuroimaging studies are funded by private sources (Caulfield and Ogbogu 2008; Nature Editorial 2008). The introduction of private funding to the research domain may be increasingly common, but it engages a number of issues related to governance and commercialization of research, concerns regarding market pressure and public trust (Caulfield and Ogbogu 2008; Kerr and Caulfield 2007), and conflicts of interest (see Cho 2002). The melding of public and private interests in research can also raise interesting questions regarding what regulations and policies apply to the work. Research conducted in the private realm may avoid many facets of governance that apply to publically funded endeavors. How this plays out in individual cases will depend largely on how the research projects are structured and who the actors are. It is an issue that bears monitoring – particularly as we see a growth in private neuroimaging (e.g., the neuromarketing arena).

A final issue worth mentioning at this point is the anticipated use of neuroimaging technologies, and fMRI in particular, in the courtroom. While a variety of potential applications are contemplated (Feigenson 2006), the introduction of fMRI for the purpose of lie detection is particularly contentious. This possibility has arisen in media reports from the United States – e.g., "MRI Lie Detection to Get First Day in Court" – (Madrigal 2009), and has received attention in bioethics and legal literature (e.g., Thompson 2007). To our knowledge, no Canadian courts have faced this issue as of yet. However, in view of leading case law regarding lie detection (*R. v. Beland*) and the *Canadian Charter of Rights and Freedoms* (sections 8, 12 and 13 in particular), it is highly questionable whether such use would in fact be permitted under Canadian law (McMonagle 2007; Downie and Murphy 2007). Nonetheless, this prospective application, and the social controversy it incites, highlights the degree to which information about research areas and their anticipated results can impact public perception of an area. It also further confirms the importance of responsible communication of research endeavors and their results (as discussed further below).

4 Research Ethics Governance in Canada

In many ways, the Canadian research ethics governance framework is both complex and fragmented. It is shaped by a number of different guiding documents and other instruments, existing at various levels including the local, national, and international (Hadskis 2007; Downie and McDonald 2004, p. 174). Providing an in-depth, comprehensive review of this framework is beyond the scope of this discussion. It will suffice for our purposes to identify relevant policies and structures, and highlight elements which may be of particular importance in the neuroimaging context.

The roots of current research ethics are widely recognized as flowing from the *Nuremberg Code* and the *Declaration of Helsinki*. The *Nuremberg Code*, which emerged from the 1947 Nuremberg trials that addressed egregious human experiments conducted during World War II, focuses on a number of key principles including the need for obtaining informed consent from research subjects. The *Declaration of Helsinki*, first developed by the World Medical Association in 1964 and updated in 2008, provides further international guidance for the conduct of ethical biomedical research on humans. Generally, it embodies three basic principles: respect for all persons, beneficence in the maximization of benefits over harms, and justice for all those who could benefit from the research (CIOMS 1993). These fundamental principles have since found their way into national and regional policies around the world including, in Canada, federal and provincial statutes, regulatory instruments, common law principles, and the civil law rules in Quebec (Hadskis 2007).

Arguably, the most important source for research ethics norms in Canada is the *Tri-Council Policy Statement: Ethical Conduct for Research Involving Humans* (TCPS 2) (CIHR 2010). The original version was issued in 1998 by the Medical Research Council [Canadian Institutes for Health Research (CIHR) today], the National Sciences and Engineering Research Council (NSERC), and the Social Sciences and Humanities Research Council (SSHRC) (together, the "Agencies"). The updated version, TCPS 2, came into effect November 29, 2010. It is the result of a lengthy revision process undertaken by the Interagency Advisory Panel on Research Ethics (PRE), and reflects the commitment of the Agencies to ensure the TCPS remains a living and evolving document.

The TCPS 2 establishes an ethical framework to which all human subject research funded by any of the Agencies must conform, or risk losing current funding and ineligibility for future funding (CIHR 2006). It has also emerged as the national norm for the evaluation of human subject research through independent Research Ethics Boards (REBs) (e.g., some professional organizations, such as the College of Physicians and Surgeons of Alberta, have voluntarily adopted its standards), and compliance is often required for publication in peer-reviewed journals (Hadskis 2007; Downie and McDonald 2004). It does not, however, have general application to research falling outside the foregoing categories. It is currently unclear exactly how much human subject research in Canada is not regulated by the TCPS 2 framework (Downie and McDonald 2004).

Respect for human dignity is the TCPS 2's key underlying value, expressed through a commitment to Respect for Persons, Concern for Welfare, and Justice (CIHR 2010, Article 1.1). As articulated in the first chapter, "[t]hese core principles transcend disciplinary boundaries and, therefore, are relevant to the full range of research covered by this Policy" (CIHR 2010, at p. 7). Other aspects of the TCPS 2 which will be particularly relevant for neuroimaging researchers include the requirements for consent (Chap. 3); fairness and equity in research participation (Chap. 4); privacy and confidentiality (Chap. 5) and conflicts of interest (Chap. 7).

Other salient points of principle and practice for researchers may be found in professional norms and codes of conduct, if the researcher involved is a member of a regulated profession (e.g., Canadian Medical Association's *Code of Ethics* (2004), and the common law. Jurisprudence addressing the issues of duty of care and standard of care in negligence cases is of particular relevance to issues relating to the conduct of human subject research (e.g., *Halushka v. Saskatchewan*; *Reibl v. Hughes*; *Weiss v. Solomon*). It is important that neuroimaging researchers be aware of and comply with their particular obligations, which will vary to some extent depending on their jurisdiction, profession, place of work, funding source and other material factors.

5 Research Ethics Issues in the Neuroimaging Context

While many of the research ethics issues engaged by neuroimaging research are not necessarily unique to this field, a number of topics are particularly relevant in this context.

5.1 Recruitment and Selection of Participants

Subject recruitment is a key step for all human subject research. In neuroimaging research, wide ranging potential applications and complex statistical analyses or pooling of collected data require special consideration of participant recruitment and selection issues. For research engaging questions of the normal human brain, healthy volunteers will be desired. In this context, recruitment may be more directly undertaken. However, comparative study of the developing, abnormal or diseased brain may require the recruitment of members of vulnerable populations, which raises additional ethical considerations.

Vulnerable populations can be defined as groups of persons who may be more easily harmed, manipulated or coerced as research participants due to their diminished competence or disadvantaged state (Rogers 1990; Sutton et al. 2003). In the context of neuroimaging research, targeted vulnerable populations may include minors (children), who can provide valuable insight into the developing brain, the mentally disabled or the mentally ill, and incarcerated individuals (e.g., for studies of violent behavior or other criminal tendencies – e.g., Raine et al. 1998), among others. Each of these groups present its own issues for recruitment and participation; for example, the unique needs and anxieties of child participants (Thomason 2009) and concerns regarding the voluntariness of consent when recruiting incarcerated individuals (Kass et al. 2007). While such groups must be protected from exploitation, principles of justice arguably require that they not be excluded from research altogether, particularly where the long-term research aims may be the development of clinical or therapeutic applications that could benefit individuals belonging to these groups (e.g., the diagnosis and monitoring of mental illness or disease - Desmond and Chen 2002).

Ultimately, in the recruitment of any individual or group for participation in neuroimaging research, fundamental human rights and respect for autonomy must always be held as paramount over scientific goals. Adhering to this approach will require a careful balancing of the relative risks of neuroimaging research against the interests of science and any potential therapeutic or clinical advancements. This balancing necessarily engages strict requirements for fully informed and voluntary consent – discussed further below (Rosen and Gur 2002). Overall, in order to preserve trust in the neuroimaging enterprise, these issues must be addressed through open dialogue and communication between researchers and potential participants (or their agents), with regard to the goals and risks associated with a given study (see Sutton et al. 2003).

5.2   Informed Consent

Informed consent is a foundational requirement for the ethical practice of research involving human participants, and neuroimaging is no exception. In this regard, neuroimaging is no exception. In substance, the informed consent of a participant to involvement in a research study is intended to represent a process of communication between the researcher and participant, and should not be reduced to a legal technicality. In the context of neuroimaging research, complying with the requirements of informed consent may be challenging in a number of different respects.

Informed consent requires that participants be aware of the material risks associated with their participation in a given study. These may include physical, psychological, emotional and social risks, among others. While the physical risks associated with neuroimaging research, and particularly fMRI scanning, may be relatively straightforward, other forms of risk may be more difficult to predict. Psychological risks may manifest in the form of discomfort or anxiety from having to remain still within the relatively small magnetic chamber (Marshall et al. 2007), or may involve unpredictable reactions to study results or incidental findings. Even more challenging are the potential social risks (e.g., stigmatization, discrimination – see Marshall et al. 2007), which may be associated with future uses of the technology; for example, behavioral diagnosis, lie detection, or insurance eligibility. The relative novelty of various neuroimaging techniques and applications also makes it challenging for researchers to adequately disclose the scientific and therapeutic limitations of some studies, and to address currently unanticipated future uses of the data. Nonetheless, the level of public excitement and hype surrounding this area of research makes it particularly important that researchers take all possible steps to ensure participants are as informed as possible about the limitations of the particular research and the technology itself.

5.3 Incidental Findings

Incidental findings in neuroimaging research can be defined as "observations of potential clinical significance unexpectedly discovered in healthy subjects or in patients recruited to brain imaging research studies and unrelated to the purpose or variables of the study" (Illes et al. 2006a, p. 783). While incidental findings are a concern that arise in other areas of human subject research (e.g., genetics), they are one of the most widely discussed research ethics issues in the context of neuroimaging (e.g., see Illes and Chin 2008). Research with neuroimaging study participants in both medical and nonmedical settings has shown that many participants expect that their scans will detect any structural brain abnormalities. This research further suggests that participants also expect that any clinically relevant information discovered will be reported to them (Kirschen et al. 2006). Accordingly, the potential discovery of such abnormalities may serve as an incentive for individuals to participate in neuroimaging research (Illes et al. 2008), regardless of whether or not such a discovery is likely (or possible) given the technology being used or the nature of the research protocol (Wolf et al. 2008). Researchers must be cognizant of and address this possible expectation both when recruiting subjects and when obtaining their consent to participate.

There is a growing consensus among commentators that researchers must develop clear and transparent procedures for dealing with incidental findings prior to embarking on the research study, and that those protocols must be discussed in detail with prospective participants so as to minimize any possibility of misunderstanding (Royal and Peterson 2008; Wolf et al. 2008). The risks of false negatives and false positives in identified incidental findings should also be considered and addressed (Royal and Peterson 2008). At a minimum, an appropriate consent procedure must be developed that includes discussion of the likelihood of an incidental finding, the duties of the researcher if one is discovered, and the potential risks and benefits associated with its disclosure to the participant (Illes et al. 2008; Wolf et al. 2008).

5.4 Other Emerging Areas of Concern

The foregoing are of course only a few of the research ethics issues implicated by neuroimaging research. Other topics of note include privacy and confidentiality of neuroimaging scans and data, communication of research results, data sharing and future use of collected data. In addition to general concerns regarding the privacy and confidentiality of research participants that are common to all areas of human subject research, and particularly health-related research, neuroimaging also brings additional unique concerns regarding brain privacy, or privacy over our thoughts and other aspects of personality (Farah et al. 2009; Alpert 2007). Whether there are real and current risks associated with these concerns is debatable at the present time. However, the issues certainly bear monitoring, particularly as current Canadian privacy legislation may prove inadequate to address these concerns.

The communication of research results is also a critical obligation for all researchers. In the context of neuroimaging research, the potential for inappropriate hype makes responsible communication of results, including their potential limitations, especially vital in this forum. The importance of this point is particularly highlighted with the current debates around the state of the science and the accuracy of conclusions that can be drawn from studies in this area (Poldrack 2006, Kulynych 2002). Guidance has been developed on the issue of science communication more generally (Bubela et al. 2009), and in the neuroscience context in particular (Illes et al. 2010), and neuroscience researchers should endeavor to incorporate these principles into their practices.

Finally, other issues that are likely to require further consideration in the future pertain to data sharing within the research community. It appears that sharing results via neuroimaging data banks and neuroinformatic platforms may be an increasing trend in this field (Racine and Illes 2007). Given that secondary uses of neuroimaging data raise additional research ethics issues (Wolf et al. 2008), these developments must be monitored to ensure the relevant issues are addressed appropriately as they arise.

6   Conclusion

Neuroimaging is unquestionably a highly intriguing area of research with significant potential both for diagnosing neural pathology and improving our understanding of how the human brain functions. Accordingly, it is not surprising that it has garnered much attention in scientific, clinical, and public domains. As the areas of inquiry for neuroimaging research, and their potential applications, continue to grow, so do the attendant ethical issues. There are a number of traditional areas of research ethics that are particularly important in the context of neuroimaging, and compliance with the associated fundamental principles can be challenging. Even where the principles are clear (e.g., necessity of informed consent), their operation in the neuroimaging context is not necessarily straightforward. Indeed, research shows significant variations in how Canadian REBs handle proposed neuroimaging research studies, even where the key principles are recognized (de Champlain and Patenaude 2006). Nonetheless, it is essential that researchers and REBs endeavor to maintain public trust and confidence with the research community by upholding fundamental ethical principles for all human subject research (McDonald 2000), including neuroimaging. This commitment will require the continued vigilance and self-monitoring of researchers and governing bodies alike to ensure that scientific and research progress develops in an ethically responsible manner.

Acknowledgments The authors thank the Canadian Institutes of Health Research for funding, our NeuroSCAN project collaborators and the University of Alberta's Health Law Institute for research and administrative support.

References

(2008) The ethical neuroscientist, Editorial Comment. Nat Neurosci 11(3):239

Alpert S (2007) Brain privacy: how can we protect it? Am J Bioeth 7(9):70–73

Bandettini P (2009) What's new in neuroimaging methods? Ann N Y Acad Sci 1156:260–293

Bartels A, Zeki S (2000) The neural basis of romantic love. Neuroreport 11(17):3829–3834

Bubela T, Nisbet MC, Borchelt R, Brunger F, Critchley C, Einsiedel E, Geller G, Gupta A, Hampel J, Hyde-Lay R, Jandciu EW, Jones SA, Kolopack P, Lane S, Lougheed T, Nerlich B, Ogbogu U, O'Riordan K, Ouellette C, Spear M, Strauss S, Thavaratnam T, Willemse S, Caulfield T (2009) Science communication reconsidered. Nat Biotechnol 27(6):514–518

Canadian Charter of Rights and Freedoms, Constitution Act, 1982, (U.K.) 1982, c. 11, Schedule B

Canadian Institutes of Health Research (2006) CIHR Procedure for Addressing Allegations of Non-Compliance with Research Policies. http://www.cihr-irsc.gc.ca/e/25178.html. Accessed 12 June 2009

Canadian Institutes of Health Research (CIHR), Natural Sciences and Engineering Research Council of Canada, Social Sciences and Humanities Research Council of Canada (2010) Tri-Council Policy Statement: Ethical Conduct for Research Involving Humans, 2nd Ed. http://www.pre.ethics.gc.ca/pdf/eng/tcps2/TCPS_2_FINAL_Web.pdf. Accessed 15 December 2010

Canadian Medical Association (2004) CMA code of ethics. http://policybase.cma.ca/PolicyPDF/PD04-06.pdf. Accessed 17 December 2010

Canli T, Amin Z (2002) Neuroimaging of emotion and personality: scientific evidence and ethical considerations. Brain Cogn 50(3):414–431

Canli T, Zhao Z, Brewer J, Grabrieli J, Cahill L (2000) Event-related activation in the human amygdala associated with later memory for individual emotional experience. J Neurosci 20 (19): RC99, 1–5

Caulfield T, Ogbogu U (2008) Biomedical research and the commercialization agenda: a review of main considerations for neuroscience. Account Res 15(4):303–320

Caulfield T, Rachul C, Zarzeczny A (2010) Neurohype and the name game: who's to blame? AJOB Neurosci 1(2):13–15

Cho MK (2002) Conflicts of interest in magnetic resonance imaging: issues in clinical practice and research. Top Magn Reson Imaging 13(2):73–38

Congdon E, Canli T (2008) A neurogenetic approach to impulsivity. J Personal 76(6):1447–1484

Council for International Organizations of Medical Sciences (CIOMS) in collaboration with the World Health Organization (WHO) (1993) International ethical guidelines for biomedical research involving human subjects. Geneva

Davatzikos C, Rapurel K, Fan Y, Shen DG, Acharyya M, Loughead JW, Gur RC, Langleben DD (2005) Classifying spatial patterns of brain activity with machine learning methods: application to lie detection. Neuroimage 28(3):663–668

de Champlain J, Patenaude J (2006) Review of a mock research protocol in functional neuroimaging by Canadian research ethics boards. J Med Ethics 32(9):530–534

Declaration of Helsinki (1964) Declaration of Helsinki with amendments in 1975, 1983, 1989, 1996, 2000 and 2008. World Medical Association, Helsinki, Finland. http://www.wma.net/e/policy/b3.htm. Accessed 26 May 2009

Desmond JE, Chen SH (2002) Ethical issues in the clinical application of fMRI: factors affecting the validity and interpretation of activations. Brain Cogn 50(3):482–497

Downie J, McDonald F (2004) Revisioning the oversight of research involving humans in Canada. Health Law J 12:159–181

Downie J, Murphy R (2007) Inadmissible, eh? Am J Bioeth 7(9):67–69

Farah M, Smith ME, Gawuga C, Lindsell D, Foster D (2009) Brain imaging and brain privacy: a realistic concern? J Cogn Neurosci 21(1):119–127

Feigenson N (2006) Brain imaging and courtroom evidence: on the admissibility and persuasiveness of fMRI. Int J Law Context:2:233–255

Green RM (2006) From genome to brainome: charting the lessons learned. In: Illes J (ed) Neuroethics: defining the issues in theory, practice, and policy. Oxford University Press, New York, pp 105–121

Hadskis M (2007) The regulation of human biomedical research in Canada. In: Downie J, Caulfield T, Flood C (eds) Canadian health law and policy, 3rd edn. Markham, LexisNexis Canada Inc., pp 257–310

Hakun JG, Ruparel K, Seelig D, Busch E, Loughead JW, Gur R, Langleben DD (2009) Towards clinical trials of lie detection with fMRI. Social Neurosci 4(6): 518–527

Halushka v. University of Saskatchewan (1965), 53 D.L.R. (2d) 437 (Sask. C.A.)

Hart A, Whalen P, Shin L, McInerney S, Fischer H, Rauch S (2000) Differential response in the human amygdale to racial outgroup vs. ingroup face stimuli. Neuroreport 11(11):2351–2355

Haynes J, Rees G (2006) Decoding mental states from brain activity in humans. Nat Rev Neurosci 7(7):523–534

Illes J (ed) (2006) Neuroethics: defining the issues in theory, practice, and policy. Oxford University Press, New York

Illes J, Bird S (2006) Neuroethics: a modern context for ethics in neuroscience. Trends Neurosci 29(9):511–517

Illes J, Chin V (2008) Bridging philosophical and practical implications of incidental findings in brain research. J Law Med Ethics 36(2):298–304

Illes J, Kirschen M, Edwards E, Stanford L, Bandettini P, Cho M, Ford P, Glover G, Kulynych J, Macklin R, Michael D, Wolf S (2006a) Incidental findings in brain imaging research. Science 311(5762):783–784

Illes J, Racine E, Kirschen M (2006b) A picture is worth a 1000 words, but which 1000? In: Illes J (ed) Neuroethics: defining the issues in theory, practice, and policy. Oxford University Press, New York, pp 149–168

Illes J, Kirschen M, Edwards E, Bandettini P, Cho MK, Ford P, Glover GH, Kulynych J, Macklin R, Michael DB, Wolf SM, Grabowski T, Seto B (2008) Practical approaches to incidental findings in brain imaging research. Neurology 70(5):384–390

Illes J, Moser MA, McCormick JB, Racine E, Blakeslee S, Caplan A, Check Hayden E, Ingram J, Lohwater T, McKnight P, Nicholson C, Phillips A, Sauve KD, Snell E, Weiss S (2010) Neurotalk: improving the communication of neuroscience research. Nat Rev Neurosci 11:61–69

Jackson G (2006) A curious consensus: brain scans prove disease? Ethical Hum Psychol Psychiatry 8(1):55–60

Kass NE, Myers R, Fuchs EJ, Carson KA, Flexner C (2007) Balancing justice and autonomy in clinical research with healthy volunteers. Nature Clin Pharmacol Ther 82(2):219–227

Kaufman M (ed) (2001) Brain imaging in substance abuse: research, clinical, and forensic applications. Humana Press Inc, Totowa

Kerr I, Caulfield T (2007) Emerging health technologies. In: Downie J, Caulfield T, Flood C (eds) Canadian health law and policy, 3rd edn. Markham, LexisNexis Canada Inc., pp 509–538

Kiehl K, Smith A, Hare R, Mendrek A, Forster B, Bring J, Liddle P (2001) Limbic abnormalities in affective processing by criminal psychopaths as revealed by functional magnetic imaging. Biol Psychiatry 50(9):677–684

Kirschen M, Jaworska A, Illes J (2006) Subjects' expectations in neuroimaging research. J Magen Reson Imaging 23(2):205–209

Kulynych J, (2002) Legal and ethical issues in neuroimaging research: human subjects protection, medical privacy, and the public communication of research results. Brain and Cognition 50:345–357

Langleben DD, Schroeder L, Maldjian JA, Gur RC, McDonald S, Ragland JD, O'Brien CP, Childress AR (2002) Brain activity during simulated deception: an event-related functional magnetic resonance study. Neuroimage 15(3):727–732

Lee T, Chan A, Raine A (2009) Hypersensitivity to threat stimuli in domestic violence offenders: a functional magnetic resonance imaging study. J Clin Psychiatry 70(1):36–45

Lieberman M, Hariri A, Jarcho J, Eisenberger N, Bookheimer S (2005) An fMRI investigation of race-related amygdala activity in African-American and Caucasian-American individuals. Nat Neurosci 8(6):720–722

Logothetis N (2008) What we can do and what we cannot do with fMRI. Nature 453 (7197):869–878

Madrigal A (16 March 2009) MRI lie detection to get first day in court, Wired Science.http://blog. wired.com/wiredscience/2009/03/noliemri.html. Accessed 16 December 2010

Marshall J, Martin T, Downie J, Malisza K (2007) A comprehensive analysis of MRI research risks: in support of full disclosure. Can J Neurol Sci 34(1):11–17

McClure S, Li J, Tomlin D, Cypert K, Montague L, Montague PR (2004) Neural correlates of behavioral preference for culturally familiar drinks. Neuron 44:379–387

McDonald M (2000) The governance of health research involving human subjects. Ottawa, Law Commission of Canada. http://www.cdha.nshealth.ca/default.aspx?page=DocumentRender&doc. Id=320. Accessed 17 December 2010

McMonagle E (2007) Functional neuroimaging and the law: a Canadian perspective. Am J Bioeth 7(9):69–70

Mobbs D, Yu R, Meyer M, Passamonti L, Seymour B, Calder A, Schweizer A, Frith C, Dalgleish T (2009) A key role for similarity in vicarious reward. Science 324(5929):900

O'Connor MF, Wellisch DK, Stanton AL, Eisenberger NI, Irwin MR, Lieberman MD (2008) Craving love? Enduring grief activates brain's reward centre. Neuroimage 42(2):969–972

Poldrack R (2006) Can cognitive processes be inferred from neuroimaging data? Trends in Cog Sci 10(2):59–63

Poldrack RA (2008) The role of fMRI in cognitive neuroscience: where do we stand? Curr Opin Neurobiol 18:223–227

R. v. Beland [1987] 2 S.C.R. 398

Racine E, Illes J (2007) Emerging ethical challenges in advanced neuroimaging research: review, recommendations and research agenda. J Empir Res Hum Res Ethics 2(2):1–10

Racine E, Bar-Ilan O, Illes J (2006) Brain imaging: a decade of coverage in the print media. Sci Commun 28(1):122–142

Racine E, Van der Loos HZA, Illes J (2007) Internet marketing of neuroproducts: new practices and healthcare policy challenges. C Q Healthcare Ethics 16(2):181–194

Raine A, Meloy JR, Bihrle S, Stoddard J, LaCasse L, Buchsbaum MS (1998) Reduced prefrontal and increased subcortical brain functioning assessed using positron emission tomography in predatory and affective murderers. Behav Sci Law 16(3):319–332

Reibl v. Hughes [1980] 2 S.C.R. 880

Rogers B (1990) Ethics and research. AAOHN J 38(12):581–585

Rosen A, Gur R (2002) Ethical considerations for neuropsychologists as functional magnetic imagers. Brain Cogn 50(3):469–481

Royal J, Peterson B (2008) The risks and benefits of searching for incidental findings in MRI research scans. J Law Med Ethics 36(2):305–314

Sutton L, Erlen J, Glad J, Siminoff L (2003) Recruiting vulnerable populations for research: revisiting the ethical issues. J Prof Nurs 19(2):106–112

Thomason M (2009) Children in non-clinical functional magnetic resonance imaging (fMRI) studies give the scan experience a "thumbs up". Am J Bioeth 9(1):25–27

Thompson S (2007) A brave new world of interrogation jurisprudence? Am J Law Med 33(2–3):341–357

Vul E, Harris C, Winkielman P, Pashler H (2009) Puzzlingly high correlations in fMRI studies of emotion, personality, and social cognition. Perspect Psychol Sci 4(3):274–290

Weisberg D, Keil F, Goodstein J, Rawson E, Gray J (2008) The seductive allure of neuroscience explanations. J Cogn Neurosci 20(3):470–477

Weiss v. Solomon, [1989] 48 C.C.L.T. 280 (Qc.C.S.)

Wolf S, Lawrenz F, Nelson C, Kahn J, Cho M, Clayton E, Fletcher J, Georgieff M, Hammerschmidt D, Hudson K, Illes J, Kapur V, Keane M, Koenig B, LeRoy B, McFarland E, Paradise J, Parker L, Terry S, Van Ness B, Wilfond B (2008) Managing incidental findings in human subjects research: analysis and recommendations. J Law Med Ethics 36(2):219–248

The Council of Europe's Next "Additional Protocol on Neuroscientific Research"?

Toward an International Regulation of Brain Imaging Research

Caroline Rödiger

Abstract Modern neurosciences are expected to be a twenty-first century challenge for manifold reasons. From the medical law perspective, opinions are highly divided concerning the optimal way to manage incidental findings in brain imaging research. Brain abnormalities place both the researcher and the volunteer in legally complex situations. Which are the duties of the researcher and which rights is the participant allowed to claim? The number of cross-national research projects is growing and adequate data sharing systems have been set up to enable the exchange more efficient, so that incidental findings have become a significant issue even from the international perspective. This remarkable development makes it necessary to envisage an instrument that has the potential to regulate neuroscientific research on international level. The Council of Europe has already presented specific approaches to the major biomedical problems of this time. The Convention on Human Rights and Biomedicine and the Additional Protocol on Biomedical Research could therefore serve as guidelines for a new document dealing with neuroscientific research.

1 Introduction

Neuroscientific research has become one of the largest research areas within the entire sphere of modern biology. Important discoveries are coming to light almost daily, improving the understanding of human brain structures and their functions and advancing medical care for patients with serious diseases. Today, brain

The author is postgraduate and research associate at the Institute of Science and Ethics, University of Bonn. She is a member of the research project "NeuroSCAN: Ethical and Legal Aspects of Norms in Neuroimaging" funded by the Federal Ministry of Education and Research.

C. Rödiger (✉)
Institute of Science and Ethics, University of Bonn, Bonner Talweg 57, 53113 Bonn, Germany
e-mail: roediger@iwe.uni-bonn.de

T.M. Spranger (ed.), *International Neurolaw*,
DOI 10.1007/978-3-642-21541-4_6, © Springer-Verlag Berlin Heidelberg 2012

imaging helps to diagnose Alzheimer in a very early stage[1] or to reduce the rate of misdiagnosis of the vegetative state[2] that has otherwise been estimated to be up to 40%.[3] The breakthroughs in neuroscientific research exert also significant influence on several other disciplines, among them law. In the last few years, a high number of research projects have been supported to investigate the impact of neuroscientific knowledge on different fields of law. One of the most prominent projects is the Law and Neurosciences Project[4] that was set up by a 3-year $10 million MacArthur Foundation grant in 2007 and aimed at examining the influence of modern neurosciences on criminal law in the U.S. justice system, for example, the use of fMRI lie detection tests in courts. Besides the application possibilities of neurosciences, the generation of neuroscientific results raises important legal questions, too. The regulatory framework which consists of generally accepted principles of biomedical research such as the informed consent, the risk-benefit analysis, or the principle of confidentiality, is particularly challenged by incidental findings. An incidental finding is a finding concerning an individual research participant that has potential health or reproductive importance and that is discovered in the course of conducting research but is beyond the aims of the study.[5] The disclosure of abnormalities can be a chance for an early treatment of brain diseases, but it can also have severe social and financial consequences. Volunteers could suffer from psychological distress and social stigma and might be discriminated against when applying for assurance and employment applications.[6] The wide range of implications depends on which procedure for handling incidental findings is undertaken by the researcher. Until now, they vary widely from one research project to another. According to a study of Brown and Hasso, 36% of researchers reported that the scans are read by a neuroradiologist and that all findings are disclosed, 47% responded that only suggestive findings are disclosed, 4% answered that disclosure depends on the type of study, and 13% reported that incidental findings are not disclosed at all.[7] Working groups, research committees, and other institutions have already published several guidelines, but they remain insufficient as they propose ethical but not legal solutions. Since this topic becomes also more and more relevant on the international level, it is high time to consider the elaboration of an international document on neuroscientific research if current treaties on biomedical research do not serve as an adequate and sufficient approach with a view to neurospecific issues.

---

[1] Butcher et al. (2009), p. 653.

[2] Monti et al. (2010), p. 588.

[3] Bosco et al. (2010), p. 88.

[4] http://www.lawneuro.org/. Accessed 30 Dec 2010.

[5] Wolf (2008), p. 219.

[6] Anonymous (2005), p. 17.

[7] Brown and Hasso (2008), p. 1426.

2   Legal Implications of Incidental Findings Raised by High-Tech Brain Imaging

Functional magnetic resonance imaging (fMRI) is one of the most popular techniques in modern brain imaging research due to its relatively low invasiveness and absence of radiation exposure. It is continuously improved in both temporal and spatial resolution, but this remarkable development also involves serious problems. Until now, high-resolution images usually need long scan time. For instance, the total acquisition time for a two-dimensional image at matrix size 1,024 needs 1,024 phase encodings in a rectangular sampling or 1,024 excitations when one phase encoding is performed per excitation, involving a total acquisition time of 102.4 s at pulse repetition time of 100 ms in T1-weighted imaging.[8] To acquire a three-dimensional image with 60 slices, the total acquisition time rises up to 102.4 min.[9] Furthermore, to visualize small anatomic structures in human brain, images with high resolution (~0.22 mm) are preferably produced with whole body 7T scanners.[10] These scanners have a bore of 60 cm and a length of 3.50 m, surrounding the participant by 425 tons of steel.[11] High-tech neuroimaging research studies necessitate careful selection of volunteers. The researcher has to make sure that they are not afflicted by anxiety or other emotional stress reactions.

Another challenge of high-field fMRI scanners is that at 7T every metal implant and even nonferromagnetic materials such as a surgical calotte fixation may lead to disturbing artifacts.[12] In addition, modern brain imaging technology has the potential to visualize smaller-sized findings[13] and thus raises the number of incidental findings. Brain abnormalities appear in nearly 40% of fMRI studies, but they are only clinically relevant in up to 8% of cases.[14] Their detection is both a blessing and a curse. On the one hand, it can be a chance for a more promising therapy and become even life-saving, on the other hand the incidental discovery can implicate a large number of social, psychic, financial, legal, and occupational consequences. This bilateral risk is going to be illustrated by the following case studies.

2.1   Case Study 1

A student volunteers for an fMRI research study and has been informed about the procedure by the researcher. The scans are not read by a neuroradiologist. A few

---

[8] Qian et al. (2010), p. 534.

[9] Qian et al. (2010), p. 534.

[10] Qian et al. (2010), p. 534.

[11] Gizewski (2010), p. 36.

[12] Gizewski (2010), p. 36.

[13] Kubicki (2009), p. 422.

[14] Illes et al. (2004), pp. 743–747.

years later, the student experiences an epileptic seizure and undergoes a neurological and neurosurgical evaluation, whereupon the physician discovers a brain tumor. The student shows a copy of the fMRI scan to the neurosurgeon. He identifies a tumor and surmises that the student could have been healed at the time of the research study.[15]

2.2   Case Study 2

A head of a family takes part in an fMRI research study. The researcher has the fMRI scan read by a neuroradiologist who discovers an incidental finding that turns out to be clinically relevant. Thereupon, the researcher informs the participant about the brain tumor. According to domestic law, the participant mentions the incidental finding to his insurance which rejects the application for occupational disablement insurance for this reason.[16]

2.3   Résumé

Both case studies show that volunteering in an fMRI study might have serious consequences depending on the proceeding. Neuroscientific research projects are becoming bigger and more international, including researchers from all over the world, so that the issue of incidental findings has become relevant even at the international level: The cross-national research studies require agreements on the protection of volunteers, not least to enable data sharing without varying degrees of confidentiality protection. So far, a treaty on neuroscientific research does not exist. However, several practical recommendations for neuroimaging research and international documents on biomedical research have already been drawn up and could serve as guidelines for a new document on neuroscientific research.

3   Regulatory Framework Applicable for Neuroscientific Research

Incidental findings are a hot topic in neuroethics and can be described as an opened "Pandora's Box"[17] with a view to their manifold legal implications. In absence of any national code or international treaty, groups of authors have published practical approaches for dealing with neuroscientific issues,[18] but for the most part, these guidelines contain simply ethical suggestions and do not envisage *legal obligations*

---

[15] Modification to a case study by Nelson (2008), p. 316.

[16] According to Anonymous (2005), p. 17.

[17] Illes and Chin (2008), p. 303.

[18] For example, Heinemann et al. (2007).

of the researcher and *enforceable rights* of the participant. Some Federal Common Rules on neuroscientific research include general stipulations such as the risk-benefit analysis, but they do not consider neurospecific issues relating to the question if the researcher is obligated to look for brain abnormalities or if s/he has to obey a precise procedure in case of incidental findings.[19]

On international level, several regulatory frameworks on or applicable to biomedical research exist such as the Nuremberg Code of 1947, the World Medical Association's Declaration of Helsinki of 1964, with its latest revision of 2008, the UNESCO's Universal Declaration on the Human Genome and Human Rights of 1997 and its Universal Declaration on Bioethics and Human Rights of 2005, and last but not least, the Council of Europe's Convention on Human Rights and Biomedicine of 1997. But with the exception of the Council of Europe's Convention, they do not have any *legally binding* force and thus do not result in legally enforceable commitments for states. Furthermore, these instruments set out legal standards such as the informed consent and other biomedical principles, but of course, they give no answer to the question of incidental findings. However, according to its statute, the Council of Europe has the competence to elaborate Additional Protocols to the Convention on Human Rights and Biomedicine that recommend specific solutions to diverse fields of science such as reproductive medicine and genetics. Thus, it is interesting to raise the question whether or not the Council also has the competence to set up similar documents on the field of neuroscience.

3.1   The Council of Europe

The Council of Europe[20] is an international organization that was founded in 1949 with the aim to protect human rights and to strengthen pluralist democracy. In the last 15 years, the Council of Europe has also made an important contribution to the development of an international biomedical law and was the first international organization to work out a *binding* international treaty concerning biomedicine.[21] The Council of Europe's Convention on Human Rights and Biomedicine (ECHRB)[22] was opened to signature in 1997 and 26 states have ratified to date.[23] Inter alia, the Convention seeks to establish a balance between the protection of volunteers and freedom of biomedical research which brings benefit to individuals suffering from diseases. Art. 15 ECHRB deals with the notion of "freedom of scientific research" and can be cited as proof that the Convention wants to take a

---

[19] Tovino (2009), p. 242.

[20] http://www.coe.int/DefaultEN.asp. Accessed 30 Dec 2010. The Council of Europe is neither identical with the European Council nor with the Council of the EU.

[21] For further information: Zilgalvis (2004), pp. 163–173.

[22] Council of Europe (1997).

[23] Eight states have signed, but not yet ratified the ECHRB.

positive approach to developments in modern medicine and biology, instead of taking the opposite strategy of issuing a series of prohibitions.[24] The ECHRB is concretized by Additional Protocols concerning different research fields. These protocols have the potential to present a specific solution to the major problems of this time. For example, the Council of Europe reacted rapidly to the production of the cloned sheep "Dolly" by working out a new protocol including ethical and legal provisions in the area of reproductive medicine. Even though the elaborated Additional Protocol on the Prohibition of Cloning Human Beings[25] left open a few questions such as the precise meaning of "human being" (or it left the precise definition to the Member States),[26] its elaboration was an effective instrument to set out legal requirements for reproductive research and to foster international discussion, last but not least because the documents of the Council of Europe have the potential to become universal instruments. The Council of Europe stresses the need for international cooperation in the field of biomedical research to extend the same protections for individuals in this field beyond its European Member States. Art. 34 ECHRB explicitly allows inviting nonmember states of the Council of Europe to sign and ratify the Convention. The wide-ranging approach of the documents of the Council of Europe has also inspired the work of other international organizations. For instance, the Preamble of the UNESCO Universal Declaration on Bioethics and Human Rights, adopted in 2005, explicitly recognizes that the European instrument is taken into account and thus underlines the global significance of the work of the Council of Europe. Therefore, the elaboration of a new Additional Protocol could do justice to the issues raised by neuroscientific research even at the international level.

3.2 A New "Additional Protocol on Neuroscientific Research"?

The goal of a new Additional Protocol dealing with neuroscientific research would consist in establishing a balance between the researcher's interests and the protection of the participant. The researcher should be able to do good scientific research in terms of Art. 15 ECHRB, but in full respect of the rights of the participant. For that purpose, the ECHRB and the Additional Protocol on Biomedical Research[27] serve as guidelines and have to be adjusted to respond to the key questions of neuroscientific research. Besides the issue of anxiety, the core problem of neuroimaging research lies in the detection and disclosure of incidental findings. Therefore,

---

[24] "Freedom of scientific research" is even constitutionally protected in some of the Member States (for example, Art. 5 German Constitution, Art. 20 Swiss Constitution).
[25] Council of Europe (1998).
[26] Andorno (2005), p. 141.
[27] Council of Europe (2005).

the new Additional Protocol should focus on the informed consent, as it represents the initial point of this challenge.[28]

3.2.1 Informed Consent

Art. 13 Para. 2 Additional Protocol on Biomedical Research contains detailed provisions for implementation of the informed consent procedure. It stipulates that "information shall cover the purpose, the overall plan and the possible risks and benefits of the research project, and include the opinion of the ethics committee. Before being asked to consent to participate in a research project, the persons concerned shall be specifically informed, according to the nature and purpose of the research:

    i. of the nature, extent and duration of the procedures involved, in particular, details of any burden imposed by the research project;
    ii. of available preventive, diagnostic, and therapeutic procedures;
    iii. of the arrangements for responding to adverse events or the concerns of research participants;
    iv. of arrangements to ensure respect for private life and ensure the confidentiality of personal data;
    v. of arrangements for access to information relevant to the participant arising from the research and to its overall results;
    vi. of the arrangements for fair compensation in the case of damage;
    vii. of any foreseen potential further uses, including commercial uses, of the research results, data or biological materials;
    viii. of the source of funding of the research project."

With a view to incidental findings, the informed consent document should address the following issues.

Nature of the Research Study

The informed consent document should explicitly refer to the nature of the neuro-scientific research project (cf. Art. 13 Para 2 Cl. 2i Additional Protocol on Biomedical Research), i.e. its *non-therapeutic aim*. According to an analysis of Kirschen et al., most of the participants expect any existing pathology to be detected and communicated to them, although subjects provide written consent to a scanning procedure for research purposes alone.[29] Therefore, it is important to underline

---

[28] Other important principles such as the principle of data protection or the principle of compensation for undue damage are going to be illuminated in the context of the informed consent, as the researcher has to provide information about appropriate arrangements.

[29] Kirschen et al. (2006), p. 207.

within the scope of the informed consent procedure that the goal of the research study is to gather data and not to provide any diagnostic or therapeutic benefit, so that brain abnormalities might remain undetected.[30] If required, the researcher has to indicate that s/he is not a medical doctor and thus not qualified to draw clinical conclusions from his/her data.[31]

Burdens of the Research Study

Incidental findings might give cause for serious concern, so that the researcher has to inform about all potential burdens (cf. Art. 13 Para. 2 Cl. 2i Additional Protocol on Biomedical Research): The detection of brain abnormalities is always psychologically distressing. In the majority of cases, brain scans contain only selected images, which are necessary for the research protocol. Consequently, the clinically significance of the incidental finding is not directly clear.[32] As they can turn out to be unimportant from a clinical point of view (false-positive findings),[33] the information about these findings generates undue anxiety. In other cases, disclosure can be a blessing thanks to early (even life-saving) treatments, but it can lead to severe psychological distress, in particular if no therapeutic assistance is available.[34] Incidental findings might influence the participant's whole future, from the family planning to his/her state of insurance and professional career.[35] As the participant might be obliged to inform his/her insurance and employer about the incidental finding according to federal insurance and labor legislation, s/he might be excluded from important private insurance policies and sustain serious discrimination on the employment market.

Furthermore, it is important to provide information about the measures that are taken in case of incidental findings.[36] Expert review of all images is neither feasible with a view to financial burdens nor advisable, because otherwise, the participants would participate in the research study with the expectation that if a brain abnormality exists, it will be discovered and reported. The research study would partially obtain a therapeutic goal. Therefore, it is more advantageous to contact a neuroradiologist for clinical review only if the researcher detects an incidental finding. Its clinical (in)significance is not deciding. In other respects, it would be difficult to determine at what point the incidental finding becomes "significant enough".

---

[30] Illes et al. (2008), p. 386.

[31] Ross (2005), p. 1118.

[32] Bernat (2010), p. 26.

[33] Illes et al. (2006), p. 783.

[34] Wolf et al. (2008), p. 374.

[35] Anonymous (2005), p. 17.

[36] Wolf et al. (2008), p. 366.

Besides, as noted above, brain abnormalities might show up as false-positive findings.

Right Not to Know

Art. 10 Para. 2 Cl. 2 ECHRB recognizes that the "wishes of individuals not to be so informed shall be observed". Disregarding a *wish* might be ethically unacceptable, though from a legal point of view, there are no sanctions. But some authors hold the view that Art. 10 Para. 2 Cl. 2 ECHRB refers to a *right* not to be informed, i.e. a right not to know,[37] so that it has to be clarified if the participant is allowed to exercise this right in the context of neuroscientific research. From a civil and criminal law perspective, the researcher cannot be held responsible for incidentally detected diseases, but s/he is obliged not to expose the participant to disproportional high risks or burdens.[38] This rule derives from the principle of non-maleficence in medical ethics and obliges the researcher to inform the participant about any incidental finding.[39] Hence, if the researcher respects the participant's right not to know, s/he violates the legal obligation to inform the volunteer about an incidental finding. If, in contrast, the researcher informs the participant about a brain abnormality against the subject's will, s/he violates his/her right not to know.[40] A certain way to avoid this dilemma is to exclude those participants from the research study who insist on their right not to know.[41] The others have to reject their right not to know within the scope of the informed consent. This prerequisite for study participation might disadvantage a small part of potential participants who would have been pleased to participate and to exercise the right not to know (approx. 3%),[42] but this approach avoids any legal dilemma ab initio. Furthermore, the participant does not dispose of a "right to participation", so that this exclusion criterion remains unproblematic from the legal point of view.

Data Protection

Neuroscientific research implies the obtainment, collection, classification, and analysis of a high number of data records. They are saved in databases, which enable their use for manifold purposes for an indefinite time and their transfer even across national borders. The researcher has to provide information about the wide

---

[37] Andorno (2004), p. 436.

[38] Heinemann et al. (2007), p. 1985.

[39] Heinemann et al. (2007), p. 1985.

[40] Rödiger (2010), p. 141.

[41] Schleim et al. (2007), p. 1043.

[42] Illes et al. (2006), p. 783.

range of temporal and spatial application possibilities. Brain data have the potential to give very important and sensitive information about the personality of the participant. If third parties become unintentionally aware of the personal information, this knowledge may lead to fatal consequences for the participant. S/he might suffer from social stigma and discrimination, in particular in research studies that examine sexual preferences including homosexual or pedophiliac addictions.[43] Furthermore, data protection is an important issue with a view to insurance and employment applications. If the assurance or the employer becomes aware of the brain abnormality, the volunteer might be excluded from important private insurance policies or sustain severe discrimination on the employment market. Therefore, the researcher has to inform precisely about the envisaged arrangements to ensure the confidentiality of personal data (cf. Art. 13 Para. 2 Cl. 2iv Additional Protocol on Biomedical Research). The creation of a pseudonym would guarantee data protection, on the one hand, and would enable the reidentification of the volunteer in case of an incidental finding on the other hand. If the participant wishes to remain completely anonymous ab initio, s/he has to be excluded from the research study. Otherwise, the researcher would get into trouble if s/he detected an incidental finding that s/he could not trace back to the participant. In this regard, the situation resembles that of the participant intending to exercise his/her right not to know.

Compensation for Undue Damage

The cases of damage are manifold and complex in neuroimaging research. To name a few, the researcher could fail to inform the participant about a brain abnormality or s/he does not provide timely information and delays the disclosure. This becomes even more precarious if the participant could have been healed. In addition, the researcher's duties are not fulfilled, once s/he communicated the incidental finding to the participant. The researcher is obliged to establish contact with a medical doctor. If s/he leaves the volunteer to his/her own devices, the researcher can be confronted with civil claims for damages and even with penal consequences. Thus, the participant has to be sufficiently informed about the arrangements for fair compensation (cf. Art. 13 Para. 2 Cl. 2vi, 31 Additional Protocol on Biomedical Research). The conclusion of a contract between the researcher and its participant is a highly desirable measure, last but not least to ensure compensation for damages. Otherwise, it could be unclear how the participant comes into fair compensation. For example, German research projects usually do not provide contracts and opinions are highly divided concerning the question of how the volunteer obtains compensation.[44]

---

[43] Ulmer et al. (2009), p. E55.
[44] Spranger (2009), p. 194.

Withdrawal of Consent

Art. 13 Para. 3 Additional Protocol on Biomedical Research provides that the person may freely withdraw consent at any time. This regulation can also be applied to neuroscientific research. It has to be specified that the legal effect of the withdrawal is just *ex nunc* (henceforward) and not *ex tunc* (retrospectively). In the other case, the researcher would be exposed to civil claims and would render himself/herself liable to prosecution. The researcher has to make clear that the volunteer is not allowed to withdraw consent s/he gave with regard to the initial rejection of the right not to know. If not, the participant could exercise the right not to know through a "back door" by withdrawing consent concerning the latest rejection.

3.2.2 Informed Consent of Vulnerable Groups

Research on neuropsychiatric disorders such as schizophrenia or Alzheimer's disease necessitates the involvement of vulnerable groups. Art. 17 Para. 1 ECHRB (Art. 15 Para. 1 Additional Protocol on Biomedical Research) stipulates inter alia that research on a person without the capacity to consent is only allowed if "the results of the research have the potential to produce real and direct benefit to his or her health". Art. 17 Para. 2 ECHRB (Art. 15 Para. 2 Additional Protocol on Biomedical Research) regulates that in exceptional cases, research is permitted subject to the following conditions:

i. the research has the aim of contributing through significant improvement in the scientific understanding of the individual's condition, disease or disorder, to the ultimate attainment of results capable of conferring benefit to the person concerned or to other persons in the same age category or afflicted with the same disease or disorder or having the same condition;
ii. the research entails only minimal risk and minimal burden for the individual concerned.

Art. 17 Para. 2 ECHRB (Art. 15 Para. 2 Additional Protocol on Biomedical Research) provides research on persons even if such research is of no immediate benefit for the individual him or herself. This regulation has met sharp criticism in some Member States,[45] because it transforms psychologically and physically disabled people into mere objects to serve other interests. Furthermore, the term of "minimal risk and minimal burden" remains unclear and ambiguous. Such a lack of precision gives too much room for abuse possibilities. Thus, neuroscientific research on vulnerable groups should only be allowed if it confers direct benefit to the participant involved in the study.

---

[45] For example, Germany has not yet ratified the ECHRB because of this regulation.

3.2.3 Informed Consent of Persons Deprived of Liberty

Neuroimaging research on persons deprived of liberty seems indispensable to analyze at what point the brains of psychopaths work differently to external stimuli and to find an adequate treatment. A famous example is that of Kent Kiehl, a neuroscientist at the University of New Mexico Albuquerque, who has used the fMRI to scan more than 1,000 inmates. He discovered that the brains of psychopaths tend to show distinct defects in the paralimbic system, which is a network of brain regions that are involved in processing emotion, inhibition, and attentional control.[46] Persons deprived of liberty are in a weak position. Therefore, it has to be paid big attention to the informed consent process. Art. 20 Additional Protocol on Biomedical Research stipulates that research studies on persons deprived of liberty that do not have the potential to produce direct benefit to their health are only allowed "if the following additional conditions are met:

i. research of comparable effectiveness cannot be carried out without the participation of persons deprived of liberty;
ii. the research has the aim of contributing to the ultimate attainment of results capable of conferring benefit to persons deprived of liberty;
iii. the research entails only minimal risk and burden."

Art. 20 Additional Protocol on Biomedical Research shows broad parallels to Art. 17 Para. 2 Additional Protocol on Biomedical Research with a view to the term of "minimal risk and burden" and the benefit which is not conferred to the participant but to the group s/he belongs to. Neuroscientific research on persons deprived of liberty should only be permitted if the study's aim is to find some personal treatment for the inmate. Otherwise, these persons would become easy prey for ambitious researchers. Moreover, it has to be made sure that the person has not any metal objects such as piercings or fragments from bullets in his/her body. Shackles have to be replaced, for example, by plastic zip ties.

4 Outlook

The Council of Europe has made substantial contributions to the promotion of biomedical research and the protection of volunteers by working out the ECHRB and the Additional Protocol on Biomedical Research. These documents serve as guidelines for a new document concerning neuroscientific research, but they have to be concretized with a view to neurospecific problems. Incidental findings represent a big issue in neuroimaging research and place both the researcher and the participants in severe situations. The initial point of this challenge is the verbal

---

[46] Hughes (2010), p. 340.

and written informed consent, which has to include detailed provisions on the nature of the research study and emerging social and financial burdens, the renouncement of the right not to know as a prerequisite for study participation, the arrangements to ensure data protection and compensation for undue damage as well as the requirements for withdrawal of consent. In addition, the researcher has to observe special rules if his/her study includes research on persons not able to consent or deprived of liberty. A new "Additional Protocol on Neuroscientific Research" would therefore have great potential to present an appropriate approach to the major issues of neuroscientific research.

Acknowledgement The author thanks the Federal Ministry of Education and Research for funding.

References

Andorno R (2004) The right not to know: an autonomy based approach. J Med Ethics 30:435–440

Andorno R (2005) The Oviedo Convention: A European Legal Framework at the Intersection of Human Rights and Health Law. J Int Biotech Law 2:133–143

Anonymous (2005) How volunteering for an MRI scan changed my life. Nature 434(7029):17

Bernat JL (2010) In Neuroimaging Trials, How Do You Handle Controls Who Are Found to Have Incidentally Discovered Abnormalities? Neurol Today 10(7):26–27

Bosco A, Lancioni GE, Belardinelli MO, Singh NN, O'Reilly MF, Sigafoos J (2010) Vegetative state: efforts to curb misdiagnosis. Cogn Process 11:87–90

Brown DA, Hasso AN (2008) Toward a Uniform Policy for Handling Incidental Findings in Neuroimaging Research. Am J Neuroradiol 29:1425–1427

Butcher JN, Mineka S, Hooley JM (2009) Klinische Psychologie. Pearson Studium, München

Council of Europe (1997) Convention on Human Rights and Biomedicine. http://conventions.coe.int/Treaty/en/Treaties/Html/164.htm. Accessed 30 December 2010

Council of Europe (1998) Additional Protocol to the Convention on Human Rights and Biomedicine concerning the Prohibition on Cloning Human Beings. http://conventions.coe.int/Treaty/en/Treaties/html/168.htm. Accessed 30 December 2010

Council of Europe (2005) Additional Protocol to the Convention on Human Rights and Biomedicine Concerning Biomedical Research. http://conventions.coe.int/treaty/en/treaties/html/195.htm. Accessed 30 December 2010

Gizewski ER (2010) High-field fMRI. In: Ulmer S, Jansen O (eds) fMRI – Basics and Clinical Applications. Springer, Berlin, pp 35–42

Heinemann T, Hoppe C, Listl S, Spickhoff A, Elger CE (2007) Incidental findings in neuroimaging: Ethical problems and solutions. Deutsches Ärzteblatt 104(27):A-1982–1987

Hughes V (2010) Science in court: Head case. Nature 464:340–342

Illes J, Kirschen MP, Karetsky K, Kelly M, Saha A, Desmond JE, Raffin TA, Glover GH, Atlas SW (2004) Discovery and disclosure of incidental findings in neuroimaging research. J Magn Reson Imaging 20:743–747

Illes J, Kirschen MP, Edwards E, Stanford LR, Bandettini P, Cho MK, Ford PJ, Glover GH, Kulynych J, Macklin R, Michael DB, Wolf SM, members of the Working Group on Incidental Findings in Brain Imaging Research (2006) Incidental Findings in Brain Imaging research. Science 311:783–784

Illes J, Chin VN (2008) Bridging Philosophical and Practical Implications of Incidental Findings in Brain Research. J Law Med Ethics 36(2):298–304

Illes J, Kirschen MP, Edwards E, Bandettini P, Cho MK, Ford PJ, Glover GH, Kulynyc J, Macklin R, Michael DB, Wolf SM, Grabowski T, Seto B (2008) Practical approaches to incidental findings in brain imaging research. Neurology 70:384–390

Kirschen MP, Jaworska A, Illes J (2006) Subjects' Expectations in Neuroimaging Research. J Magn Reson Imaging 23:205–209

Kubicki H (2009) I still haven't found what I'm looking for, but I may have found something else: non-physician researchers and incidental findings in magnetic resonance imaging. Saint Louis Univ J Health Law Policy 2:413–430

Monti MM, Vanhaudenhuyse A, Coleman MR, Boly M, Pickard JD, Tshibanda L, Owen AM, Laureys S (2010) Willful Modulation of Brain Activity in Disorders of Consciousness. New Engl J Med 362:579–589

Nelson CA (2008) Incidental findings in Magnetic Resonance Imaging (MRI) Brain Research. J Law Med Ethics 36(2):315–319

Qian Y, Zhao T, Hue YK, Ibrahim TS, Boada FE (2010) High-resolution spiral imaging on a whole-body 7T scanner with minimized image blurring. Magn Reson Med 63:543–552

Rödiger C (2010) Ein internationales Abkommen zur Regelung bildgebender Verfahren? Zufallsfunde als Herausforderung für ein künftiges internationales Instrument. In: Spranger (ed) Aktuelle Herausforderungen der Life Sciences. Lit-Verlag, Münster, pp 129–146

Ross K (2005) When volunteers are not healthy. Eur Mol Biol Org Rep 6(12):1116–1119

Schleim S, Spranger TM, Urbach H, Walter H (2007) Zufallsfunde in der bildgebenden Hirnforschung. Empirische, rechtliche und ethische Aspekte, Nervenheilkunde 26:1041–1045

Spranger TM (2009) Rechtliche Implikationen der Generierung und Verwendung neurowis-senschaftlicher Erkenntnisse. In: Schleim, Spranger, Walter (eds) Von der Neuroethik zum Neurorecht? Vandenhoeck & Ruprecht, Göttingen, pp 191–213

Tovino SA (2009) Incidental Findings A Common Law Approach. Account Res 15(4):242–261

Ulmer S, Jensen UR, Jansen O, Mehdorn HM, Schaub J, Deuschl G, Siebner HR (2009) Impact of Incidental Findings on Neuroimaging Research Using Functional MR Imaging. Am J Neuroradiol 30:E55

Wolf SM (2008) Introduction: The Challenge of Incidental Findings. J Law Med Ethics 36(2): 216–218

Wolf SM, Paradise J, Caga-anan C (2008) The Law of Incidental Findings in Human Subjects Research: Establishing Researchers' Duties. J Law Med Ethics 36(2):361–383

Zilgalvis P (2004) European law and biomedical research. In: Council of Europe publishing (ed) Ethical eye – biomedical research. Koelblin-Fortuna-Druck, Baden-Baden, pp 163–173

Legal Landscape of Neuroscientific Research and Its Applications in Finland

Salla Silvola

Abstract This article focuses on the regulation of research and use of neuroscientific knowledge in Finland. As no separate regulation on neuroscientific research exists, legislation on medical research has been taken as a starting point for the legal analysis. The recently extended scope of application of the Medical Research Act has both positive and negative effects to multidisciplinary research projects such as neuroscientific research. Although the Act now takes better account of novel research areas outside the scope of traditional medical science, relevant expertise in the ethics committees may be difficult to find. Generally, the Finnish legislation responds reasonably well to the particularities of neuroscientific research such as incidental findings. As the application of neuroscientific knowledge in many areas is still rather sporadic, there has not been sufficient incentive to introduce legislation in the area. The only recorded demands for guidance in the area involve the use of polygraphs in criminal investigation and court proceedings.

1 Legal Landscape of Neuroscientific Research

1.1 Relationship Between Medical and Neuroscientific Research

The legal landscape of neuroscientific research is tightly connected with the nature of neuroscientific research. Neuroscientific research often combines elements from many different disciplines ranging from technological sciences to

Salla Silvola, LLD (Helsinki), MA (London), Docent in Medical and Bio Law, University of Helsinki, Finland; supported by the project "NeuroSCAN: Ethical Concept and Norms", Academy of Finland (SA 1124638).

S. Silvola (✉)
University of Helsinki, Yliopistonkatu 3, P.O. Box 400014 Helsinki, Finland
e-mail: salla.silvola@helsinki.fi

humanities.[1] Its basis lies, nevertheless, in medical research. Medical research is the sole area of research that is extensively regulated by law in Finland. The legal definition of medical research has recently been extended.

Intervention in the integrity of a person is the basic condition for the Medical Research Act[2] to be applicable. As the name of the Act indicates, the Act is applied to research, which aims to increase knowledge on health or on the causes, symptoms, diagnosis, treatment, and prevention of diseases or the nature of disease in general. The Act is also applied if human embryos or foetuses are investigated for the purposes stated above.[3]

In 2009, the National Advisory Board on Research Ethics published guidelines on the ethical principles on research in humanities, social and behavioural sciences and on organising its ethical review.[4] Broadening the scope of the Medical Research Act was one of the recommendations put forward in the guidelines. In 2010, the scope of the Medical Research Act was extended to include not only medical research in the strict sense, but also research in nursing and health studies, as well as studies in sports and nutrition.[5] All of these research areas share a common denominator. All involve interventions in the physical (and mental) integrity of the research participants for the purpose of increasing knowledge on human health.

Although neuroscientific research often involves elements from, for example, behavioural sciences, it is very likely that any neuroscientific study that also involves elements of medical procedures or the use of medical equipment falls under the scope of application of the Medical Research Act. This was the case before the scope of the Medical Research Act was broadened and it is even more likely now.[6] Therefore, even if one part of a multidisciplinary research project involves an intervention into the physical integrity of a person in a health-related research project, the whole protocol shall be scrutinised by an ethics committee according to the Medical Research Act.[7] Hence, in the following, I will focus on the law of medical research, in particular on the provisions of the Medical Research Act. In addition, the general legal rules on protection of personal data are referred to, as these provisions apply to any activity where handling of personal data is involved, including neuroscientific research.

---

[1] Revonsuo (2009).

[2] Nr. 488 of 1999.

[3] Medical Research Act, Section 2.

[4] National Advisory Board on Research Ethics (2009).

[5] Act on the Amendment of the Medical Research Act Nr. 794 of 2010.

[6] Lötjönen (2009b), p. 216.

[7] Another question is whether all parts of the research protocol should be scrutinised against the standards stated in the Medical Research Act, or only the part that falls under the broadened scope of the Act.

1.2   Constitutional and Human Rights Aspects

As in many constitutions and human rights conventions, freedom of science is also protected in the Constitution of Finland.[8] According to the Constitution, "the freedom of science, the arts and higher education is guaranteed".[9] The interpretation of this provision is not well reported in the preparatory materials.[10] However, it is stated that this provision is closely connected to the provision on freedom of speech and that freedom of science includes the right of the scientist to choose the theme of his or her research and the methods of its investigation.[11]

Scientific freedom is restricted by other constitutional rights such as a person's right to private life and personal integrity.[12] Self-determination is a basic element of both provisions. The rights of research participants cover not only the right to physical integrity against any unauthorised touching, but also the right to privacy in the meaning of protection of personal data. Confidentiality and protection of private life become central topics, for example with regard to any incidental findings that may be discovered during the conduct of a neuroscientific investigation.

The same dichotomy between freedom of scientific research and protection of persons undergoing biomedical research can also be seen at the international level. Finland ratified the Council of Europe Convention on Human Rights and Biomedicine and two of its additional protocols in 2009.[13] According to Article 15 of the Convention, scientific research in the field of biology and medicine shall be carried out freely, albeit subject to the provisions ensuring the protection of the human being. Article 16 sets the general conditions for protection of persons undergoing research and Article 10 states, *inter alia*, that everyone is entitled to know any information collected about his or her health. However, the wishes of individuals not to be so informed shall be observed. In exceptional cases, the patient's right to know may be restricted in the interests of the patient.

1.3   Legal Limits of Neuroscientific Research

The legal preconditions for medical research according to the Medical Research Act can roughly be grouped into two categories: the organisational requirements and the rights of research participants. The organisational requirements include such conditions as a compulsory review by an ethics committee or the qualifications of

---

[8] Act Nr. 731 of 1999.

[9] Section 16, para 3.

[10] Bill Nr. 1 of 1998 and Bill Nr. 309 of 1993.

[11] Bill Nr. 309 of 1993, p. 64. See also Miettinen (2001).

[12] Sections 7 and 10 of the Constitution of Finland.

[13] Council of Europe (1997, 1998, 2002).

the person in charge of the research project. The rights of the research participants are triggered by the actual conduct of the study. Informed consent and the problem of incidental findings belong to this category of legal preconditions.

1.3.1 Organisational Preconditions for Neuroscientific Research

Ethics Committee Review

Medical research protocols have been reviewed in ethics committees in Finland since the end of 1960s. At first, ethics committee review was regulated at a low normative level.[14] It was not until the year 1999 that ethics review became a clear legal requirement along with the adoption of the Medical Research Act. Before 1999, ethics committees functioned at the institutional and departmental level.[15] Many of the researchers were themselves members in the ethics committees where their own research plans were evaluated. Although the researchers could not take part in the handling of their own research protocol, the review by an ethics committee was more akin to guidance by colleagues than neutral and independent evaluation of the research plan.

The Medical Research Act brought along change in the organisation of the ethics committees. The legal requirement was that there ought to be at least one research ethics committee per hospital district, and the ethics committees had to fulfil the criteria for objectivity and lay representation set by law.[16] There are 20 hospital districts in Finland. Some hospital districts with most active research centres founded more than one ethics committee. For the first 10 years after the Medical Research Act entered into force, ethics committee review was conducted by more than 25 regional ethics committees spread across the country. A national research ethics committee was founded to review international multicentre clinical trials on pharmaceuticals and to monitor and guide the regional ethics committees.[17]

In 2010, the system of ethics committee review was changed again. It became obvious that the standards of ethics committees varied depending on the volume of research protocols they reviewed. The functioning of ethics committees was most efficiently organised and the staff was most experienced in larger research centres such as university hospitals.[18] The Medical Research Act was amended to appreciate this. At present, all medical research protocols in the country are reviewed by

---

[14] Immonen (1992), p. 26. See also Lötjönen (1997), p. 874.

[15] Alone in the connection of the Helsinki University Central Hospital, more than 32 ethics committees were in operation in 1986. Penttilä (1992), p. 81.

[16] Sections 18 and 19.

[17] The committee's official name (translated from Finnish) was Sub-Committee on Medical Research of the National Advisory Board on Health Care. It was founded by Decree on the National Advisory Board on Health Care, Nr. 494 of 1998.

[18] Bill nr. 65 of 2010, p. 12.

ethics committees established by five hospital districts with university hospitals. Some hospital districts have founded more than one ethics committee and the number of ethics committees totals nine at present. In addition, the national research ethics committee was reorganised and relocated as part of the organisation of the National Supervisory Authority for Welfare and Health.[19] It is still mandated to review international multicentre clinical trials on pharmaceuticals and to monitor and guide the university hospital ethics committees, but as the number of regional ethics committees has considerably decreased, the national ethics committee is expected to take a larger share of reviewing actual research protocols.

Along with the decrease in the number of ethics committees and the increasing volume of their work, it is inevitable that the expertise and the amount of time available for each research protocol in the ethics committees will become more limited. This is particularly challenging with multidisciplinary research protocols handling novel research areas such as neuroscientific research. Some hospital districts have solved this problem by establishing multiple committees or permanent sub-committees to assist with protocols that require expert knowledge.[20] In addition, all ethics committees have the opportunity to request additional *ad hoc* expert opinions, if necessary. Balancing between objectivity and sufficient expertise can be particularly demanding in a small country.

Person in Charge of the Research Project

The Medical Research Act also requires every research project to have a person in charge of the project. The person in charge of the research shall ensure that there are competent staff and suitable tools and equipment available for the research and that the research is otherwise conducted under safe conditions. It is for the person in charge to ensure that the research is conducted in accordance with the provisions of the Medical Research Act, the international obligations concerning the status of research participants, and the rules and guidelines that govern research.[21]

The person in charge of the research also has the responsibility to suspend the research immediately where required for the safety of the research participant. If new knowledge related to the carrying out of the research or the medicinal product used in the research that has a bearing on the safety of the research participants emerges in the course of the research, the person in charge of the research shall immediately undertake precautionary measures to protect the research participants together with the commissioning party to the research (if a separate commissioning

---

[19] The name and organisation of the national ethics committee was changed by an amendment to the Medical Research Act, Nr. 794 of 2010, and by Decree on National Committee on Medical Research Ethics, Nr. 820 of 2010.

[20] The Hospital District of Helsinki and Uusimaa is an example of the first option, and the ethics committee of the Hospital District of Southwest Finland is an example of the second approach.

[21] Medical Research Act, Section 5, Subsection 2.

party exists). The commissioning party (or, such lacking, the person in charge of the research) shall immediately inform the ethics committee of such new knowledge and of the measures undertaken based thereon. In addition, information and measures regarding clinical trials on medicinal products shall be immediately communicated to the Finnish Medicines Agency.[22]

Until 2010, the professional qualifications of the person in charge of the research project were strictly limited to medical doctors or dentists. However, together with the broader scope of application of the Medical Research Act, more flexibility had to be introduced. At present, the person in charge does not necessarily need to be a medical doctor, unless the research involves a clinical study on pharmaceuticals. In other research projects, the sufficiency of professional and scientific qualifications required from the person in charge of the research are evaluated by the ethics committee on an individual basis in relation to the research project in question.[23] This amendment may well prove useful in multidisciplinary projects such as neuroscientific research.

Other Organisational Aspects

If a neuroscientific research project involves *medicinal products*, the legal criteria for conducting the study become slightly more complicated. Clinical trials on medicinal products are not only regulated at the national level but also at the level of the European Union. The so-called Clinical Trials Directive[24] sets additional conditions that the EU member states have implemented in their national legislation. One of them is that not only does a research project involving medicinal products need to be reviewed by the ethics committee prior to its commencement, but the sponsor shall also be required to submit a valid request for authorisation to the competent authority of the Member State in which the sponsor plans to conduct the clinical trial.[25]

In Finland, the competent authority on medicinal products is the Finnish Medicines Agency (Fimea). According to the directive, the competent authority has 60 days to react to the request by requiring additional information or by approving or rejecting the request. If the competent authority has not reacted in 60 days, the study may go ahead. If the study involves medicinal products for gene therapy, somatic cell therapy or genetically modified organisms, the time period has been extended to 90 days.[26] When *genetically modified organisms* are involved in

---

[22] Medical Research Act, Section 5, Subsection 3.

[23] Medical Research Act, Section 5, Subsection 1 and Bill Nr. 65 of 2010, p. 19.

[24] European Union (2001).

[25] Article 9, para 2.

[26] In the case of xenogenic cell therapy, there is no time limit to the authorisation period. See Article 9, paras 4 and 6 of the Directive. In Finland, the requirements of the Clinical Trials Directive have been implemented in the Medical Research Act and the Act on the Finnish Medicines Agency Nr. 593 of 2009 (which replaced the earlier Act on the National Agency of Medicines Nr. 35 of 1993).

experimental gene therapy or xenogenic cell therapy, a licence by the Board for Gene Technology is also required according to the Gene Technology Act.[27]

Moreover, if the neuroscientific research project involves *medical devices* or their accessories, yet another control authority enters into play. A notification is to be made to the National Supervisory Authority for Welfare and Health, if studies are to be conducted with medical devices without CE marking, if the device is used in the study contrary to the purpose or the instructions defined by the manufacturer, or if the device is an active implantable device as defined in Annex IX of the Council Directive 93/42/EEC of 14 June 1993 concerning medical devices.[28] To use devices that generate *ionising radiation*, a safety licence is required from the Radiation and Nuclear Safety Authority in accordance with the Radiation Act.[29]

According to Section 13 of the Act on Status and Rights of Patients,[30] the health care professionals or other persons working in a health care unit cannot give confidential information contained in *patient records* to outsiders without a written consent by the patient without a specific exception provided in law. In research projects that involve active participation from research participants, acquiring written consent is normally not an issue.[31] Consent for accessing patient records and for handling the patient's health-related data during the research project is acquired at the same time and using the same consent form as for participation in the research project itself. If the participant is not capable of giving informed consent, consent is provided by his or her proxy according to the Medical Research Act.[32]

One final organisational feature to note is that in order to conduct a study in a public health care institution or any other facility, permission from the central administration of that organisation is also required.

---

[27] Nr. 377 of 1995. See also Lehtonen (2006), pp. 246–249.

[28] European Union (1993). See also the Regulation on Clinical Investigation on Medical Devices and their Accessories, National Supervisory Authority for Welfare and Health (2010). The National Supervisory Authority for Welfare and Health is also responsible for issuing licenses for research on embryos and for research on human tissues and cells, if the consent of the tissue or cell donor cannot be acquired. See Medical Research Act, Section 11, and Act on the Medical Use of Human Organs and Tissues, Nr. 101 of 2001, Sections 19 and 20.

[29] Nr. 592 of 1991.

[30] Nr. 785 of 1992.

[31] In other types of research it could be possible, by exception, to get access to patient records without individual written consent. According to Section 28 of the Act on the Openness of Government Activities (Nr. 621 of 1999), the public authority (e.g. a health care centre or a hospital district), in whose care the documents are, may in individual cases give access to patient documents for the purpose of scientific research, if it is obvious that the giving of the information does not violate the interests for the protection of which the secrecy obligation has been prescribed. When considering an application for permission, it must been taken care that the freedom of scientific research is secured. If documents are needed from more than one health care unit, may the National Institute for Health and Welfare grant the permission on behalf of individual health care units (Act on the Status and Rights of Patients, Section 13, Subsection 4).

[32] Lötjönen (2008, 2009a) and Nieminen (2009).

1.3.2   Rights of Research Participants

Weighing Risks and Benefits

All of the organizational requirements listed above are geared towards the protection of research participants. However, until the research project has received approval from a research ethics committee and it has passed the conditions set by e.g. the supervisory bodies for medicinal products and/or medical devices, the research participants are rarely involved in any way.

According to the Medical Research Act, the interests and well-being of research participants must always be put before any benefits to science or society. Measures are to be taken to prevent any risks or harmful effects to the research participant as far as possible, and the research participants may be exposed only to measures where the expected health or scientific benefit is unequivocally greater than the potential risks or harm to the research participant.[33]

Although the interests and well-being of research participants is always to be put before benefits to science or society, this legal condition does not require that there *always* ought to be a health benefit to the participants. This is clear from the further criterion, according to which the participants may be exposed only to measures where the expected health *or scientific benefit* ought to be greater than the potential risks or harm to the research participant. This is crucial to neuroscientific research, as in many neuroscientific research projects the science is not yet sufficiently advanced to promise any health benefits. If there were therapeutic benefits to offer, greater risks could be afforded. However, a research project that does not involve any direct health benefits to the participants will be scrutinised very carefully in terms of the risks involved.

Informed Consent

Formal Requirements

Once the research project has been approved by the ethics committee and all the other organisational requirements have been met, the recruitment of research participants begins. One very concrete legal condition to medical research is the requirement of informed consent. According to the Medical Research Act, medical research cannot be conducted without the research participant's informed consent in writing. Consent in writing also fulfils the condition of express consent required for the processing of sensitive data according to the Personal Data Act.[34] There are exceptions to the rule of written informed consent where consent cannot be obtained owing to the urgency of the matter (research into emergency conditions)

---

[33] Medical Research Act, Section 4.

[34] Personal Data Act, Section 12.

or when consent in writing would be contrary to the interests of the research participant (e.g. studies on drug addicts, where written evidence of consent would deter participants from taking part in the study). Neither of these exceptions, however, applies to neuroscientific research in its usual forms.

However, neuroscientific research may well be conducted on a research participant who is not able to write due to a neurological condition. In this case, provided that the potential research participant has the legal capacity to consent, the research participant can consent orally in the presence of at least one witness who is not dependent on the research.[35]

Informing the Research Participants

According to the Medical Research Act, the research participants must have their rights, the purpose and nature of the research and the procedures it involves properly explained to them. The potential risks and harm must also be properly explained. The Personal Data Act requires that information is given on the purpose of the processing, if and where additional information is collected, how the data is protected and who will have access to the data.[36] All the information above shall be given in a manner that research participants are aware of all issues connected with the research that have a bearing on their decision-making.

As neuroscientific research is still developing and new information appears rapidly, the requirement of informing the research participants fully of *all* the risks involved may be a challenge. In practice, one principle applies: the riskier the research, the more information is required. Although the researchers may not know exactly what kind of information can be extrapolated e.g. from the scanned images of the research participant's brain, they will be able to explain the nature of the research, what kind of procedures and equipment will be involved, how invasive the procedures will be and the physical risks involved in them. Research, by its nature, is inquisitive, and the purpose of it is to gather new scientific information. It should be explained to the participants that unexpected issues may arise.

Incidental Findings

One particular issue that has caused worry in neuroscientific research is the problem of incidental findings.[37] While scanning for neurological activity for research

---

[35] Additional conditions apply to research on minors or adults who do not have the legal capacity to consent themselves. In these cases, the consent of the representative of the participant is required and the level of risk to the participant is regulated more closely. On the problems involved with deteriorating legal capacity and a comparative account on respecting the self-determination of persons with cognitive impairment, see Lötjönen (2009a).

[36] Personal Data Act, Section 26. See also the Medical Research Decree Nr. 986 of 1999, Section 3.

[37] On the ethical dimensions of incidental findings in neuroscientific research, see Takala (2009).

purposes, the images can show signs of a medical condition (e.g. a tumour or multiple sclerosis). As incidental findings are by now a known risk, the research participants should be informed of this risk as well. It certainly has a bearing to their decision-making, and is therefore required by the Medical Research Act. Contrary to the purely therapeutic circumstances, the Medical Research Act does not contain a provision, according to which the research participants would have a right to refuse this information prior to giving consent. An interesting question is, however, whether after receiving the information on the risk of an incidental finding, the research participant can instruct the researchers not to disclose the finding if the risk materialises.

It is relatively clear that unless the research team has been so instructed, they would generally have a legal duty to disclose it. After all, according to the provisions on the duties of the persons in charge of the research, they have a legal duty to protect the research participants if new knowledge related to the carrying out of the research emerges in the course of the research that has a bearing on the safety of the research participants.[38] In addition, patients are entitled to the health and medical care that is required by their state of health according to the Act on the Status and Rights of Patients,[39] and all health care professionals have a statutory general duty to promote and maintain health, to prevent illness and to cure those who are ill according to the Act on Health Care Professionals.[40] Based on the above, it is clear that the research team has a legal duty to disclose their findings, if the information can be used to benefit the research participant (the condition is at least partly treatable) and the participant has not specifically forbidden the disclosure.

However, if the person in charge of the research project has a legal duty to protect the research participants, how should one understand the content of this duty? Does "protection" always mean disclosure of information or could it in some cases also be understood to require withholding information? If the participant has given an informed refusal on the disclosure, this could also be understood as protecting the research participant from unwanted information. Although the Medical Research Act itself does not regulate this issue, both the Personal Data Act and the Act on the Status and Rights of Patients emphasise the patient's (or the data subject's) right to self-determination also when access to information is concerned.[41] These provisions should be applied in the lack of particular direction from the Medical Research Act.

---

[38] Medical Research Act, Section 5, Subsection 3. This specific duty exists independent of the Sections 26 and 27 of the Personal Data Act, according to which the researchers are not legally obliged to give research-related information to the research participants, if the data in the file are used solely for scientific research.

[39] Section 3.

[40] Act on Health Care Professionals, Section 15.

[41] The relevant provisions of the Personal Data Act (Sections 27–29) assume that the data subject has initiated the request for access to information and the Act on the Status and Rights of Patients states clearly that information shall not be given against the will of the patient (Section 5).

In cases where the revealed condition is treatable or at least partly treatable, it may be difficult to understand why the research participant has wished to refuse the information. In these cases, the researcher acts wisely if he or she makes sure that the refusal is indeed a fully informed decision and that the participant has the legal capacity to make it in order to free him or herself from the legal duty to protect. If the researcher is still uncomfortable with the participant's decision, he or she can refuse to accept the person as a research participant. Unlike with right to medical treatment, nobody has a right to be included in a research project as a research participant.

If the revealed condition is untreatable, it may be in the participant's interest not to know also from an objective point of view. As there is no therapeutic incentive to inform the patient, the emphasis should lie in this case heavily with the wishes of the fully informed participants. If the participant's wishes are not known, justification for not revealing potentially harmful information can be found from the personal data legislation and from the Act on the Status and Rights of Patients. Both contain exceptions on withholding data from data subjects in circumstances when disclosure would potentially cause serious danger to the health of the data subject.[42]

Withdrawal of Consent

Another legal criterion with informed consent under the Medical Research Act is that research participants are entitled to withdraw their consent at any point prior to the completion of the research. They must be informed of this right before the start of the research. Withdrawal of consent must not involve any negative consequences for the research participant.[43]

The discussion here concerns the point of time when the research is completed in the sense that the research participant still has the right to withdraw his or her consent without any negative consequences. Does the "completion of research" mean the end of the actual research procedures or some point after it, e.g. during the process of analysing the research results before publication?

When thinking about the stereotypical medical research situations, the intention of the law-maker must have been on the physical research procedures. However, in neuroscientific research, the reason for the participant's withdrawal of consent may not be the pain or inconvenience of the research procedure itself, but, for example, worry about the use of the sensitive information gathered from the scanned images. If this information is identifiable to a certain research participant and the participant wishes to withdraw his or her consent to the processing of data, the provisions of the Personal Data Act apply. According to Section 12 of the Act, the data shall be

---

[42] Personal Data Act, Section 27, and Act on the Status and Rights of Patients, Section 5.

[43] Medical Research Act, Section 6.

erased from the data file immediately, as withdrawal of consent means that there no longer is a justification for its use in the study.[44]

2 Application of Neuroscientific Knowledge

The protection of the integrity of the person sets boundaries to the application of neuroscientific knowledge in the same way as in research into neuroscientific knowledge. The purpose of the application and its justification play an important role here. Albeit very few specific legal rules exist in this area, some clear limits can be pointed out while in some areas the limits have to be constructed from the more general principles.

2.1 Therapeutic Medicine

The application of neuroscientific knowledge that is probably the easiest to justify is its use in therapeutic medicine. Developing safer and better treatment methods is, after all, the underlying purpose of the majority of medical research projects. There are, however, areas in therapeutic medicine, which cause discussion with regard to neuroscientific knowledge. One of these areas is the use of neuroscientific procedures in the treatment of some mental illnesses, such as severe depression. In the Current Care guideline on depression prepared by the Finnish Medical Society Duodecim, treatment using electricity and transcranial magnetic stimulation treatment (TMS) are mentioned as treatments with strong evidence-based foundation, whereas vagal nerve stimulation (VNS) and deep brain stimulation (DBS) are new treatment methods, the status of which is still being investigated. [45]

The discussion in this area concerns not so much violation of legal boundaries but the lack of sufficient medical evidence of the efficiency of the treatment. Because of the potential risks involved, the capacity of the person consenting to the procedure is emphasized.[46]

---

[44] Personal Data Act, Section 12, Subsection 2.

[45] Working group set by the Finnish Medical Society Duodecim and the Finnish Society of Psychiatrists (2010). Current Care guideline on Depression. Current Care guidelines are evidence-based treatment guidelines prepared by the Current Care working groups (including approximately 700 volunteer health care top professionals from a range of fields across Finland) in cooperation with Current Care editors employed by the Finnish Medical Society and relevant interest groups in support of decision making in health care.

[46] Soini (2009).

The legal capacity of a person suffering from depression or another mental illness has rarely come into evaluation by the courts in Finland.[47] The general rule is that despite a diagnosed mental illness, the person remains legally capable of deciding on his or her own treatment. This is not altered by e.g. a court decision to appoint a legal guardian to handle the person's financial affairs or even when the indications for involuntary treatment are fulfilled. In order for the legal guardian to have any competence on the patient's personal affairs, it shall be specifically mentioned in the court order and even then the patient can decide on his or her own treatment, if he or she is actually capable of doing so.[48]

When treating a patient's mental illness in involuntary treatment, a patient must be cared for, as far as possible, in mutual understanding with the patient. The treatment and examination of the patient is restricted to medically acceptable methods. Involuntary treatment is only justified to the extent that failure to administer the treatment would seriously jeopardise the health and safety of the patient or others. Psychosurgical or other treatments that seriously or irreversibly affect the patient's integrity may only be given with the written consent of an adult patient, unless it is question of a measure that is necessary to avert a danger to the patient's life.[49]

Hence, it can be said that even with patients diagnosed with severe depression, the patient remains the primary person to decide over his or her treatment provided. If experimental methods of therapy are proposed, they cannot be administered against the will of the patient in involuntary treatment. Outside involuntary care, the adult patient who does not have the capacity to decide on the treatment may be treated in mutual understanding with his or her proxy.[50] In these circumstances, it is possible for the proxy to consent even to experimental therapy. However, in order for the experimental therapy to be medically indicated, all other traditional methods of treatment must have been exhausted.

2.2 Medicinal Products for Enhancement

Medical knowledge is increasingly used not only to cure illness but also to enhance qualities that are considered normal from the medical point of view. Typical

---

[47] See, however, the decision by Vantaa District Court Nr. 7869 (00/6566, 3 November 2000) in Mäki-Petäjä-Leinonen (2003), p. 263.

[48] According to Section 23, Subsection 2 of the Guardianship Services Act (Nr. 442 of 1999), a person who has been declared incompetent may self decide on matters pertaining to his or her person, if he or she understands the significance of the matter. According to Section 29, Subsection 2 of the Act, if the ward cannot understand the significance of the matter, the guardian shall be competent to represent the ward also in matters pertaining to his or her person, if the court has so ordered.

[49] Mental Health Act (Nr. 1116 of 1990), Section 22 b.

[50] According to Section 6 of the Act on the Status and Rights of Patients, possible proxies for an adult patient are his or her legal representative (by court order or by authorisation), a family member, or a person otherwise close to the patient (including a cohabitation partner).

examples of this are the use of pharmaceuticals to enhance memory or concentration or to control anxiety. The pharmaceuticals used for these purposes are not necessarily forbidden substances, but they can have been approved for the treatment of an unrelated illness or a related condition, which is medically relevant. The legal issue involved in this area is whether the physician, who the patient approaches, acts against his professional duties, if he or she prescribes the pharmaceutical for other than medically indicated purposes.

The primary legal rule for this can be found from the Act on Health Care Professionals, which states that "the aim of the professional activities of health care professionals is to promote and maintain health, to prevent illness, to cure those who are ill and to alleviate their suffering. In their professional activities, health care professionals must employ generally accepted, empirically justified methods, in accordance with their training, which should be continually supplemented".[51] Licensed physicians or dentists are entitled to prescribe medicines from a pharmacy.[52] Prescription of medicines is regulated in more detail in the Decree on Prescribing Medicinal Products.[53] For example, Section 10 of the Decree states that the person prescribing medicines can only make a prescription to a person whose need for medication he or she has ascertained by personal investigation or by other reliable means. Section 11 further states that particular care and caution shall be applied when prescribing medicinal products that may be misused.

The prescription of medicinal products that affect the central nervous system is monitored closely. If a suspicion arises on the physician misusing his or her right to prescribe these products, the pharmacy can inform either the National Supervisory Authority for Welfare and Health or the local Regional State Administrative Agency of the suspected misuse. The National Supervisory Authority for Welfare and Health is authorised to use a variety of administrative measures to discipline for the misconduct, including the right to cancel the right of a health care professional to practise his or her profession, or to impose restrictions on the right to practise professional activity as a licensed professional for a fixed period or until further notice.[54] Generally speaking, the former sanction has mainly been used in cases where the health care professionals have been confirmed having an abuse problem and the latter when the physicians have misused their right to prescribe medicinal products to others.

According to the summarised decisions published by the National Supervisory Authority for Welfare and Health, these disciplinary measures have been used mainly in cases where the abuse of prescription rights was used for acquiring

---

[51] Act nr. 559 of 1994, Section 15.

[52] Section 22.

[53] A new Decree (Nr. 1088 of 2010) on Prescribing Medicinal Products by the Ministry of Social Affairs and Health will come into force 1 January 2011.

[54] Act on Health Care Professionals, Sections 26–33.

intoxicants or their replacement.[55] However, a further inquiry to the Authority revealed isolated cases where the physician had been found to misuse his or her prescription rights for enhancement purposes; however, these cases have been very rare.

2.3   Legal Proceedings

One of the most controversial applications of neuroscientific knowledge is the use of polygraph test, also known as a lie detector test, in the context of police investigations and court proceedings. The Finnish National Bureau of Investigation has had access to polygraph equipment since 1995.[56] The equipment measures electronic activity on the skin (electrodermal activity) and pulse.[57]

The polygraph tests currently performed in Finland use the method of "guilty knowledge testing"[58] (also known as "concealed information testing"), rather than the more traditional method of "control question testing". The difference between the two tests is that in the control question testing the person's physiological reactions are measured against questions on his or her guilt, whereas in the concealed information testing the person's reactions are measured against some facts that only the guilty person (in addition to the investigators) can have knowledge of. In the latter case, the questions may involve, for example, a murder weapon, clothes that the victim was wearing, or items that the assailant left at the crime scene. The guilty knowledge test is said to be preferred to the control question test because the percentage of false positive results (i.e. the proportion of innocent persons receiving a guilty-result) is somewhat lower than with the control question testing. On the other hand, the problem with the guilty knowledge test is that it provides more false negative results (i.e. the guilty person may pass the test without being detected) than the control question test.[59]

Without taking a stand on whether either of these tests is a scientifically valid method for criminal investigation, it is noteworthy that in Finnish law there are no legal provisions concerning the use of polygraph tests either during police investigation or as evidence in court proceedings. Because all coercive measures against

---

[55] Summaries of the decisions are published by the National Supervisory Authority of Welfare and Health on their website (National Supervisory Authority of Welfare and Health 2010).

[56] Klami (1996), p. 209.

[57] Hamilo (2010).

[58] The guilty knowledge test was developed already in the 1950s by David Lykken (Lykken 1959).

[59] Jokinen (2005), p. 73. See also Iacono and Lykken (1997), who asked the members of Society for Psychophysiological Research and American Psychological Association about their opinion on the two tests. Seventy-seven percent of those who replied estimated the guilty knowledge test to be based on scientifically valid principles compared to 36%, who said the same about the control question test.

the suspect must be regulated by law, the result is that the use of polygraph tests is voluntary. Refusing to undergo the test cannot be used as evidence against the party to the proceedings.[60] The test is mainly used to assist police in its investigation, but it has also been used as evidence in court proceedings.[61] According to the Finnish Code of Judicial Procedure, the court can evaluate all evidence freely unless there is a special provision in law on its significance.[62] No cases have been reported on any conviction that has solely been based on the polygraph. Representatives from the police and the attorneys representing clients in court have, however, expressed an opinion that guidance should be introduced in this area.[63]

2.4 Insurance Companies

As far as it has been possible to establish, neither the polygraph test nor any other neuroscientific applications have been used by insurance companies in Finland. The only polygraph for professional use lies at the premises of the National Bureau of Investigation.[64] As no activity has been reported in this area, drafting legislation on the issue has not been called for. It is, however, common practise that while applying for an insurance to cover the expenses for a future illness, the applicant consents the insurance companies to have access to the relevant patient documentation. Therefore, if the patient has undergone some neuroscientific tests in the past in connection to his or her medical treatment, this information will also be revealed to the insurance companies.

2.5 Employment

It is increasingly common practice for the employers to require psychological testing in the connection of certain job application processes. As the quality and reliability of these tests varies, it was considered necessary to adopt legislation to control the area. For this purpose (*inter alia*), the Act on the Protection of Privacy in

---

[60] Klami (1996), p. 215.

[61] According to the news article published in connection of a recent murder case where guilty knowledge test had been applied, hundreds of persons have been tested using the polygraph test, but the results have been used in court only in around ten times (Talja 2010).

[62] Code of Judicial Procedure, Chapter 17, Section 2.

[63] Juhani Sirén quotes Mr. Jorma Hautala, who is one of the police officers conducting polygraph tests in Finland in his article, Sirén (2003). See also Klami (1996), p. 216, who comments the need for legislation from the point of view of an attorney representing his clients.

[64] Jokinen (2005), p. 12.

Working Life[65] was adopted in 2004. According to the Act, personality and aptitude assessments can only be performed with the employee's consent. The employer has the duty to ensure that the assessment methods used are reliable, the persons conducting the assessment are experts, and the findings of the assessment are free from error.[66] Upon request, the employer or an assessor designated by the employer must provide the employee with a written statement on the assessment of the employee's personality or aptitude free of charge. If the employer has received the statement orally, the employee must be informed of its content.[67]

The Act is also applied e.g. for the use of health care services and processing information on drug use by the employer, camera surveillance in the workplace and retrieving and opening electronic mail messages belonging to the employer. For example, when carrying out employee health examinations and tests, professionals in health care, properly trained laboratory personnel and health care services must be used.[68] Genetic testing on the request of the employer during recruitment or during employment relationship is not permitted. The employer has no right to know whether or not the employee has previously taken part in such testing either.[69]

As far as can be estimated from the data available, neuroscientific testing is not used in recruitment or during employment in Finland. However, if this kind of testing was to be performed, it would have to fulfil the requirements set out in the Act on the Protection of Privacy in Working Life. Although the provisions on personality and aptitude testing were drafted for the purpose of ensuring the quality and reliability of psychological tests, there is nothing in the formulation of the provisions, which would prevent their application also to neuroscientific testing.

References

Council of Europe (1997) Convention for the protection of human rights and dignity of the human being with regard to the application of biology and medicine: convention on human rights and biomedicine. Eur Treaty Ser 164

Council of Europe (1998) Additional protocol to the convention for the protection of human rights and dignity of the human being with regard to the application of biology and medicine, on the prohibition of cloning human beings. Eur Treaty Ser 168

Council of Europe (2002) Additional protocol to the convention on human rights and biomedicine concerning transplantation of organs and tissues of human origin. Eur Treaty Ser 186

European Union (1993) Council Directive 93/42/EEC of 14 June 1993 concerning medical devices. L 169/1, 12.7.1993

---

[65] Nr. 759 of 2004.

[66] In order to ensure the reliability of the tests and the qualifications of the experts, the employers work together with the professional unions such as the Finnish Psychological Association.

[67] Section 13.

[68] Section 14.

[69] Section 15.

European Union (2001) Directive 2001/20/EC of the European Parliament and of the Council of 4 April 2001 on the approximation of the laws, regulations and administrative provisions of the Member States relating to the implementation of good clinical practice in the conduct of clinical trials on medicinal products for human use. L 121/34, 1.5.2001

Government Bill to the Parliament for Amending the Provisions on Constitutional Rights in the Constitutional Acts (Bill Nr. 309 of 1993)

Government Bill to the Parliament for a New Constitution Act of Finland (Bill Nr. 1 of 1998)

Government Bill to the Parliament for Acts on the Amendment of the Medical Research Act, Section 13 of the Act on the Status and Rights of Patients and Section 18 of the Act on the Status and Rights of the Clients of Social Welfare (Bill Nr. 65 of 2010)

Hamilo M (2010) Valpastuminen paljastaa syyllisen. Tiede Magazine. http://www.tiede.fi/artikkeli/1316/valpastuminen_paljastaa_syyllisen. Accessed 29 Dec 2010

Iacono WG, Lykken DT (1997) The validity of the lie detector: two surveys of scientific opinion. J Appl Psychol 82:426–433

Immonen T (1992) Ihmiseen kohdistuva lääketieteellinen tutkimus: eettisten toimikuntien asema. In: Lahti R (ed) Biolääketiede ja laki, Sosiaali –ja terveyshallituksen raportteja 54, Helsinki, pp 19–78

Jokinen A (2005) Rikos jää tekijän mieleen: Muistijälkitesti rikostutkintamenetelmänä. Poliisiammattikorkeakoulun oppikirjoja 12. Edita, Helsinki

Klami HT (1996) Valheenpaljastuskoe Suomen oikeudessa. Defensor Legis 2:209–216

Lehtonen L (2006) Biolääketieteelliseen tutkimukseen tarvittavat luvat ja tutkimustoiminnan valvonta. In: Lehtonen L (ed) Bio-oikeus lääketieteessä. Edita, Helsinki, pp 239–256

Lykken DT (1959) The GSR in the detection of guilt. J Appl Psychol 43:385–388

Lötjönen S (1997) Ihmiseen kohdistuva lääketieteellinen tutkimustoiminta ja siihen soveltuvat oikeussäännöt. Lakimies 6:856–879

Lötjönen S (2008) Medical research on minors in Finland. Eur J Health Law 15:135–144

Lötjönen S (2009a) Autonomy and dignity in clinical medical research on adults with cognitive impairment. In: Aasen HS, Halvorsen R, da Silva AB (eds) Human rights, dignity and autonomy in health care and social services: nordic perspectives. Intersentia, Antwerp, pp 161–175

Lötjönen S (2009b) Neurotieteellisen tutkimuksen oikeussääntely Suomessa. In: Launis V (ed) Neuroetiikan hyvä ja paha. UNIpress, Kuopio, pp 215–226

Miettinen T (2001) Tieteen vapaus. Kauppakaari. Lakimiesliiton Kustannus, Helsinki

Mäki-Petäjä-Leinonen A (2003) Dementoituvan henkilön oikeudellinen asema. Suomalaisen Lakimiesyhdistyksen julkaisuja A-sarja N:o 241, Suomalainen Lakimiesyhdistys, Helsinki

National Advisory Board on Research Ethics (2009) Ethical principles of research in the humanities and social and behavioural sciences and proposals for ethical review. http://www.tenk.fi/ENG/ethicalreview.htm. Accessed 29 Dec 2010

National Supervisory Authority of Welfare and Health (2010) Kanteluratkaisujen lyhennelmät http://www.valvira.fi/tietopankki/ratkaisulyhennelmat. Accessed 29 Dec 2010

National Supervisory Authority for Welfare and Health (2010) Regulation Nr. 3 on clinical investigation on medical devices and their accessories

Nieminen L (2009) Lapset tutkimuskohteena: kuka päättää lapsen osallistumisesta tutkimukseen? Lakimies 2:226–253

Penttilä E (1992) Ihmiseen kohdistuva lääketieteellinen tutkimus: koetoiminta kyselytutkimuksen valossa. In: Lahti R (ed) Biolääketiede ja laki, Sosiaali –ja terveyshallituksen raportteja 54, Helsinki, pp 79–109

Revonsuo A (2009) Neurotiede ja filosofia ajan hermolla. In: Launis V (ed) Neuroetiikan hyvä ja paha. UNIpress, Kuopio, pp 13–40

Sirén J (2003) Valehtelijan painajainen, City magazine 2. http://www.city.fi/artikkeli/Valehtelijan+painajainen/714/. Accessed 29 Dec 2010

Soini M (2009) Neurotiede ja tietoon perustuva suostumus. In: Launis V (ed) Neuroetiikan hyvä ja paha. UNIpress, Kuopio, pp 104–126

Takala T (2009) Aivokuvantamisen eettiset ulottuvuudet ja säätelylliset haasteet. In: Launis V (ed) Neuroetiikan hyvä ja paha. UNIpress, Kuopio, pp 127–141

Talja P (2010) Muistijälkitestin mukaan Ulvilan puuttuva murha-ase olisi kirves (21 April 2010). http://www.mtv3.fi/uutiset/rikos.shtml/2010/04/1105364/muistijalkitestin-mukaan-ulvilan-puuttuva-murha-ase-olisi-kirves. Accessed 29 Dec 2010

Working group set by the Finnish Medical Society Duodecim and the Finnish Society of Psychiatrists (2010) Depression. Current Care guideline (latest update 21 October 2010). www.kaypahoito.fi. Accessed 29 Dec 2010

The Obtainment and Use of Neuroscientific Knowledge in France

Legal Requirements and Implications

Caroline Rödiger

Abstract As a reaction to the rapid developments in modern neurosciences, the French legislator proposed the implementation of neuroscientific rules in the French Law on Bioethics in January 2010. Neuroscientific research has indeed not yet been covered by any national code or international treaty and opinions are highly divided with a view to the legal framework for neuroscientific research. Notably incidental findings in brain imaging research pose a big challenge for both the researcher and the participant. They might cause psychological distress, social stigma, and severe financial burdens on the participant's side and the researcher might be confronted with civil claims for damages or even render himself liable to prosecution. Therefore, it is going to be analyzed in a first step if the insertion of neurospecific rules in the Law on Bioethics could shed light on the management of incidental findings and the regulation of neuroscientific research. Another important issue in neurolaw is that of the recent application possibilities of neuroimaging techniques such as the use of fMRI scanners as lie detectors in courts or as communication methods for vegetative state patients. Even if these procedures are not yet a daily occurrence in France, it is an important time to consider the civil, criminal, and constitutional consequences in a second step.

The author is postgraduate and research associate at the Institute of Science and Ethics, University of Bonn. She is a member of the research project "NeuroSCAN: Ethical and Legal Aspects of Norms in Neuroimaging" funded by the Federal Ministry of Education and Research.

C. Rödiger (✉)
Institute of Science and Ethics, University of Bonn, Bonner Talweg 57, 53113 Bonn, Germany
e-mail: roediger@iwe.uni-bonn.de

T.M. Spranger (ed.), *International Neurolaw*,
DOI 10.1007/978-3-642-21541-4_8, © Springer-Verlag Berlin Heidelberg 2012

1   Introduction

Since the first successful fMRI study in 1991,[1] brain imaging research has progressed rapidly and the neurosciences have become one of the most widely supported research areas worldwide. Today, fMRI systems have up to 9.4 T high field magnets, which permit to visualize minimal brain structures and thus to make new discoveries. This development not only increases the efficacy of medical treatment, but also affects several other disciplines such as ethics, economy, sociology, politics, or even theology. In the last few years, another "neuro field" has emerged: Neurolaw (Neuroloi). The French Centre d'analyse stratégique, which stands under the direction of the French Prime Minister and fulfills the functions of monitoring, expertise and decision-making assistance in implementing and carrying out public policies, defines this item as follows: "The notion 'Neurolaw' describes the ensemble of neuroscientific research among the results can – with a view to diverse measures from pharmacology to neuropsychology and functional brain imaging – contribute to the illumination of legal and juridical procedures."[2] From the medical law perspective, this definition raises the foregoing question of the legal framework for "the ensemble of neuroscientific research". Until now, research studies in neurosciences have not yet been covered by any national code or international treaty. By customary law, researchers provide an informed consent document that includes details about the research study such as its procedure, duration, and purpose, but it is highly controversial how to deal with "incidental findings", which are accidentally detected abnormalities during a research study that appear in almost 40% of fMRI studies, but that are only clinically relevant in up to 8% of cases.[3] The ethical and legal implications of such findings are various. They can cause fears and psychic problems on the participants' and their relatives' side, but even more relevant from a legal point of view is that they may lead to disadvantages or rejection of insurance and employment applications.[4] Initial points of this discussion are the informed consent and other biomedical principles, which need a specification. Therefore, it has to be examined in a first step, if any national biomedical regulations can be applied – with or without any modification – for neuroscientific research in due consideration of the issue of incidental findings. Second, the definition of the Centre d'analyse stratégique raises the question of the precise "legal and juridical procedures" that profit by neuroscientific knowledge. The idea of using fMRI scanners in courts as lie-detection tests has spread widely and raises several questions in criminal

---

[1] See Belliveau et al. (1991), pp. 716–719.

[2] Free translation of Centre d'analyse stratégique (2009), No. 159, p. 1: "Le terme «neuroloi» désigne de manière générique l'ensemble des travaux en neurosciences dont les résultats peuvent – à divers échelles allant de la pharmacologie à la neuropsychologie en passant par l'imagerie cérébrale – participer à l'éclairage des procédures légales et judiciaires."

[3] Illes et al. (2004), p. 743.

[4] Anonymous (2005), p. 17.

procedure and constitutional law. Lesser known but more relevant to the actual praxis is the use of brain imaging methods to help corporal disabled persons make their own medical and civil decisions without any assistance of custodians. Hence, possible application fields of neuroscientific knowledge in due consideration of legal implications are going to be illuminated in a second step.

2 Legal Framework for Neuroscientific Research

In biomedical research, a number of principles have been established to assure a balance between the protection of the proband on the one hand and the researcher's interests on the other hand[5]: The principle of human dignity (principe de dignité humaine)[6] does not include a subjective right (droit subjectif) of the participant, but an interdiction of third persons to endanger the body of others.[7] Also applicable to biomedical research in general is the risk-benefit analysis[8] (analyse bénéfice risque), which puts on the scale the risks and potential benefits of biomedical progress: The researcher has to assure that the amount of benefit clearly outweighs the amount of risk. Furthermore, a principle of big significance for the researcher is that of freedom of research (liberté de la recherche). It can be traced back to Art. 4 Declaration of the Rights of Man and of the Citizen (Déclaration des droits de l'Homme et du Citoyen) of 1789.[9] Its constitutional value has been determined by decision of the Constitutional Court (Conseil constitutionnel) of July 29, 1994,[10] but of course, it has its limitations in fundamental rights of the proband, considerably in the principle of human dignity. The mentioned principles do not raise major concerns with a view to incidental findings. The researcher neither violates the principle of human dignity nor augments the medical risk by simply disclosing a

---

[5] The most important biomedical principles are listed in the Additional Protocol to the Convention on Human Rights and Biomedicine concerning Biomedical Research of the Council of Europe, which is not yet legally binding for France.

[6] The principle of human dignity is neither expressly mentioned in the French Constitution of October 4, 1958 nor in the post-war Constitution of October 27, 1946, even though a preliminary draft contained the notion of *dignité de la personne humaine*. By decision of July 24, 1997, the French Supreme Court declared for the first time that the principle of human dignity is a principle of the French Constitution and accepted it as a constitutional category.

[7] Laude et al. (2009), p. 658.

[8] Art. L1121-2 Para. 1 PHC, Art. 16 ii Convention on Human Rights and Biomedicine, Art. 6 Para. 1 Additional Protocol to the Convention on Human Rights and Biomedicine concerning Biomedical Research.

[9] Art. 4 Declaration of the Rights of Man and of the Citizen: "Liberty consists in the freedom to do everything which injures no one else; hence the exercise of the natural rights of each man has no limits except those which assure to the other members of the society the enjoyment of the same rights. These limits can only be determined by law."

[10] Cour Const (1994).

brain abnormality, which has already existed before.[11] However, the content of other biomedical principles such as the informed consent, the right (not) to know, the principle of data protection and the right to compensation is far from being definitive in neuroscientific research. Therefore, it has to be analyzed if any written national rules could be applied for neuroscientific research and thus help to specify its legal requirements.

2.1 Application of the Law on Bioethics?

The French Law on Bioethics (Loi de bioéthique), also known as Huriet's Law (Loi de Huriet)[12] and now part of the Public Health Code (Code de la santé publique),[13] sets out the legal conditions concerning the protection of individuals involved in biomedical research. Initially, Huriet's Law included only regulations concerning cloning, genes, and organ donation. Later, it has strongly been influenced by the European Directive 2001/20/CE on the approximation of laws, regulations and administrative provisions of the Member States relating to the implementation of good clinical practice in the conduct of clinical trials on medicinal products for human use.[14] The inserted dispositions entailed a more participant orientated protection and engendered a multiplication of administrative forms for the sponsors. As a reaction to the fast-growing knowledge in biomedical research, the French legislator determined a rotational revision of the Law on Bioethics. In January 2010, the question of implementing neurospecific rules in the Law on Bioethics was raised for the first time.[15]

Art. L1121-1 Para. 1 Public Health Code (PHC) defines its scope of application as all biomedical research involving human beings, with the aim of increasing biological or medical knowledge.[16] Therefore, Art. L1121 PHC et seq. can be applied for neuroscientific research in due consideration of Art. L1121-2 PHC, which states that no "biomedical research shall be carried out on a human being, if it is not based on the latest state of scientific knowledge and on sufficient preclinical

---

[11] The disclosure of a brain abnormality does not become critical until the researcher delays the notification of the incidental finding or violates another legal duty.

[12] The Law on Bioethics (2010).

[13] The Public Health Code (2010).

[14] A directive is a legislative act of the European Union which is binding with respect to the end to be achieved while leaving some option as to form and method of execution. The European Directive (2001).

[15] Assemblée nationale (2010).

[16] Free translation of Art. L1121-1 Para. 1 PHC: "Les recherches organisées et pratiquées sur l'être humain en vue du développement des connaissances biologiques ou médicales sont autorisées dans les conditions prévues au présent livre et sont désignées ci-après par les termes 'recherche biomédicale'."

experimentation, if the predictable risk incurred by the persons who lend themselves to the research is out of proportion to the benefit expected for these persons or the interest of this research [risk-benefit analysis], if it does not aim at widening the scientific knowledge and the means susceptible to improve the human condition, if the biomedical research was not conceived by such way as to reduce at least the pain, the inconveniences, the fear and other predictable inconvenience connected to disease or to the research, by taking into account particularly the degree of maturity of minors and the capacity of understanding of adults unable to express an informed consent. The interest of the persons who lend themselves to biomedical research always prevails over the only interests of science and society. The biomedical research shall begin only if all these conditions are fulfilled. Their respect must be constantly maintained."[17] With a view to the issue of incidental findings, the informed consent of healthy persons and vulnerable groups, the right to know and not to know, the principle of data protection, and the right to compensation are going to be illuminated in the following.

2.1.1 Informed Consent

The informed consent, one of the most important principles in biomedical research, stipulates that the participant has to be specifically informed about the research project before being asked to consent to participate in the research study. According to Art. L1122-1 Para. 1 PHC, the researcher or a physician representing him/her informs the participant about objective, methods and duration of the research (No. 1), foreseen benefits, obligations, and predictable risks, inclusively in case of abandoning the research before its aim (No. 2), potential medical alternatives (No. 3), options of medical supervision at the end of the research, if such a supervision is required, in case of premature termination of the research and in case of exclusion of the research (No. 4) and the opinion of the committee cited in Art. L1123-1 PHC and the authorization of the competent authority mentioned in Art. L1123-12 No. 5 Cl. 1 PHC. Furthermore, the researcher has to inform him/her

---

[17] Free translation of Art. L1121-2 PHC: "Aucune recherche biomédicale ne peut être effectuée sur l'être humain: si elle ne se fonde pas sur le dernier état des connaissances scientifiques et sur une expérimentation préclinique suffisante; si le risque prévisible encouru par les personnes qui se prêtent à la recherche est hors de proportion avec le bénéfice escompté pour ces personnes ou l'intérêt de cette recherche; si elle ne vise pas à étendre la connaissance scientifique de l'être humain et les moyens susceptibles d'améliorer sa condition; si la recherche biomédicale n'a pas été conçue de telle façon que soient réduits au minimum la douleur, les désagréments, la peur et tout autre inconvénient prévisible lié à la maladie ou à la recherche, en tenant compte particulièrement du degré de maturité pour les mineurs et de la capacité de compréhension pour les majeurs hors d'état d'exprimer leur consentement. L'intérêt des personnes qui se prêtent à une recherche biomédicale prime toujours les seuls intérêts de la science et de la société. La recherche biomédicale ne peut débuter que si l'ensemble de ces conditions sont remplies. Leur respect doit être constamment maintenu."

about his/her right to communication, during or at the end of the research, and about obtained information concerning his/her health (No. 5 Cl. 2). Where appropriate, s/he has to be informed about the interdiction of simultaneous participation to another research study or about the period of provided exclusion by the protocol and the inscription in the national file according to Art. L1121-16 PHC (No. 6).[18]

These regulations are generally applicable to neuroscientific research, but they have to be concretized and extended in a manner that does justice to the problem of incidental findings: Within the scope of "predictable risks" (No. 2), it has to be clarified to what extent abnormalities can create burdens. The researcher should indicate that – apart from their clinical (in)significance – incidental findings can cause psychic problems and may have a negative effect on the state of insurance and employment conditions, at worst disability insurances/occupational disablement insurances and contracts of employment could be refused.[19] In addition, it is all the more required to provide information about the *arrangements for responding* to adverse events or concerns of research participants. It is important to emphasize that the research study does not pursue any diagnostic aim, so that abnormalities are likely to remain undiscovered. But if the researcher discloses an incidental finding, s/he should contact a neuroradiologist immediately. As an element of the right to communication (No. 5 Cl. 2) information about the *arrangements for access* to information relevant to the participant arising from research is also required. Furthermore, Art. L1121-1 PHC does not mention *arrangements* to ensure respect for private life and confidentiality of personal data as well as *arrangements* for fair compensation in case of damages. The researcher should optionally inform about any potential further (non-)commercial use of the research results, the data and the source of funding of the research project.

Art. L1122-1 Para. 2 PHC stipulates that the participant has the right to withdraw his/her consent at any time without taking any responsibility or facing any prejudice.[20] The term "at any time" is capable of being misunderstood. Of course, the proband is allowed to withdraw his/her consent whenever s/he likes, but the *legal*

---

[18] Free translation of Art. L1122-1 Para. 1 PHC: "Préalablement à la réalisation d'une recherche biomédicale sur une personne, l'investigateur, ou un médecin qui le représente, lui fait connaître notamment: 1° L'objectif, la méthodologie et la durée de la recherche; 2° Les bénéfices attendus, les contraintes et les risques prévisibles, y compris en cas d'arrêt de la recherche avant son terme; 3° Les éventuelles alternatives médicales; 4° Les modalités de prise en charge médicale prévues en fin de recherche, si une telle prise en charge est nécessaire, en cas d'arrêt prématuré de la recherche, et en cas d'exclusion de la recherche; 5° L'avis du comité mentionné à l'article L. 1123-1 et l'autorisation de l'autorité compétente mentionnée à l'article L. 1123-12. Il l'informe également de son droit d'avoir communication, au cours ou à l'issue de la recherche, des informations concernant sa santé, qu'il détient; 6° Le cas échéant, l'interdiction de participer simultanément à une autre recherche ou la période d'exclusion prévues par le protocole et son inscription dans le fichier national prévu à l'article L. 1121-16."

[19] Anonymous (2005), p. 17.

[20] Free translation of Art. L1122-1 Para. 2 PHC: "Il informe la personne dont le consentement est sollicité de son droit de refuser de participer à une recherche ou de retirer son consentement à tout moment sans encourir aucune responsabilité ni aucun préjudice de ce fait."

effect of his/her withdrawal occurs only *ex nunc* (henceforward). If its effect was *ex tunc* (retroactively), the researcher would be exposed to civil claims and would render himself/herself to liable to prosecution.

2.1.2 Informed Consent of Vulnerable Groups

Numerous serious and still not sufficiently treatable diseases such as dementia, apoplexia, or coma require neuroscientific research on vulnerable populations. The PHC envisages four vulnerable groups: First, pregnant, parturient, and nursing woman (Art. L1121-5 PHC), second, persons deprived of their freedom by a judicial or administrative decision, hospitalized persons without consent and persons in a medical or social establishment (Art. L1121-6 PHC), third, minors (Art. L1121-7 PHC), and finally incapacitated adults (Art. L1121-8 PHC). According to Art. L1121-5–L1121-8 PHC research on vulnerable persons is allowed provided that there is a direct benefit to the individual and that the risk outweighs the benefit, or that research is of value to other persons of the same vulnerable group and risks and exposure are of minimal character.[21]

The fact that research is permitted even if it does not benefit its participant, but contributes to the health benefit of other persons of the same vulnerable group, is giving cause for concern. At worst, this provision could leave free hand to abusive ambitions on the part of the researcher. In addition, the terms "risks and exposure (...) of minimal character" and "minimal risk and burden" are – irrespective of attempts to give definition – very vague. To guarantee the best level of protection of vulnerable people, neuroscientific research on them should only be allowed if it is with direct benefit to the individual.[22]

2.1.3 Right to Know and Not to Know

The right to know and not to know is an extension to the principle of autonomy (principe d'autonomie) and determines that the proband can decide to be or not to be informed about the study results concerning his/her health. Most participants want to exercise their right to know, because the detection of a brain abnormality can be a chance for an early and more promising therapy and can thus become life-saving. According to Art. L1122-1 No. 5 Cl. 2 PHC the participant has the right to know all information concerning his/her health, including the right to information concerning costs (Art. L1111-3 Para. 1 Cl. 1 PHC) and concerning the medical file (Art. L1111-7 Para. 7 PHC). These regulations apply to neuroscientific research without further modification.

---

[21] A parallel can be drawn to the analogous wording of Art. 17 Para. 2 Convention on Human Rights and Biomedicine.

[22] Germany has not yet ratified the Convention on Human Rights and Biomedicine for this reason.

However, a very few number of probands (approx. 3%)[23] do not want to be informed about any incidental finding and exercise their right not to know for manifold reasons. Some prefer to take the risk of an open though uncertain future, others fear the trembling uncertainty about the clinical (in)significance of an abnormality. The detection of brain tumors also involves serious legal problems that can cause unpredictable harm. Since the proband might be obliged to inform insurances or employers about the incidental finding, s/he might be excluded from important private insurance policies or sustain severe discrimination on the job market.

The chapter of the PHC dealing with biomedical research does not refer to the right not to know, but the precedent chapter which covers the rights of an invalid and of users of the health system includes the following regulation: Art. L1111-2 Para. 4 PHC stipulates that the will of a patient not to know a diagnosis or a prognosis must be respected except when third parties are exposed to a risk of infection.[24] Furthermore, in case of the detection of a grave abnormality in the context of genetic tests, the PHC provides in Art. L1131-1 Para. 3 Cl. 1 that the physician has to inform the patient about the risks s/he would expose to his/her family members if s/he kept silent.[25] According to Laude et al., the latter stipulation involves that the right not to know has not to be respected, once family members are potentially concerned.[26] Their interest predominates the right not to know (and the individual right of data protection).[27]

Of course, the idea of exposing third parties to risks is also applicable in neuroscientific research: For instance, a school bus driver takes part in an fMRI research study and claims his right not to know. Thus, the researcher does not inform the participant about the incidental finding. As a consequence of the tumor, the bus driver collapses and causes a fatal accident. Related problems arise in the modified case: A head of a family takes part in an fMRI study. S/he makes use of his/her right not to know and the researcher detects a hereditary abnormality such as Neurofibromatosis 2. At the time of research participation, his/her children, who have inherited the disease, could have been treated.

In addition to the numerous cases that refer to third parties, the right not to know clashes with the legal obligation of the researcher to inform the participant about an incidental finding. The legal duty has its origin in medical ethics and derives from

---

[23] Illes et al. (2006), p. 783.

[24] Free translation of Art. L1111-2 Para. 4 PHC: "La volonté d'une personne d'être tenue dans l'ignorance d'un diagnostic ou d'un pronostic doit être respectée, sauf lorsque des tiers sont exposés à un risque de transmission."

[25] Free translation of Art. L1131-1 Para. 3 Cl. 1 PHC: "En cas de diagnostic d'une anomalie génétique grave posé lors de l'examen des caractéristiques génétiques d'une personne, le médecin informe la personne ou son représentant légal des risques que son silence ferait courir aux membres de sa famille potentiellement concernés dès lors que des mesures de prévention ou de soins peuvent être proposées à ceux-ci."

[26] Laude et al. (2009), p. 692.

[27] Laude et al. (2009), p. 693.

the ethical principle of nonmaleficence. Although the researcher cannot be held responsible for incidentally detected diseases, s/he is obliged not to expose the proband to a disproportional high risk or burden and to provide an appropriate compensation in case of damages resulting from research. Hence, the following dilemma is apparently inevitable, in particular in case of clinically relevant findings: If the researcher respects the exercise of the right not to know on the participant's side, s/he violates the legal obligation to inform the subject about incidental findings. If, in contrast, the researcher informs the participant about an brain abnormality against his/her will according to the legal duty, s/he violates the proband's right not to know. A certain way to escape from the dilemma is the following: As the participant is allowed to reject the right not to know, it is imaginable to grant access to the study, under the condition that the participant expressly rejects the exercise of the right not to know in advance. Regardless whether interests of third parties are concerned or not, this provision avoids any collision between the legal duty to inform and the right not to know *ab initio*, without raising any concerns from a legal point of view, not least because the participant does not dispose of a "right to participation". The participant should subscribe a corresponding waiver declaration, e.g., within the scope of the informed consent document.

2.1.4 Data Protection

The protection of personal data is an important principle with regard to incidental findings. It has to be guaranteed that third parties such as employers and assurances only take note of brain abnormalities if law provides for it, otherwise participants might become victims of discriminatory measures. The principle of nondiscrimination (principe de non-discrimination) is laid down in Art. L1132-1 Labour Code (Code du travail)[28] and has a vast scope of application. It forbids discrimination in particular against employees, direct or indirect, on the grounds of origin, sex, habits, sexual orientation, age, family situation or pregnancy, genetic characteristics, membership or nonmembership real or believed of an ethnic group, a nation or a race, political opinions, union activities, religious convictions, physical appearance, family name, state of health, or disability.[29] Art. 16–13 Civil

---

[28] Ex-Art. 122-45 Labour Code.

[29] Free translation of Art. 1132-1 Labour Code: "Aucune personne ne peut être écartée d'une procédure de recrutement ou de l'accès à un stage ou à une période de formation en entreprise, aucun salarié ne peut être sanctionné, licencié ou faire l'objet d'une mesure discriminatoire, directe ou indirecte, telle que définie à l'article 1er de la loi n° 2008-496 du 27 mai 2008 portant diverses dispositions d'adaptation au droit communautaire dans le domaine de la lutte contre les discriminations, notamment en matière de rémunération, au sens de l'article L. 3221-3, de mesures d'intéressement ou de distribution d'actions, de formation, de reclassement, d'affectation, de qualification, de classification, de promotion professionnelle, de mutation ou de renouvellement de contrat en raison de son origine, de son sexe, de ses mœurs, de son orientation sexuelle, de son

Code (Code civil) stipulates that no one may be discriminated against on the basis of his genetic features.[30]

It should be emphasized that the principle of nondiscrimination not only applies to genetic characteristics and other above-mentioned features, but also to all kinds of brain abnormalities (which can show genetic characteristics such as Neurofibromatosis 2 at the same time).

Participants could not only suffer from severe discrimination on the job market, but also be excluded from important insurance policies. Art. L133-1 Insurance Code (Code des assurances) regulates that access to life or disablement insurance is guaranteed according to the terms provided for in the hereafter reproduced Art. L1141-1–L1114-3 of the PHC.[31] Under the terms of Art. L1114-1 Cl. 1 PHC, firms and organizations that propose a life or disablement insurance shall not take into account the results of genetic characteristics of a person who desires to benefit from such insurance policy even when the said results are transmitted to them by that person or with his/her consent.[32] According to Art. L1114-1 Cl. 2 PHC, these firms and organizations should additionally be allowed neither to ask any questions relating to genetic tests and results, nor to subject a person to pass genetic tests before the conclusion of the contract and during the performance of said contract.[33]

On the one hand, this stipulation protects persons which suffer from genetic diseases to a high extent, but others who have also contracted a severe nongenetic disease are excluded.[34] This leads to a substantial discrimination, which can hardly be justified, particularly with a view to malignant tumors. Therefore, a corresponding regulation, which explicitly alludes to (clinically relevant) incidental findings would be desirable.

The chapter about biomedical research does not envisage any general provision concerning data protection. However, the precedent title including the rights of a

---

âge, de sa situation de famille ou de sa grossesse, de ses caractéristiques génétiques, de son appartenance ou de sa non-appartenance, vraie ou supposée, à une ethnie, une nation ou une race, de ses opinions politiques, de ses activités syndicales ou mutualistes, de ses convictions religieuses, de son apparence physique, de son nom de famille ou en raison de son état de santé ou de son handicap."

[30] Free translation of Art. 16-13 Civil Code: "Nul ne peut faire l'objet de discriminations en raison de ses caractéristiques génétiques."

[31] Free translation of Art. L133-1 Insurance Code: "L'accès à l'assurance contre les risques d'invalidité ou de décès est garanti dans les conditions fixées par les Articles L. 1141-1 à L. 1141-3 du code de la santé publique ci-après reproduits."

[32] Free translation of Art. L1114-1 Cl. 1 PHC: "Les entreprises et organismes qui proposent une garantie des risques d'invalidité ou de décès ne doivent pas tenir compte des résultats de l'examen des caractéristiques génétiques d'une personne demandant à bénéficier de cette garantie, même si ceux-ci leur sont transmis par la personne concernée ou avec son accord."

[33] Free translation of Art. L1114-1 Cl. 2 PHC: "En outre, ils ne peuvent poser aucune question relative aux tests génétiques et à leurs résultats, ni demander à une personne de se soumettre à des tests génétiques avant que ne soit conclu le contrat et pendant toute la durée de celui-ci."

[34] For example, the Court of Cassation (Cour de cassation) determined in the judgment of October 7, 1998 that a HIV-infected person has to inform his/her insurance.

diseased person and the users of the health system stipulates in Art. L1110-4 Para. 1 PHC that every patient has the right to respect for his/her privacy and the right to keep data concerning his/her person secret.[35] Therefore, any personal information collected during fMRI studies should be considered and treated as confidential, for example by creating a pseudonym for the brain images. This procedure would guarantee the protection of personal data even if the researcher asks a third person (e.g., a neuroradiologist) for advice in case of a probable brain abnormality.

2.1.5   Right to Compensation

In the context of neuroscientific research, several imaginable events of damage exist. In case of faulty data security practices, for example, third parties such as insurances or employers could become aware of brain abnormalities against the will of the insurant or employee. This, in turn, could lead to fatal consequences concerning the participant's insurance status and working condition. At worst, as already indicated, disability insurances/occupational disablement insurances and contracts of employment could be refused. Furthermore, it is possible that the researcher does not handle incidental findings in an appropriate manner. S/he neglects to inform the participant about an incidental finding or s/he delays the notification, just to name two possibilities.

The PHC has a few compensation rules that apply, regardless of any potential contractual responsibility. The sponsor who does not need to be identical with the researcher assumes the compensation for the proband in line with Art. L1121-10 Para. 1 PHC. However, s/he can exculpate himself/herself from compensation if s/he has not caused the damage or if the participant has withdrawn his/her consent.[36] If the sponsor does not bear any responsibility, the victims can be indemnified under the conditions of Art. L1142-3 PHC according to Art. L1121-10 Para. 2 PHC.[37] Art. L1126-7 PHC stipulates that only the District Court (Tribunal de Grande Instance) is competent for judgments concerning compensation resulting from biomedical research.

---

[35] Free translation of Art. L1110-4 Para. 1 PHC: "Toute personne prise en charge par un professionnel, un établissement, un réseau de santé ou tout autre organisme participant à la prévention et aux soins a droit au respect de sa vie privée et du secret des informations la concernant."

[36] Free translation of Art. L1121-10 Para. 1 PHC: "Le promoteur assume l'indemnisation des conséquences dommageables de la recherche biomédicale pour la personne qui s'y prête et celle de ses ayants droit, sauf preuve à sa charge que le dommage n'est pas imputable à sa faute ou à celle de tout intervenant sans que puisse être opposé le fait d'un tiers ou le retrait volontaire de la personne qui avait initialement consenti à se prêter à la recherche."

[37] Free translation of Art. L1121-10 Para. 2 PHC: "Lorsque la responsabilité du promoteur n'est pas engagée, les victimes peuvent être indemnisées dans les conditions prévues à l'article L. 1142-3."

The informed consent document should expressly refer to the compensation rules and to the proceedings which have to be taken in case of any complaints and issues of exculpation and responsibility.

Due to the principle of noncommercialization of the human body (principe de non-commercialité du corps humain), the participant does not benefit from any further compensation.[38] Art. L1121-11 Para. 1 Cl. 1 PHC only authorizes compensation for the expenditures that are determined in an objective manner (e.g., travel expenses).[39]

This regulation can be adopted in neuroscientific research without any modification, because it avoids that persons of lower social ranks would be tempted to participate for mere financial reasons. Moreover, a prior medical examination such as that in terms of Art. L1121-11 Para. 3 PHC is not required and would indirectly inure to the benefit of the participant, so that the research study would become a self-serving character.[40]

3 Application of Neuroscientific Techniques

The knowledge obtained by neuroscientific research has widespread fields of application that involve a large number of legal issues. Actually, fMRI is primarily used in the medical context to reveal mental diseases and to adjust medical attendance. Brain imaging also helps to reduce the rate of misdiagnosis, for example that of the vegetative state, which has otherwise been estimated to be up to 40%:[41] In the famous case of Rom Houben, an fMRI brain scan revealed a false diagnosis of the patient's vegetative state, leading physicians to conclude that the patient had in fact been awake for 23 years. During this time, Rom Houben had been unable to communicate due to simple muscle paralysis. Apart from the medical application field, fMRI examinations help independent reviewers in law procedures to reveal mental disorders which have a wide influence on the culpability of the accused. According to Art. 122–1 Para. 1 Criminal Code (Code pénal), a person is not criminally liable who, when the act was committed, was suffering from a psychological or neuropsychological disorder, which destroyed his/her discernment or his/her ability to control his/her actions.[42] Of course, the same

---

[38] Laude et al. (2009), p. 662.

[39] Free translation of Art. L1121-11 Para. 1 Cl. 1 PHC: "La recherche biomédicale ne donne lieu à aucune contrepartie financière directe ou indirecte pour les personnes qui s'y prêtent, hormis le remboursement des frais exposés et, le cas échéant, l'indemnité en compensation des contraintes subies versée par le promoteur."

[40] As mentioned above (2.1.1), the research study does not pursue any diagnostic aim.

[41] Bosco et al. (2010), p. 88.

[42] Free translation of Art. 122-1 Para. 1 Criminal Code: "N'est pas pénalement responsable la personne qui était atteinte, au moment des faits, d'un trouble psychique ou neuropsychique ayant aboli son discernement ou le contrôle de ses actes."

applies to civil law. For instance, contracts could turn out to be invalid according to Art. 1123 Civil Code, which determines that any person may enter into a contract, unless s/he has been declared incapable of it by law. On the level of civil procedure law, the capacity to take legal action could be restrained or excluded.

But much more legally relevant and exciting are "new" application fields such as the use of fMRI scanners as lie detection tests in courtrooms or as communication methods for coma patients.

3.1 Use of fMRI Scanners as Lie Detection Tests in Courtrooms

Since the nineties, the question of legitimacy of lie-detection tests has been a heated public debate. "The Polygraph" (Le Polygraphe), a Canadian–French–German film about a man who undergoes a lie-detection test in order to clarify whether he is guilty or not, called big attention in 1996 and described a scenario which has become true today: In some countries such as in the USA, in Singapore, Israel,[43] and India,[44] the use of lie detectors in courtrooms is already a daily occurrence. For instance, the Brain Electrical Oscillations Signature (BEOS) Test, a brain-based lie detection test, has been used in more than 300 Indian cases since 2003.[45] In 2008, this test was even used to convict a 24 year old of murder.[46] The fMRI scanner has not yet been used as a lie detection test in French courts. The Centre d'analyse stratégique argues that lie-detection tests are against human dignity.[47] Furthermore, French courts have already dealt with related subject matters: The French Supreme Court (Cour de Cassation) judged in 2000[48] and 2001[49] that hearings conducted under hypnosis were irregular and compromised the rights of defense, because hypnosis is against the defendant's consent and free will. The Criminal Court of Seine (Tribunal Criminel de Seine) forbade the use of narcoanalysis ("truth serum") by decision of February 23, 1949.

Furthermore, the use of fMRI scanners as lie-detection tests in courtrooms violates the principle of liberty of evidence (principe de liberté de la preuve) in

---

[43] Centre d'analyse stratégique (2009), No. 128, p. 5.

[44] See Rödiger (2011).

[45] Puranik et al. (2009), p. 817.

[46] State of Maharashtra v Aditi Baldev Sharma and Pravin Premswarup Khandelwal, Sessions Court Pune June 12, 2008, Sessions Case No. 508/07.

[47] Centre d'analyse stratégique (2009), No. 159, p. 7.

[48] Cass. Crim. (2000).

[49] Cass. Crim. (2001).

criminal procedure law. Art. 427 Para. 1 Criminal Procedure Code (Code de procedure pénale) stipulates that except where the law otherwise provides, offences may be proved by any mode of evidence and the judge decides according to his/her innermost conviction.[50] However, the principle of liberty of evidence has its limits, where *reasonable doubts* exist as to the validity of scientific evidence.[51] One of the most known selling brain scan services, No Lie MRI, promotes on his website that the current accuracy is more than 90% (and is estimated to be 99% once the product development is complete).[52] Therefore, the use of fMRI scanners as lie-detection tests is doubtable evidence and cannot be invoked in court.

3.2   Use of fMRI Scanners as Communication Methods for Vegetative State Patients?

Moreover, brain imaging methods have the potential to communicate not only with locked-in persons, but also with true vegetative state patients. For example, the neuroscientist Adrian M. Owen used the fMRI scanner to demonstrate that one of his vegetative state patients was able to give yes and no answers to question stimuli.[53] This fast-paced development and application of neuroscientific knowledge has the potential not only to improve patient well-being and medical treatment upon a diagnosis of vegetative state, but may also provide greater patient autonomy in medical decision making. The ability of patients in vegetative state to make personal informed decisions raises important legal concerns. All decisions have to be made in accordance with the will of the patient. If the patient is fully conscious and able to express his/her will via fMRI techniques, s/he can make his/her own decisions and does not need a custodian to represent his/her interests in medical decisions. In the same sense, facilitating patient communication through neuroscientific techniques may also affect other areas of civil law, such as contractual capacity and all cases in which personal action and intention are required, for instance in the declaration of marriage or the last will.

---

[50] Free translation of Art. 427 Para. 1 Criminal Procedure Code: "Hors les cas où la loi en dispose autrement, les infractions peuvent être établies par tout mode de preuve et le juge décide d'après son intime conviction."

[51] Other legal systems have less severe limits such as Germany where *improper* evidence is excluded according to Art. 244 Para. 3 Cl. 2 German Criminal Procedure Code.

[52] The website of No Lie MRI can be found at: http://noliemri.com/products/Overview.htm. Accessed 30 Dec 2010.

[53] The Washington Post (2010).

4 Outlook

The analysis of different biomedical principles in the first part shows that existing national rules are suitable for orientation, but they are far away from being sufficient, in particular with a view to the challenges raised by incidental findings. Incorporating neurospecific rules in the Law on Bioethics would be a convenient way to face the issues. This solution would be in accordance with the proposition of Jean Leonetti, secretary of the parliamentary commission for revising the bioethics laws, who suggested the incorporation of some neuroscientific rules in the Law on Bioethics in January 2010.[54] The Council of State (Conseil d'Etat), the Agence de la biomédecine, the Office parlementaire d'évaluation des choix scientifiques et technologiques and the National Consultative Ethics Committee on Life and Health Sciences (Comité consultative national d'éthique pour les sciences de la vie et de la santé) were asked to make concrete propositions concerning the implementation of neuroscientific rules in the Law on Bioethics. The main question was whether it should be a law providing guidelines, i.e., a more flexible and responsive law[55] or rather, a detailed law which will turn out to be incomplete. In the outcome, the majority voted for a law providing guidelines and suggested broadening the scope of responsibility of the agency and conferring upon them neuroscience issues.[56] This solution has a few drawbacks, because the guidelines would not make a big difference to the existing rules concerning biomedical research and could even become redundant. By contrast, a detailed law would contain a first practical approach for researchers how to deal with incidental findings. Of course, it would turn out to be incomplete, but this lies in the nature of neuroscientific research that poses almost daily new challenges to neuroscientists, ethicists, and lawyers and is therefore both a blessing and a curse.

With a view to the legal use of neuroscientific knowledge, France drags far behind other countries such as the USA and India. The use of fMRI scanners as lie-detection tests in French courts is still a science fiction story, but their use in the medical context seems to gain in importance. fMRI techniques are likely continue to shed light on patient decision-making processes and expressions of will. It is therefore important to keep an eye on the legal consequences of these remarkable developments.

Acknowledgement The author thanks the Federal Ministry of Education and Research for funding.

---

[54] Assemblée nationale (2010).

[55] Centre d'analyse stratégique (2009), No. 128, p. 6.

[56] Centre d'analyse stratégique (2009), No. 128, p. 7.

References

Anonymous (2005) How volunteering for an MRI scan changed my life. Nature 434(7029):17

Assemblée nationale (2010) Mission d'information sur la révision des lois bioéthiques: Rapport de Jean Leonetti – mercredi 20 janvier 2010. http://www.assemblee-nationale.fr/presse/communiques/20100120-02.asp. Accessed 30 Dec 2010

Belliveau JW, Kennedy DN, McKinstry RC, Buchbinder BR, Weisskoff RM, Cohen MS, Vevea JM, Brady TJ, Rosen BR (1991) Functional mapping of the human visual cortex by magnetic resonance imaging. Science 254:716–719

Bosco A, Lancioni GE, Belardinelli MO, Singh NN, O'Reilly MF, Sigafoos J (2010) Vegetative state: efforts to curb misdiagnosis. Cogn Processing 11:87–90

Cass. Crim. (2001) November 28, 2001: Bull. Crim., No. 247

Cass. Crim. (2000) December 12, 2000: Bull. Crim., No. 369

Centre d'analyse stratégique (2009) Analyse Perspectives scientifiques et éthiques de l'utilisation des neurosciences dans le cadre des procédures judiciairies, La note de veille, No. 159, p 1. http://www.strategie.gouv.fr/IMG/pdf/NoteVeille159.pdf. Accessed 30 Dec 2010

Centre d'analyse stratégique (2009) Analyse. Impact des neurosciences: quels enjeux éthiques pour quelles regulations?, La note de veille, No. 128, p 6. http://www.strategie.gouv.fr/IMG/pdf/NoteVeille128.pdf. Accessed 30 Dec 2010

Convention on Human Rights and Biomedicine (2010) http://conventions.coe.int/Treaty/EN/Treaties/Html/164.html. Accessed 30 Dec 2010

Cour Const (1994) July 29, 1994, No. 94–345 DC, RJC 1–595

European directive 2001/20/CE: http://ec.europa.eu/health/files/eudralex/vol-1/dir_2001_20/dir_2001_20_en.pdf. Accessed 30 Dec 2010

Illes J, Kirschen MP, Karetsky K, Kelly M, Saha A, Desmond JE, Raffin TA, Glover GH, Atlas SW (2004) Discovery and disclosure of incidental findings in neuroimaging research. J Magn Reson Imaging 20:743–747

Illes J, Kirschen MP, Edwards E, Stanford LR, Bandettini P, Cho MK, Ford PJ, Glover GH, Kulynych J, Macklin R, Michael DB, Wolf SM, members of the Working Group on Incidental Findings in Brain Imaging Research (2006) Incidental findings in brain imaging research. Science 311:783–784

Laude A, Mathieu B, Tabuteau D (2009) Droit de la santé. Presses Universitaires de France, Paris

Law on Bioethics (2010) http://ec.europa.eu/research/biosociety/pdf/french_law.pdf. Accessed 30 Dec 2010

Public Health Code (2010) http://www.legifrance.gouv.fr/. Accessed 30 Dec 2010

Puranik DA, Joseph SK, Daundkar BB, Garad MV (2009) Brain signature profiling in India: it's status as an aid in investigation and as corroborative evidence – as seen from judgments. In: Proceedings of XX All India Forensic Science conference, Jaipur, pp 815–822. http://www.axxonet.com/cms-filesystem-action/publications/beos_in_india.pdf. Accessed 30 Dec 2010

Rödiger C (2011) Das Ende des BEOS-Tests? Zum jüngsten Lügendetektor-Urteil des Supreme Court of India. Nervenheilkunde 30:74–79

State of Maharashtra v Aditi Baldev Sharma and Pravin Premswarup Khandelwal, Sessions Court Pune June 12, 2008, Sessions Case No. 508/07: http://lawandbiosciences.files.wordpress.com/2008/12/beosruling2.pdf. Accessed 30 Dec 2010

The Washington Post (2010) In 'vegetative state' patients, brain scanners show some alert minds: http://www.washingtonpost.com/wp-dyn/content/article/2010/02/03/AR2010020302887.html. Accessed 30 Dec 2010

Legal Implications of Neuroscientific Instruments with Special Regard to the German Constitutional Order

Tade Matthias Spranger

Abstract Whereas the past few years have repeatedly been entitled as the "era of biotechnology," most recently one has to get the impression that at least the same degree of intention is being paid to the latest developments in the field of neurosciences. It is by now nearly impossible to oversee the number of research projects dealing with the functionality of the brain – for instance concerning the organizational structure of the brain – or projects dealing with the topic of mind reading. Massive efforts have also been taken in the field of prediction; for instance, it is possible to analyze certain structures and thereby presume the research paticipant's decision before he/she has ever told it.

The concept of neurosciences covers a wide range of different scientific branches, all of which investigate the structure and functioning of the nervous system. As far as research in connection to the human brain is concerned, the term of brain research is nearly used as a synonym. The concerned research projects do not only cover experimental basic research, but also search for therapeutic methods of treating nervous diseases.

1 Introduction

The neurobiological research of the brain is carried out on three different levels. The first level deals with the functioning of greater brain areas, for instance special tasks of the cerebral cortex, such as the amygdale or the basal ganglia. The mid-level describes actions within communities of thousands of different cells. And the lower level covers actions of singular cells and molecules. Whereas there have been

T.M. Spranger (✉)
Institute of Science and Ethics, University of Bonn, Bonner Talweg 57, 53113 Bonn, Germany
e-mail: spranger@iwe.uni-bonn.de

T.M. Spranger (ed.), *International Neurolaw*,
DOI 10.1007/978-3-642-21541-4_9, © Springer-Verlag Berlin Heidelberg 2012

significant improvements on the first and the third level, there have not yet been any improvements on the mid-level.[1]

The methods of brain research show many differences and follow diverse methods: image-guided procedures such as the Positron Emission Tomography (PET) and the functional Magnetical Resonance Tomography (fMRT) evaluate the energy resources needed in different areas of the brain and show a very precise resolving power up to millimeters. In a temporal sense however, they are at least seconds behind the actual effects.

The more classical electroencephalography (EEG) shows the electrical activity of neural united cell structures in real time, but does not give definite information about the area of action. To gain more accuracy, one should use the Magnetical Encephalography, which is able to show changes of magnetic fields as precise as milliseconds.[2]

From the view of the arts, brain research is especially of utmost interest, because it shows possible solutions for the debates on consciousness, mind, soul, intellect, or emotions. Whereas in the philosophical branch the discussion is already at a high level concerning the range and the methods, the legal discussion[3] on the topic mainly focuses on consequences for the criminal law or the criminal procedure law, with special regard to the debate on free will.[4] Some contributions from the field of civil law furthermore discuss questions of liability in connection to image-guided procedures and questions of employment and insurance law in connection with so-called incidental findings. The Public Law and especially the Constitutional Law, however, are really yet starting to discuss the topic.[5] Thus, this paper tries to point out connections between the current discussion and the Constitutional Law and especially to evaluate the problem from the perspective of the Basic Rights. Although some methods might be seen as future-scenarios, a closer glance shows that the main part of concerned methods is in need of a legal discussion already today.

2 Constitutional Landscape

In particular for historical reasons, the constitutional dimension enshrined in the so-called Basic Law plays a pivotal role in the legal assessment of latest technological developments in Germany. Within the Basic Law, which ties the legislative process

---

[1] See on this and on the following aspects: Das Manifest – Elf führende Neurowissenschaftler über Gegenwart und Zukunft der Hirnforschung, in: Gehirn & Geist 2004, 30.

[2] Das Manifest – Elf führende Neurowissenschaftler über Gegenwart und Zukunft der Hirnforschung, in: Gehirn & Geist 2004, 30 et seq.

[3] The international literature does not ignore the described limitation in content, see: Garland and Frankel (2004); O´Hara (2004); Goodenough (2004); Corvat (2004); Busey (2006); Santosuosso (2009).

[4] Critical about this: Fischer, Strafgesetzbuch, 55. Aufl. 2008, Sect. 13 Rn. 9a.

[5] General aspects on the demand for new regulations as a consequence of developments in life-sciences: Simitis (2008).

to the constitutional order and binds state administration to uphold the law, the basic rights constitute the most important provisions. Covering such manifold aspects as human dignity (Art. 1), equality before the law (Art. 3), freedom of faith, conscience, and creed (Art. 4), school education (Art. 7) or freedom of assembly (Art. 8), the basic rights' tremendous effects depend on four different dimensions:

- As individual rights to defend oneself against the State (so-called Abwehrrechte/ subjektive Rechte; human dignity as primus inter pares)
- As claim to active support by the State (so-called Anspruchs- oder Leistungsrechte, only accepted as rare exceptions, cf. Art. 6 para. 4: "Every mother shall be entitled to the protection and care of the community.")
- As normative basis for the State's duty to protect the individual (so-called staatliche Schutzpflichten, in particular enshrined in Art. 1 para. 1 and Art. 2 para. 2)
- As normative basis for the so-called third-party effect ("mittelbare Grundrechts-drittwirkung"): Basic Rights as a normative "lighthouse" for the State's legal assessment of subconstitutional matters

2.1 Human Dignity

Once again, the Basic Right of human dignity of Art. 1 para. 1 German Basic Law can be determined as the initial and crucial point of the legal discussion.[6] Above all, one has to raise the essential question, whether the common image of a human being is going to be changed by neuroscientifical developments or even turned upside down. In a second step, one has to determine, which dangers are to be feared for the "old" or also the "new" understanding of human dignity in connection with neuroscientific instruments. Finally, despite all possible dangers connected to the new developments, one must not forget that sometimes technical improvement is just the condition for living with human dignity.

2.1.1 General Understanding of Human Dignity

In a contribution named manifest, dealing with current and future developments of brain research, 11 leading researchers have formulated a precise demand some years ago: "The Arts and Neurosciences will have to lead an intensive discussion in order to design a new image of human beings."[7]

---

[6] Also see Heun (2005).

[7] Das Manifest – Elf führende Neurowissenschaftler über Gegenwart und Zukunft der Hirnforschung, in: Gehirn & Geist 2004, 30 (37). Furthermore: Schleim (2008), p. 146 et seq.

In fact, especially the relevance of the debate on free will cannot be denied for a just application of Basic Rights. That is why one can state that only the recognition of the individual as free in will and reasonable will lead to individual freedom guaranteed by the Basic Rights.[8]

The current debate on the question of free will goes back to the so-called Libet-Experiment. In this experiment, the physiologist Benjamin Libet evaluated in 1979, in which ways the temporal order of a concrete will to act is connected to its motor implementation. Within the experiment, the probands were asked to choose an optional point of time for raising the right hand and to remember the time of day of that action. In some cases, they were asked to act most spontaneously, in other cases they were asked to plan their action about one second before carrying it out. Evaluating the results, Libet found out that the subjective will to act comes in any case after the point of time, in which the motor cortex begins to prepare the move. If this leads to the conclusion that the will to act cannot cause the initialization of the motor cortex, one could principally criticize the figures of free will and liability of the human being.

However, Libet's experiment has ever since been subject to strong criticism, which was not only focused to possible methodological mistakes. However, recently there has been a significant amount of opinions stating that they have come to the same conclusions, but using more modern and more specific methods.[9] Those opinions follow the results of Libet and try to show the falsity of the concept of a free will of human beings. For instance, Gerhard Roth states that one has to call the feeling of subjective freedom of decision and action an illusion.[10]

In certain areas of the Criminal Sciences, the debate about free will is focused on the call for a new scheme of "Betterment Criminal Law"[11] or "Protective Criminal Law"[12] instead of the old "Debt Criminal Law", all based on the new results of neurobiology. The question now is in which way this Criminal Law discussion takes over to the Constitutional debate. As the jurisdiction of the Federal Constitutional Court shows a strong connection between the principle of personal debt and the guarantee of human dignity,[13] the more Criminal Law-based debate on free will is widely regarded as a major challenge for the understanding of the principle of human dignity, stated in Art. 1 para. 1 of the German Basic Law.[14] However, this argumentation does not yet take into account the quality of the concerned intervention.

---

[8] Möllers (2008).

[9] Libet (2004).

[10] In this way Roth (2003).

[11] Herdegen (2006).

[12] Hillenkamp (2005), 313 (317).

[13] Heun (2005). Schreiber (2006).

[14] In this direction also see Hillenkamp (2005), 313 (317).

The Principle of Debt and Human Dignity

As far as the Federal Constitutional Court links the penalty with the provision of individual debt, this kind of guarantee has traditionally been linked with Art. 2 para. 1 Basic Law rather than to the human dignity.[15] Although the quotation of the perpetrator, who must not be reduced to a pure object of governmental crime prevention while denying his constitutionally protected value and esteem[16] is well known, it does not affect the individual debt, but rather the practical dealing with governmental punishment.

As for this point, the Federal Constitutional Court states:

> It follows directly out of Art. 1 para. 1 Basic Law in connection to the Principle of the Social State, that the state is demanded to secure a minimum of existence in order to secure a life in human dignity. Understanding human dignity in that sense, makes it inacceptable to let the state take one's individual freedom without giving him the chance to gain it back some time.[17]

Taking these considerations into account, the principle of human dignity is not that much important for the question "if" the state punishes, but rather for the question "how" the state punishes. It was not until the famous "East German Border Guard Decision" that the Federal Constitutional Court left this line of argumentation and took a more constitutional approach, now also taking into account the human dignity in a more specific way:

> In terms of care for the Criminal Law, Art. 1 para. 1 Basic Law determines the character of penalty and the relationship between debt and expiation. The basic principle 'no penalty without debt' has the quality of a constitutional principle and derives from the human dignity and Art. 2 para. 1 Basic Law in connection with the Principle of the Legal State.[18]

Intensity of the Threatening Intervention

A negation of individual debt based on neuroscientific results would consequently open up the two following options to act: One possibility would be to determine individual debt as a social construction, which is widely independent from neuroscientific findings. The other possibility would be to completely renew the term of debt and adjust it to the new framing conditions. Collisions with the principle of human dignity, however, only arise in the fictional constellation that neuroscientific

---

[15] Decision of the Federal Constitutional Court, in Germany cited as: BVerfGE 20, 323 (331).

[16] Decision oft he Federal Constitutional Court, in Germany cited as: BVerfGE 45, 187 (228).

[17] See decision of the Federal Constitutional Court, in Germany cited as: BVerfGE 45, 187 (228 f.). Further decisions point out the importance of human dignity as well: BVerfGE 6, 389 (439) sees the reason for a reasonable punishment in the Principle of the Legal State. In this direction also BVerfGE 20, 323 (331).

[18] Decision of the Federal Constitutional Court, in Germany cited as: BVerfGE 95, 96 (140).

findings make the original term of debt inapplicable, but the state still sticks to the old concept of debt.

It is important to note that this concept would not question the original image of the human being, which is the basis for the definition of human dignity. To the contrary, in such a case the existing understanding of human beings and their dignity would have to be strengthened, because otherwise the thesis of free will would lack a connection to the Basic Rights. Despite these concepts, it is to be pointed out that the principle of debt is at least also based on human dignity and that not the principle of human dignity is based on the principle of debt. Consequently, any recalibration of this constellation must take into account the frame of Art. 1 Basic Law, which is not violated by such modifications. Therefore, the whole discussion is only linked with constitutional matters in that way, that any penalty without debt would indicate an intervention into human dignity *in a legal sense*. But so far neuroscientific findings do not have the necessary power to overcome the old principle of debt.

Taking a closer look, one can see that punishment is first of all based on the actions of the individual and his or her responsibility for that. In other words, debt is the subjective attribution of actions deflecting from the law. This means that the perpetrator is held responsible in the sense of a social construction,[19] if he could have acted in another way. However, the possibility of acting in another way is not to be understood in an interdeterministic sense of personal freedom.[20]

As far as the Criminal Law Debt Decision means a difference between the actions of the individual and the actions to be expected, this evaluation means a general view of things inherently. If a perpetrator "really" is guilty cannot be stated definitely. Thus, criminal responsibility is only an analogy to "real" debt.[21]

What is also important to note in this connection is that none of the ambassadors of a hard determinism[22] was able to find an acceptable definition of institutions such as "reasonability" or "evaluation".[23] To the contrary, all hints point at the conclusion that the human being is not being determined neuronally, genetically, or in any other way, but that on some levels and in different ways, there can be found psychological, genetic, or neuronal soft determination. Beside this, experiences, environmental influences, and the education play a decisive role for the development of every human being and his existence.[24]

---

[19] Jakobs (2009).

[20] Schreiber (2006), (1074 et seq.).

[21] Schreiber (2006), (1076 et seq.). Also see: Fischer, Strafgesetzbuch, 55. Aufl. 2008, vor Sect. 13 Rn. 8.

[22] The interesting question, if and in how far a lack of interdeterministic freedom could cause a state of missing determination, shall not be subject to this paper.

[23] For the neuroscientific debate, see: Schreiber (2006), p. 1069 (1074).

[24] Schreiber (2006), p. 1069 (1076); Spranger (2007b), 161 (176).

Thus, a triumph of neuronal reductionism is not to be expected.[25] The neuroscientific debate on determinism is rather – beyond the theological, psychological and physical determinism – just one more field of the well-known and general debate on determinism.[26]

Therefore, the problems of freedom or not freedom of the human will in some areas of brain research will not be able to seriously attack the basis or the functioning of the criminal law definition of debt, so that there will not be any problems with the constitutional understanding of the human being. Apart from that, human dignity is one of the essential aspects of the individual debt, whereas the principle of personal debt is vice versa no essential content of human dignity or the image of the human it is based on. From a neuroscientific point of view, the human dignity does not have to fear any consequences. Any other opinions cannot be proofed by facts or arguments.

Further developments, however, have already left the level of a rather academic discussion and thus give more motivation for a constitutional assessment. As far as methods of destroying one's will or massive interventions into the personal identity and integrity are seen as "typical interventions"[27] into Art. 1 para. 1 Basic Law, this holds true especially for neuroscientific measures in a procedural way.

2.1.2 Dangers for Human Dignity

Procedural Use

In the USA companies such as Cephos[28] and NO Lie MRI[29] canvass the usage of image-guided procedures – especially the functional magnetic resonance tomography – in court trials and other legal areas. Also other countries show a strong development of comparable methods: In India, two states with together about 160 million citizens use the so-called Brain Electrical Oscillations Signature (BEOS) Test, which is initially based on the EEG. The activity measured by the EEG is evaluated by a specific software. An interpretation of the data by the investigator is not planned.[30] In 2008, the decision of an Indian court became famous, basing its conviction essentially on the result of a BEOS-Test and convicting the concerned woman to a life-long sentence.

The advocacies of such methods claim that such methods are able to detect lies, but are technical improvements to the common lie-detector. Legally speaking, this

---

[25] Das Manifest – Elf führende Neurowissenschaftler über Gegenwart und Zukunft der Hirnforschung, in: Gehirn & Geist 2004, p. 30 (37).

[26] In this direction probably also: Heun (2005), 853 (857); Schreiber (2006), p. 1069 (1071).

[27] Pieroth/Schlink, Grundrechte, 24. Aufl. 2008, Rn. 361.

[28] http://www.cephoscorp.com (12 February 2010).

[29] http://noliemri.com (12 February 2010).

[30] Spranger (2009b).

makes it possible to compare the new technology to the standards set up for the usage of polygraphs. However, this comparison brings several points of criticism. First of all, it is pointed out that image-guided procedures do not – in contrast to the polygraph – measure physical reactions, which hint at excitement, fear or nervousness, but only show the brain activity indirectly. Thus, the usage of image-guided procedures is not aiming at a conclusion from physical reactions to a feeling to truth or untruth, but from the brain function to the quality of a thought. Image-guided procedures would thus show a "weaker indirectness".[31]

Apart from the fact that the degree of directness or indirectness cannot be seen as a definite parameter for the suitability of a method of proof, also the claim for a "weaker indirectness" seems not to be forcing in any way, as the interpretation of the result is much more important in connection to the usage of image-guided procedures than in connection to a polygraph.[32] The first problem already arises when it comes to the question, in which temporal field the brain activity is to be measured. This brings forward the task to correctly determine the time, which the proband needs to read and understand the questions.

What is even more important is the fact that, at least according to the current state of the art, there cannot be determined certain areas in the brain for "lie" or "truth". Therefore, the usage of image-guided procedures leads to a process which is very intense in time and work, and which can possibly lead to a statement about the general quality of a thought. Whereas the sequence of a polygraph is "physical function – feeling – truth/lie", the sequence of image-guided procedures must consequently be "brain function – image-interpretation for the evaluation of the quality of a thought – truth/lie". The thesis of a "weaker indirectness" is not conforming to this result.[33]

Following this line of argumentation, a comparison to the existing jurisdiction on polygraphs is acceptable when answering the question by which constitutional provisos the usage of image-guided procedures is legitimated in German courts, as both methods are aiming at providing evidence for the truth or untruth of certain statements by means of measuring physical or physiological actions.

The Federal Constitutional Court does not see the human dignity, but rather the general right of privacy[34] as the relevant basis:

> The usage of the lie-detector is aiming at finding out involuntary and invisible physical reactions by technical means, in order to draw conclusions concerning the truth or untruth of certain statements. Such kinds of 'personal screening,' which deny the fact that any statement is an essentially personal effort and see the individual as a mere object of the procedure, interfere into the right to personal privacy in a non-acceptable way.[35]

---

[31] Beck (2006).

[32] Spranger (2007b), 161 (164 et seq.).

[33] Spranger (2007b), 161 (164 et seq.).

[34] In Germany: Allgemeines Persönlichkeitsrecht (APR).

[35] Decision of the Federal Constitutional Court, in Germany cited as: BVerfG, NJW 1982, 375.

In contrast to this, the Federal Supreme Court[36] pointed out in an earlier decision:

> The usage of the answers within the interrogation with the polygraph of the accused person as measures of proof, was improper according to Art. 1 para. 1 Basic Law and section 136a of the Code of Criminal Procedure, regardless of his consent to this.
>
> The polygraph aims at gaining more and different statements of the accused person, among these statements which are made automatically and could not be made without the polygraph. Beyond the conscious and willing answer, the polygraph "answers", without giving the accused person the chance to prevent this. Such an insight in the inner sphere of the accused person violates the freedom of will (section 136a Code of Criminal Procedure) and is improper in a Criminal Court Trial.

In 1998, this opinion was adjusted as follows:

> The points of criticism, which have so far been brought forward against the usage of the polygraph, have no validity any more, if the concerned person has worked together with the interrogators voluntarily. The senate thus does not see any violation of the human dignity of the concerned person in cases, where the state has demanded the usage of such instruments.
>
> A more different approach, taking into account the consent of the concerned person, better suites the scope of Art. 1 para. 1 Basic Law. It is the declared scope of Art. 1 para. 1 Basic Law not to restrict, but to protect human dignity, which contains the principal freedom to decide on one's own fate.

Following this jurisdiction, a violation of human dignity is at least not given, when the concerned person has consented to the procedure. This result can only be held by those who state that human dignity does not have to be protected against the will of the concerned person and thus demands a protection against one own. The dogmatic relevance and acceptance of this option show the need for a closer look. At this point, one could name the jurisdiction on illustrious topics such as peep shows, the arguable pleasure of dwarf-throwing or various technical death-games.

If one argues that the ability to determine what is protected by human dignity should stay with the concerned person, one has to forbid the protection of the individual against himself. An intervention covered by the informed consent of the concerned person thus would not even mean a violation of human dignity, if it would mean a total "screening" of the concerned person or his brain functions. On the other side, the governmental force on using such methods would not fit the constitutional understanding of human dignity.

"Mind Reading"

More general than the usage of lie-detection are all neuroscientific projects which are pooled under the terms of "mind reading" or "brain reading".[37] Such projects do

---

[36] In Germany: Bundesgerichtshof, BGH.

[37] On this: Schleim (2006); Schleim (2008).

not only try to find out if a statement is true or false, but also aim at the screening of certain and current contents of thoughts.

As for the state, such methods are of special interest for secret intelligences. For instance, at least in the USA possible fields of application in connection to the fight against terrorism are being discussed and applicants for the military forces have to undergo a polygraph test.[38] The legal problems in this context show strong connections to the discussions concerning state torture or the legal assessment of the German Federal Aviation Security Act.[39]

Constitutional aspects have to be considered with regard to certain options of application in the private field. This holds true for the uncontrolled usage by companies, which could be able to try and gain certain information about contenders, customers, or simply market actors.[40] However, the main focus of the current possible problems lies on possible violations of right to data protection and the constitutional right to determine the usage of one's own personal data, and not that much on problems in connection to human dignity.

2.1.3   Enabling of Humane Existence

Whereas the discussion has so far mainly been focused on potential risks, other fields show options where one should think of fields of application which are able to guarantee a humane existence. This does especially apply for elements of neuroprosthetics.

The general aim of neuroprosthetics is to substitute disabled nervous systems by means of smallest technical instruments. For instance, deaf patients are given a new sense of hearing via electrostimulation through a cochlea implant and deep brain stimulation shall help patients with Morbus Parkinson. Furthermore, gateways between the brain and the computer, so-called "Brain-Computer Interfaces" (BCIs), help patients, who cannot move anymore (locked-in syndrome), to communicate to their environment.[41]

Furthermore, the methods of measurement used with neuroprosthetics, use signals of the body and transform them into electric signals, which are able to stimulate muscles or nervous or artificial systems, such as an artificial hand (actorics). The digitally strengthened, processed, and filtered schemes of innervations become signals, which then are able to move prosthetics.

Also, feedbacks are generally possible: Sensors being implemented into an artificial hand can measure the temperature of an object and send this to a microprocessor being implemented into the lower arm, so that the concerned brain areas

---

[38] See interview with John-Dylan Haynes, in: Gehirn & Geist 2006, 63.

[39] In Germany: Luftsicherheitsgesetz.

[40] "Neuromarketing" and "Neuroeconomy".

[41] http://www.imtek.de/bmt/index.php?page=http://www.imtek.de/bmt/content/neuroprothetik. php (12 February 2010).

are able to feel heat. Within the nearer future, it shall be possible for patients with palsy of arms or legs, to control "via thoughts" roboter arms or electro wheelchairs.[42]

Interaction as a Basis for a Humane Life

Such an improvement of options to interaction, which are sometimes even reduced to sero with challenged persons, may affect on the objective-legal content of human dignity, as it covers also personal identity and personal integrity.

In this context, the terms of identity and integrity first of all mean a state of highest possible self-autonomy.[43] As far as the term of identity shows an inner and an outer connection, the usage of neuroprosthetics especially affects the outer connection, which is the human being, who is free, but social, and who can present himself by the help of other human beings.[44]

The factual ability to interact is of utmost importance in order to concretize the content of the image of a human being.[45] This provision is best reflected in the area of custodianship law, where it is the declared aim to suit the wishes of the concerned person while respecting his autonomy.[46] This aim shall be accomplished by only allowing custodianship in rare exceptions and respecting the will of the concerned person.

Relevance for the Initiation of Legal Custodianship

Not only when it comes to the point, where we have to ask about the ability of personal wishes and declaration of disabled persons, but also when it comes to the creation of a custodianship, neuroscientific findings will become more and more important. If a legal custodianship shall be established due to a mental disease, and the common methods could be improved via neuroscientific findings, the law of custodianship has to make use of such methods with regard to the principle of human dignity in cases where a certain improvement can be expected.

---

[42] VDE Verband der Elektrotechnik Elektronik Informationstechnik e.V. (ed.), VDE-Studie zum Anwendungsfeld Neuroprothetik: Mikrosysteme in der Medizin, Anwendung - Technologie - Vergütung, 2005.

[43] Höfling, in: Sachs (Hrsg.), Grundgesetz, 4. Aufl. 2007, Art. 1 Rn. 35 with further references.

[44] Höfling, in: Sachs (Hrsg.), Grundgesetz, 4. Aufl. 2007, Art. 1 Rn. 36.

[45] Decision of the Federal Constitutional Court, in Germany cited as: BVerfGE 4, 7 (15 et seq.).

[46] See Palandt/Diederichsen, BGB, 68. Aufl. 2009, Sect. 1896 Rn. 4.

2.2 Physical Integrity

Neuroscientific findings and methods affect the principle of human dignity in many different ways. But also further guarantees of protection, such as the protection of life and personal integrity have to be discussed with regard to the developments of brain research.

2.2.1 Right to Defense

As far as Art. 2 para. 2 Basic Law also contains the protection of psychological integrity,[47] one has to point out the discussion on the so-called "thought-manipulation". If indeed thoughts would be manipulated or deleted by the state in a harmful way, this would certainly mean a violation of the personal integrity. But if the treatment rather aims at deleting bad thoughts with the consent of the concerned person, this treatment could well be seen as a medical treatment.

2.2.2 Duty to Protect

Furthermore, Art. 2 para. 2 Basic Law establishes a common duty to protect, which is important to deal with. This holds true especially with the predictive potential of neuroscientific measures. According to a strong opinion in literature, the fight against violence is not only one of the essential governmental tasks, but also demanded due to Art. 2 para. 2 Basic Law. The state has to provide the highest possible level of security to the citizens and is entitled to use all existing methods to defend an attack against the personal integrity or the life of an individual.[48]

As neuroscientific research projects on aggression and prevention against aggression show,[49] the current discussion very much focuses on preventive measures, which have to be taken very early in order to prevent from aggression.[50] In some ways, such efforts can be compared to the science-fiction movie "Minority Report" or with the novel the film is based on by the American author Philip K. Dick of 1965.[51] The story deals with the apartment precrime within the New York

---

[47] Pieroth/Schlink, Grundrechte, 24. Aufl. 2008, Rn. 393.

[48] Murswiek, in: Sachs (ed), Grundgesetz, 4. Aufl. 2007, Art. 2 Rn. 196. On Criminal Law aspects: Federal Supreme Court, in Germany cited as: BGH, NJW 2004, 237 et seq. In this direction the decision of the European Court of Human Rights points, too: EGMR, NJW 2003, 3259 et seq.; Furthermore: Meyer-Ladewig, Kommentar zur Europäischen Menschenrechtskonvention, 2. Aufl. 2006, Art. 2 Rn. 7c.

[49] See for instance Tremblay (2008).

[50] Das Manifest – Elf führende Neurowissenschaftler über Gegenwart und Zukunft der Hirnforschung, in: Gehirn & Geist 2004, 30 (36).

[51] Also see: Schneider and Merkle (2007) p. 227 et seq.

Police Department, which aims at preventing future crimes by means of pre-cognition. The (future) perpetrators are imprisoned and brought to so-called caveat – an artificial state of steady unconsciousness – without any court trial.

Whereas such a comparison might well enhance the media interest in the topic, it also holds the threat that neuroscientific research is seen as a dangerous and rather fantasy-like research branch. What is true is that many research projects, on the national level and on the international level, deal with aspects of human aggression, which could at least bring new results concerning the prognosis of the concerned person. Already today, the area of brain research shows interesting answers in connection to the following questions[52]:

- Are there any neurogenetic constellations, in which there is a higher risk of crime?
- Is the emotional cognition disturbed?
- Are there psychopathic character traits?
- What does the driving structure of a person look like?
- Have there been successful therapeutic measures?
- Is there a difference between the statements of a person and his inner feelings?

In the relevant context, it is not necessary to develop new criteria concerning the differentiation between the prognosis and the term used in section 66 Criminal Code, which is "affinity to severe crimes".[53] Nevertheless, the relevance of such research methods becomes obvious already today in connection with actions according to the laws on mentally ill people. If, and as far as, such methods are able to improve the prognosis on potential risks, the state has to fulfill its duty of protection by using such methods. However, the application of such methods also aims at the protection of the concerned persons themselves. Not only the impor-tance of possible treatment according to the Mentally Ill Act, but also the quantity demands a brought usage of the new methods.

The practical side of such duties of protection does not bring an unlimited usage of neuroscientific methods, however. In general, the corridor of evaluation is very wide[54] for the state and only has to pay attention to the prohibition of insufficient means on the one hand and to the prohibition of disproportionate means on the other hand. Thus, a definite duty of application cannot be easily established. On the other hand, the concerned developments must not be ignored, which means that courts or authorities have to consider the use the earlier the better they suit the state of the art.

---

[52] Examples according to Markowitsch (2009) with further references.

[53] Fischer, Strafgesetzbuch, 55. Aufl. 2008, Sect. 66 Rn. 2 f.

[54] Lang, in: Epping/Hillgruber, BeckOK GG, of: 01.10.2008, Art. 2 Rn. 59 f.

2.3   General Right of the Personality

The General Right of the Personality, deriving from Art. 2 para. 1 in connection with Art. 1 para. 1 Basic Law gains special importance, when it comes questions of self-determination or self-reservation. Alike DNA-samples of identification have to be protected according to the individual's right to determine the usage of his own personal data, and the coding part is part of the absolutely protected core of personality,[55] also individual brain data are protected by the constitution.

What is of interest, are especially the consequences deriving from this definition for the design of legal relationships between private persons, which become even more important in neuroscientific research projects between the researcher and the proband. Especially the problem of so-called incidental findings is strongly connected to the General Right of the Personality.

With image-guided procedures, up to 40% of the probands show anomalies, of which about 8% are of relevance and thus can be called incidental findings.[56] The consequences of such an incidental finding are countless and reach from questions of liability in case of an insufficient, delayed, or not existing illumination up to a possible degradation in insurance.[57]

Dealing with these relationships has to consider the importance and scope of the General Right of Personality and the right to determine the usage of one's own personal data, as well as the existing norm-legal regulations of data protection. But even more, one has to take into account conflicts of personal rights deriving from the prior-claimed "right not to know" and the existing duty of the researcher to state the findings, also with special regard to the relatives of the proband.

According to this discussion, at the moment there would only be one way to solve the problem: the proband has to be informed that incidental findings might occur during the study, that the study does not have any diagnostic aims and that it might also be that anomalies are not detected. Furthermore, the proband has to be informed about possible negative effects of the incidental finding for third parties or relatives.

The consent of the proband then has to explicitly state that he consents to the illumination about possible incidental findings. Probands refusing to consent under this condition have to be excluded from the study.[58]

---

[55] Decision of the Federal Constitutional Court, cited in Germany as: BVerfGE 103, 21 (32 et seq.).

[56] General: Kim et al. (2002); Illes et al. (2004), 743 et seq.

[57] On this and the following aspects: Schleim et al. (2007b), 1041 et seq.

[58] Extensively Spranger (2009c), (196 et seq.).

2.4   Prohibition of Discrimination

Besides the discussion about rights to freedom, the current developments also have relevance for rights to equality. The initial point would be the Basic Right of Art. 3 para. 3 sentence 2 Basic Law, which constitutes a general prohibition of discrimination against disabled persons. Whereas this right does not constitute any claim for benefits, it strengthens the legal position of disabled persons in the form of so-called derivative benefit rights.[59] Above all one can mention the demand for construction without any barriers.[60]

Furthermore, one has to consider the high relevance of the prohibition of discrimination for Civil Law relationships.[61] For instance, neuroscientific methods can have effects for the ability to testify or the ability to give legally binding declarations. As for the ability to testify, the Federal Constitutional Court has stated:

> Persons, who are intellectually speaking able to form their last will, must not be detained from forming their last will solely due to their physically limited abilities.[62]

At this point of the discussion, once again the field of application mentioned in connection to the Brain–Computer Interfaces is of special interest. Such instruments do not only enable people to live in human dignity, but also secure the demand of Art. 3 para. 3 sentence 2 Basic Law and enable disabled persons to form their last will, which they are guaranteed to by Art. 14 para. 1 sentence 1 Basic Law.[63]

2.5   Ability to Vote

At the end of this "tour de raison" one has to point out that beyond the mentioned, rather classical fields of application, there have to be considered further constellations, in which the constitutional background becomes only obvious at second site. For instance, the relevant literature shows first discussions on the question, which consequences of neuroscientific developments have to be expected in connection to the ability to vote according to Art. 38 para. 2 Basic Law. Whereas some opinions state that there is no relevance for the right to vote, because this right solely demands a certain level of development from the concerned person,[64] other

---

[59] Osterloh, in: Sachs (ed.), Grundgesetz, 4. Aufl. 2007, Art. 3 Rn. 305 et seq.

[60] On this: Dageförde (2007), 1036 et seq.; Braun, Das Gesetz zur Gleichstellung behinderter Menschen im Bereich des Bundes, in: RiA 2002, 177 et seq.

[61] Osterloh, in: Sachs (Hrsg.), Grundgesetz, 4. Aufl. 2007, Art. 3 Rn. 307.

[62] Decision of the Federal Constitutional Court, in Germany cited as: BVerfGE 99, 341 et seq.

[63] Wendt, in: Sachs (Hrsg.), Grundgesetz, 4. Aufl. 2007, Art. 14 Rn. 194.

[64] Heun (2005), 853 (854).

authors hold the view that the ability to vote is not given due to the missing ability of legal action.[65]

What seems to be important to note in this context is, that alike within the discussion on criminal responsibility, also the right to vote is not more than a social construction. And, alike within the field of criminal responsibility, there is not to be expected any major consequences for the right to vote from neuroscientific developments.

3 Beyond Constitutional Law: Problems Connected to the Generation of Knowledge

3.1 Legal Conflicts

Due to the differences of a researcher–proband relationship to a physician–patient relationship, there is regularly no contract of medical treatment which would stipulate any basis for the further legal assessment. However, this does not at all mean that there would be no legal problems in this context. To the contrary, risks of liability and responsibility become even more important in noncontractual relationships, in cases of a proband's damnification or a damnification to a third party. Consequently, first of all one should answer the question, which possible constellations may cause a damnification. An opinion in literature sees the following constellations as relevant: report on an incidental finding against the declared will of the proband or without his consent, becoming of a necessary treatment in form of therapy with the risk of adverse reactions, change for the worse in the assurance status, the evaluation of the images is not carried out properly or with a delay, the information about the incidental finding is deferred self-responsible with the effect of avoidable consequential damages, the form of information leaves the proband without any help (lack of a personal and temporally appropriate dialogue), omission of further diagnostic measures, wrong advice for necessity of further diagnostic measures.

This enumeration has been highly criticized, for those constellations would only have a higher likelihood of turning into damnification, which does not mean that a legally relevant damage and consequently a claim for damages have to evolve from the constellations in any case. Nevertheless, the enumeration helps to show the great variety of possible fields of conflicts. Neither can the risk of a responsibility according to Criminal Law (especially with regard to the delicts of assault and battery) be excluded by this, nor can the possibility of claims for damages according to Civil Law be excluded. The following few examples shall stress some further legal implications:

---

[65] Schreiber (2006), p. 1069 (1070).

Example 1: A pilot takes part in an fMRT-study. In the course of the project, an incidental finding occurs, which is not picked up as an issue in the following due to a lack of professional competence. The pilot's affection becomes acute a short time later during a flight, which causes the crash of the airplane. The relatives of the killed passengers and the airline want to hold the researcher responsible for the accident, because he has not initialized a further treatment despite his knowledge of the incidental finding.

Example 2: With regard to a certain project, the probands are only informed about the technical procedures. Incidental findings are only mentioned as common possibility. In the course of the research on a female student of medicines, an inoperable anomaly is detected, which can cause a serious danger to her life under certain circumstances. A few years later the proband wants to safeguard the buying of her house by a life insurance. Does she show her knowledge about the finding, no insurance company will ever contract with her; does she not show her knowledge, the contract will be vulnerable due to fraud.

3.2 Practical Consequences

Speaking from a practical point of view, the problem is to handle most sensitive questions without prohibiting the described research projects in general. On the other hand, it has to be clear that proper and broad information of the probands will probably complicate finding new probands.

Concerning the written information of the concerned person, he has to be informed about the possibility of incidental findings, which includes a variety of information, such as about the fact that the project does not have any diagnostic aims and that possible anomalies might also not be detected, or about possible negative consequences of an incidental finding for himself or for a third party, for instance with regard to the insurance status.

The proband's consent, which has to be based on the described information, has to stretch on the will to be informed about incidental findings. Otherwise, the risk of liability would be shifted to the researcher in an unfair way. This would mean that he bears the risk of liability only because he has special knowledge, but is held off from carrying out indicated measures of treatment by the concerned person. Any probands refusing to be informed about these facts should consequently be excluded from the project by the researcher.

With regard to the researcher's best own will, a growing opinion in literature states that the researcher should not explicitly take responsibility for the evaluation of clinical relevance. The better way for the researcher himself would be to make sure the finding is carefully evaluated by a competent professional (e.g. a neuroradiologist) before the decision about the information.

Other opinions in literature state the case that a competent expertise of a neuroradiologist is only "preferable" and that it is sufficient for the patient to be informed about the incompetent evaluation of the images by the researcher.

However, this does not really solve the problem of liability, but rather transfers the problem into the stadium of evaluation.

Furthermore, the information of the concerned person is not able to meet legal standards of informed consent, if it is based on wrong facts. Therefore, it is necessary to guide the concerned person and not to leave him alone with the wide field of possible legal problems he cannot know about. Still, this does not mean that any research project has to be guided by a competent person for the evaluation of the findings. This procedure is only relevant for cases in which an anomaly is detected.

3.3   Temporary Result

In many clinics, the above-mentioned solution is already being carried out. However, the major problem is not the fact that the clinics do not want to meet the new standards, but that they cannot meet certain standards due to financial problems. Therefore, in the future, researchers should claim further financial support for such procedures already in their application forms.

4   Problems of Application

Whereas the legal framework of neuroscientific studies is already adequately being discussed today, the discussion about the possible usage of generated findings mainly focuses on future developments. It seems to be absolutely necessary to speed the discussion up, in order to overcome the law's general character as a reactive measure and to avoid the legislative measures falling way behind the technical developments.

4.1   Criminal Law and Criminal Procedure

With regard to Criminal Law and Criminal Procedure, three main aspects have to be discussed from the legal point of view.

4.1.1   Criminal Responsibility

Enormous effects can be caused for the evaluation of the criminal responsibility. In this context, the German Federal Supreme Court has already set up important standards more than 50 years ago. In its decision the court stated that any punishment demands responsibility. The reasons for the call for responsibility should be the free, self-responsible and moral self-determination of a human being, which gives him the ability to choose the right instead of illegality and to avoid legal prohibitions. Following this decision, criminal responsibility is seen as the reason and the degree of punishment.

Within the German Criminal Code, section 20 deals with the responsibility and states that a person who is unable to foresee the injustice of his activities due to a clinical mental health problem, is not held responsible for his activities. If the ability described in section 20 is diminished, the punishment of the perpetrator can be adjusted according to section 21. Neuroscientific methods could well help specialists to determine a missing responsibility according to sections 20 and 21 of the Criminal Code.

4.1.2   Prevention of Criminal Actions

Some opinions in literature are afraid that the neuroscientific development could cause a shift from a Criminal Law of responsibility to a Criminal Law of prevention, in which the assumed dangerousness of the perpetrator determines the degree of punishment. However, also within the current legal situation the prognosis on the perpetrator's dangerousness affects the degree of punishment in many ways, so that the above-mentioned opinion does not really stress any new relevant points.

The legal difficulties are not so much about the evaluation of the perpetrator's character after a criminal activity, but rather about the prediction on future criminal activities. As a consequence, not only the risk of repetition would have to be taken into account, but also the risk of a risk of first offense. According to the individual political opinions, some people see the risk of imprisoning disposed persons, while other people see the chance of strengthening the public security by measures which do not necessarily have to be sanctions.

4.1.3   The Future of a Neuroscientific Lie-Detector

The question if, and if yes how, neuroscientific methods can be used in a criminal procedure as a kind of polygraph, is currently discussed controversially. Commonly known as "lie-detector", the polygraph has been subject to court decisions many times. In 1954, the Federal Supreme Court for Criminal Law first dealt with the topic. The decision stated that using a polygraph in a court trial violates section 136a of the Code of Criminal Procedure and Art. 1 of the Basic Law, namely the human dignity. The acceptance of the measure does not depend on its positive effects or on the consent of the accused person. To the contrary, the nonacceptance derives directly from the written law. The violation of human dignity was explained by the sentence that the accused person is participant in the court trial and not subject to the court trial. Also the decision shows an extensive part on the technical acceptance of the polygraph, it clearly states that this aspect is not decisive for the decision. The key-sentence of the decision is the violation of human dignity. In contrast to other Basic Rights of the German Constitution, the right of human dignity cannot be compared to any other right and every intervention into the human dignity is at the same time a violation. From the Court's point of view, it is furthermore not decisive whether or not the accused person has consented to the usage of the polygraph. This opinion is based on the older common opinion that nobody, not even the concerned person himself, can abdicate the right of human dignity. Today, this opinion is widely overcome, because it does not take into account that the free decision of the individual is constitutive for the interpretation of personal freedom and that a forced freedom turns freedom into its opposite.

Therefore, the Federal Supreme Court reacted to the upcoming new opinions and changed its opinion at least in parts. The decision makes clear that in cases where the accused person works together with the court voluntarily, the common arguments against the usage of a polygraph are not relevant. Under those circumstances, the Court did not see a violation of human dignity or of section 136a of the Code of Criminal Procedure. Although the polygraph evaluates physical data, it does not allow the Court a view into the soul of the accused person. This is stressed by the fact that there could not be determined any direct connection between a lie and the connected physical reaction (no specific lie response). On the other side, any extra-corporeal circumstances can cause physical reactions. Therefore, it cannot even be proofed if the accused person tells the truth. The dignity of the concerned person is not violated because of the fact that he is connected to technical devices throughout the process. To this he has consented. However, the Court does not accept the usage of the polygraph, because it is a totally inappropriate measure of proof according to section 244 para. 3 of the Code of Criminal Procedure. This section states that a method of proof has to be declined if it is obviously not appropriate and able to give proof.

As a consequence of this decision, technical measures intervening deeply into the concerned person's integrity shall be acceptable, at least as long as he or she has given informed consent and the relevant method is able to give definite proof.

For the question whether or not this decision can be transferred on methods of neuroimaging, it is after all not decisive, which intensity the intervention shows. If the concerned person has consented to the usage of the method, it can at least not violate the personal rights of the concerned person.

Therefore, the decisive question is rather, whether or not the chosen method is capable or – like the polygraph – totally incapable to give proof. According to the steady jurisdiction, a method of proof is incapable, if the common life experience shows that this method cannot help to clear up the case and therefore the usage of the method would be a mere deceleration of the court trial. It is not yet foreseeable if and if yes, when or to which extend future neuroscientific procedures will be able to fulfill the criteria of appropriate proof.

With this as a background, it is today of interest, if the Court could decline a specialist, basing his evaluation on neuroscientific measures. In former decisions, the Federal Supreme Court has stated that a specialist is totally inappropriate according to section 244 para. 3 of the Code of Criminal Procedure, if his evaluation is based on methods which cannot be controlled or which are not yet completely tested.

According to this jurisdiction, the capability of the evaluation will depend on the degree to which the specialist has based his overall-result on the critical method. If this method is only one measure among others, to stress the overall-result, the Court will probably not decline the specialist's evaluation.

4.2 Challenged Persons

Despite the fact that still many legal assessments merely stress the inherent risks of new neuroscientific procedures, it cannot be denied that there are also possible constellations, in which the application of neuroscientific measures could even help to fulfill state duties. Most of all, this applies for the protection of disabled persons. Article 3 para. 3 sentence 2 of the German Basic Law determines that no person shall be discriminated because of his or her disability. This Basic Right is not only to protect the concerned person against interventions, but also gives a legal assignment to the legislator, to pay attention to the protection of disabled persons whenever enacting a new law. This development can be seen best in the field of building law, where the Federal States have now enacted certain regulations which are to ensure that public buildings are build in an appropriate way which allows disabled persons to use the building without barriers. From a constitutional point of view, the state is committed by Art. 3 para. 3 sentence 2 of the Basic Law to carefully assess new methods of neuroscientific measures which are likely to give benefits, and to decide whether or not they may be used.

4.2.1 Daily Routine

Many disabled persons have problems coping with the most common activities of everyday life. In this field, it is of special interest to look at the neuroscientific usage of thoughts in order to control computers or machines. For some people, this would simply mean that some everyday tasks could be fulfilled by thoughts. In other cases, where people are captured in a kind of locked-in state, it would possibly mean the only chance for them to communicate with other people. Several research projects are currently dealing with these questions.

4.2.2 Improvement of the Legal Situation

But it is not only the field of everyday tasks, which could be arranged in a completely new way by measures of neuroscientific developments. Even when it comes to the legally binding declaration of a person's will, one can foresee some important improvements of the status quo.

Especially some of the most important instruments of Civil Law – such as buying contract, contract of work, contract of personal services – cannot be used by disabled persons due to the fact that they are seen as incapable of contracting in a legal sense. According to section 104 no. 2 of the German Civil Code, a person is incapable of contracting, if he is in a state of mental disease of the brain, if this state is not just momentarily. If an incapable person declares his will, it is invalid according to section 105 para. 1 of the Civil Code. Consequently, the concerned person is not able to give a declaration of his will in a legally binding way and thus cannot be party to a contract. Only in cases of everyday affairs, the contract becomes valid, when it is completely fulfilled according to section 105a of the German Civil Code.

If there is no everyday legal affair at stake, fundamental rights or duties of the concerned person might be affected, which can best be illustrated at the example of limited abilities of establishing a testament.

According to section 2231 Civil Code, a testament can be established in two different ways, i.e. by notarial recording or by a declaration of the testator according to section 2247 Civil Code. This norm constitutes that the declaration has to be in a self-handwritten form and has to contain the personal signature. Furthermore, it shall contain the time and date of the declaration.

This shows that the law sets as a pre-condition the testator's ability to physically establish his own testament. Only to some extent the form of notarial recording can solve this problem. In this context, section 2232 Civil Code determines that the testator can also declare the will to the notary or give a written document to the notary, which he must not necessarily have written on his own. In addition, section 2233 para. 2 Civil Code makes clear that a testator who is unable to read written documents can only establish his testament by declaration to the notary.

In all cases the law sets up special conditions for the establishment of a testament – such as the ability to give an oral or written declaration, the ability to read – which many disabled persons are not able to fulfill. Also the so-called emergency testaments – before the mayor according to section 2249, before three witnesses according to section 2250 or on sea according to section 2251 – do not solve the problem, but rather deal with the situation that it is, due to an emergency, not possible to establish the testament before a notary.

As a consequence of those regulations, many disabilities lead to the lack of ability to establish one's own testament, according to section 2229 para. 4 Civil Code. This section determines that a person who is, due to a mental disease, unable to see the meaning of his own declaration is incapable of establishing a testament.

4.2.3 Jurisdiction of the Federal Constitutional Court

Some 10 years ago the Federal Constitutional Court has made some adjustments to this very strict scheme and has decided that a general exclusion of the ability to establish a testament for mutes unable to write would violate the Basic Rights of the guarantee of succession according to Art. 14 para. 1 sentence 1 Basic Law and the prohibition of discrimination according to Art. 3 para. 1 and para. 3 sentence 2 Basic Law.

In the relevant literature, some authors suggest to transfer this decision to comparable constellations, for instance the situation of mute blind persons who are incapable of the embossed printing or for barren illiterates. Nevertheless, the principle of the Federal Constitutional Court is still valid that the testator must have the ability to somehow declare his own will to another person in order to be able to establish his testament. This result is also to the benefit of the concerned person, because he must be protected against assumed wills, which are then interpreted as his own.

At this point, the discussion about an improvement of the situation for disabled persons by neuroscientific methods has its initial point. If neuroscientific methods can create new ways of communication for disabled persons, these persons are very close to being able to give legally binding declarations.

This would mean an enormous range of legal opportunities: Apart from the limitation of possible mentoring relationships, the chance of giving informed consent in the physician–patient relationship or the possibility of an advance directive there can be thought of various situations, in which disabled persons could gain back a lot of their personal autonomy. In contrast to the current discussion, whether or not neuroscientific findings could prohibit a free will, the above-mentioned discussion is rather about the possibility to improve current situations of downgrading by establishing new methods of declaring a person's will.

4.3 Manipulation of Thoughts

A negative connotation is linked with the wording "thought manipulation". However, it is rather decisive that the voluntary element plays a role of utmost importance here. If thoughts shall be deleted with a bad intention, this would violate several elements of crime and cause possible damage claims. On the other hand, a deletion of bad thoughts, which could eventually also lead to a pathological state, could possibly be seen as a medical treatment.

4.4 Further Fields of Application

Especially in the field of thought-reading, there are various possible scenarios which can be thought of: There are discussions about the abusive use of thoughts by promotion companies as well as discussion about the fight against terrorism with brain-scanning machines at airports, which could then possibly convert terrorists. Also in the field of the employment sector, neuroscientific methods could be helpful for the employer to get to know more about the employee's motivation and character.

At least within the German legal system such scenarios could be eased by means of constitutional principles. If state authorities intervene into Basic Rights of citizens – in this case especially into the individual's right to determine the use of one's own personal data (Art. 2 para. 1 in connection with Art. 1 para. 1 Basic Law) – this intervention has to fulfill the constitutional requirements and has to be appropriate. The evaluation of appropriateness is task both of the legal sciences and of the jurisdiction.

If the intervention is not caused by the state, but by private persons, like in the constellation of promotion agencies, a direct violation of Basic Rights is not possible, because only state authorities are directly beholden by the Basic Rights. But if it came to a court trial before a civil or criminal court, the court would have to consider the importance of the Basic Rights according to the principles of the indirect third-party-effect of Basic Rights.

On a nonconstitutional level, the generated findings would regularly be considered personal data. Consequently, special regulations on data protection would have to be taken into account, which are especially strict if the collection, usage and storage is carried out by a state authority. In this context, one would have to consider stricter regulations on data protection also for private organizations and companies.

5   Conclusion

The increasing number of neuroscientific findings being generated has to be carried out with respect to the binding law system and according to the principles of the Good Clinical Practice. As far as the methods of neuroimaging are concerned, those methods lead to a great variety of legal questions, which can only be dealt with when paying attention to certain preconditions beforehand the research project.

Anyway, it cannot be called sufficient to only claim the missing responsibility of the researcher because of a lack of a physician–patient relationship or to stress an incompetent evaluation of findings. The future usage of generated or future results should be the initial point to a "pro-active" discussion about the legal framework. Problems of Constitutional or Civil Law, which are currently only rarely discussed, will thus have to be declared the centre point of a neurolegal debate.

References

Aholt A, Neuhaus C, Teichert T, Weber B, Elger CE (2007) Neurowissenschaftliche Analyse des Regret-Effektes und der Beeinflussbarkeit der Kaufentscheidungszufriedenheit. NeuroPsychoEconomics 2(1):76–91
Beck S (2006) Unterstützung der Strafermittlung durch die Neurowissenschaften? Juristische Rundschau 81(4):146–150
Birbaumer N, Strehl U, Hinterberger T (2004) Brain-computer interfaces for verbal communication. In: Dhillon G, Horch K (eds) Neuroprosthetics: theory and practice. World Scientific, New Jersey, pp 1146–1157

Blankertz B, Losch F, Krauledat M, Dornhege G, Curio G, Müller K-R (2008) The Berlin brain-computer interface: accurate perfomance from first-session in BCI-naive subjects. IEEE Transactions on Biomed Eng 55(10):2452–2462

Boujong K (2003) § 136a. In: Pfeiffer G (Hrsg) Karlsruher Kommentar zur StPO (5. Aufl.), (Rn. 34.). Beck, München

Busey, Loftus (2006) Cognitive science and the law. Trends Cogn Sci, 111 et seq

Corvat, McCabe (2004) The brain and the law. Philos Trans R Soc B 1727 et seq

Dageförde H-J (2007) Barrierefreies Bauen nach der neuen Bauordnung für Berlin. Grundeigentum 2007, 1036 ff

Garland, Frankel (Hrsg) (2004) Neuroscience and the law – brain, mind, and the scales of justice.

Goodenough, Prehn (2004) A neuroscientific approach to normative judgment in law and justice. Philos Trans R Soc B 1709 et seq.

Günther K (2006) Hirnforschung und strafrechtlicher Schuldbegriff. Kritische Justiz 38(2):116–133

Hager (2002) Zur Neufassung der Regelungen über das barrierefreie Bauen. In: Verwaltungsblätter für Baden-Württemberg 2002, 71 f

Heinemann T, Hoppe C, Listl S, Spickhoff A, Elger CE (2007) Zufallsbefunde bei bildgebenden Verfahren in der Hirnforschung. Dtsch Ärztebl 104(27):A1982–A1987

Herdegen (2006) Schuld und Willensfreiheit. In: Kempf, Jansen, Müller (ed) Verstehen und Widerstehen, Festschrift für Christian Richter II. p 233 (244)

Heun (2005) Die grundgesetzliche Autonomie des Einzelnen im Lichte der Neurowissenschaften. JZ 853 (855)

Hillenkamp T (2005) Strafrecht ohne Willensfreiheit? Eine Antwort auf die Hirnforschung. Juristenzeitung 60(7):313–320

Hübner G (2005) Die neue Sächsische Bauordnung. Sächsische Verwaltungsblätter 13(9):213–223

Illes J, Kirschen MP, Karetsky K, Kelly M, Saha A, Desmond JE, Raffin TA, Glover GH, Atlas SW (2004) Discovery and disclosure of incidental findings in neuroimaging research. J Magn Reson Imaging 20:743–747

Jäde H (2003) Barrierefreies Bauen in der Bayerischen Bauordnung. In: KommunalPraxis BY 2003, 364ff

Jakobs (2009) Strafrechtliche Schuld als gesellschaftliche Konstruktion. Ein Beitrag zum Verhältnis von Hirnforschung und Strafrechtswissenschaft. In: Schleim, Spranger, Walter (ed) Von der Neuroethik zum Neurorecht? p 244 et seq

Kim, Illes, Kaplan, Reiss, Atlas (2002) Incidental findings on pediatric MR images of the brain. Am J Neuroradiol 1674 et seq

Lampe, Pauen, Roth (eds) (2008) Willensfreiheit und rechtliche Ordnung

Libet (2004) Haben wir einen freien Willen? In: Geyer (ed) Hirnforschung und Willensfreiheit. Zur Deutung der neuesten Experimente. p 268 et seq

Löwith K (1928) Das Individuum in der Rolle des Mitmenschen. Drei Masken Verlag, München

Markowitsch (2009) Mind reading? – Gutachten vor Gericht. In: Schleim, Spranger, Walter (ed) Von der Neuroethik zum Neurorecht? p 133 (142 et seq.)

Möll T (2007) Messung und Wirkung von Markenemotionen – Neuromarketing als neuer verhaltenswissenschaftlicher Ansatz. Gabler, Wiesbaden

Möllers (2008) Willensfreiheit durch Verfassungsrecht. In: Lampe, Pauen, Roth (ed) Willensfreiheit und rechtliche Ordnung. 250 (258, 260)

o.V. (2006) Is this the bionic man? Nature 442:109

O´ Hara (2004) How neuroscience might advance the law. Philos Trans R Soc B 1677 et seq

Pfeiffer G (2005) Strafprozessordnung (5. Aufl.). Beck, München

Putallaz F-X, Schumacher BN (2008) Für eine Metaphysik der Person. In: Putallaz F-X, Schumacher BN (eds) Der Mensch und die Person. Wissenschaftliche Buchgesellschaft, Darmstadt p 9–14

Roth (2003) Willensfreiheit, Verantwortlichkeit und Verhaltensautonomie des Menschen aus Sicht der Hirnforschung. In: Dölling (ed) Jus humanum. Grundlagen des Rechts und Strafrecht. Festschrift für Ernst-Joachim Lampe zum 70. Geburtstag. p 43 (56)

Santosuosso (Hrsg) (2009) Neuroscience and the Law.

Schleim (2006) Zeig mir dein Hirn – und ich sag dir, was du denkst1. Gehirn & Geist, 60 et seq

Schleim S (2008) Gedankenlesen – Pionierarbeit der Hirnforschung. Heise, Hannover

Schleim S, Spranger TM, Walter H (2007a) Von der Neuroethik zum Neurorecht ? Der Beginn einer neuen Debatte. Nervenheilkunde 26(9):813–817

Schleim S, Spranger TM, Urbach H, Walter H (2007b) Zufallsfunde in der bildgebenden Hirnforschung. Empirische, rechtliche und ethische Aspekte. Nervenheilkunde 26(11): 1041–1045

Schneider K, Merkle C (2007) "Minority report" – fiction or reality?: legal questions raised by neuroscience with particular emphasis on German criminal law. J Int Biotechnol Law 4(6):227–232

Schreiber (2006) Ist der Mensch für sein Verhalten rechtlich verantwortlich? In: Kern, Wadle, Schroeder, Katzenmeier (ed) Humaniora. Medizin – Recht – Geschichte, Festschrift für Adolf Laufs zum 70. Geburtstag. p 1069

Schüssler M (2002) Polygraphie im deutschen Strafverfahren. Lang, Frankfurt a.M

Simitis (2008) Biowissenschaften und Biotechnologie – Perspektiven, Dilemmata und Grenzen einer notwendigen rechtlichen Regelung. JZ 693 et seq

Spranger TM (1999) Testiermöglichkeit schreib- und sprechunfähiger Personen. Sozialrecht und Praxis 9(4):219–220

Spranger TM (2000) Extremisten – Zur Einstellung von politisch Radikalen in den Öffentlichen Dienst. In: Menzel J (Hrsg) Verfassungsrechtsprechung. Hundert Entscheidungen des Bundesverfassungsgerichts in Retrospektive. Mohr-Siebeck, Tübingen, pp 254–259

Spranger TM (2007a) Der freie Wille von Menschen mit Behinderung. Sozialrecht und Praxis 12(9):547–550

Spranger TM (2007b) Neurowissenschaften und Recht. In: Sturma D (Hrsg) Jahrbuch für Wissenschaft und Ethik. de Gruyter, Berlin, pp 161–178

Spranger TM (2009) Der Einsatz neurowissenschaftlicher Instrumente im Lichte der Grundrechtsordnung. JuristenZeitung 17, JZ 2009, Heft 21, 1033–1040

Spranger (2009) Neurowissenschaftliche Tests im indischen Strafprozess. Zur Relevanz des BEOS-Tests in der Neurorechtsdebatte. Nervenheilkunde 150

Spranger (2009) Rechtliche Implikationen der Generierung und Verwendung neurowissenschaftlicher Erkenntnisse. In: Schleim, Spranger, Walter (ed) Von der Neuroethik zum Neurorecht? p 194

Sturma D (ed) (2006) Philosophie und Neurowissenschaften. Suhrkamp, Frankfurt a.M

Tremblay (2008) Understanding development and prevention of chronic physical aggression: towards experimental epigenetic studies. Philos Trans R Soc B 2613 et seq

Truccolo W, Friehs GM, Donoghue JP, Hochberg LR (2008) Primary motor cortex tuning to intended movement kinematics in humans with tetraplegia. J Neurosci 28(5):1163–1178

Wolpaw JR, Birbaumer N (2006) Brain-computer interfaces for communication and control. In: Selzer ME (ed) Textbook of neural repair and rehabilitation. Cambridge University Press, Cambridge, pp 602–614

Yoon C, Gutchess AH, Feiberg F, Polk TA (2006) A functional magnetic resonance imaging study of neural dissociations between brand and person judgments. J Consum Res 33:31–41

Neurolaw in Greece: An Overview

Takis Vidalis and Georgia-Martha Gkotsi

Abstract Given the rapid advancements in neuroscience and its growing involvement in legal proceedings, in this paper, we aim to address the question of whether and to which extent Greek legislation could be revisited in the light of the most recent neuroscientific discoveries. "Reading" the human cognitive and emotional functions by modern neuroscientific technology is relevant to a number of Greek legal provisions, an overview of which is presented in this paper. In the first chapter, we describe the general framework that governs an adult's capacity of will, taking into consideration the constitutional aspect, the civil law's approach and some special topics concerning medical law and research, with the aim to examine how these issues could be illuminated with a neuroscientific perspective. The second chapter is exclusively dedicated to criminal law, in an effort to evaluate the potential influence of neuroscience on the Greek criminal justice system. The penal legislation concerning the assessment of criminal responsibility, the evaluation of the sentence, and the admissibility of neuroscientific techniques in criminal Courts, as well as some special issues concerning juvenile offenders and crime prevention are presented. Finally, a unique Greek case where the use of a lie detector was permitted in the context of a criminal trial is cited and briefly analysed.

This overview leads to conclude that although the Greek legal system refers extensively to situations of interest for neurolaw, the acceptance of neuroscientific methods for determining the cognitive or mental status of persons involved in civil, medical and criminal relationships is rarely considered as important. However, the

Takis Vidalis, Hellenic National Bioethics Commission, PGP in Bioethics, University of Crete
Georgia-Martha Gkotsi, Ph.D candidate, University of Lausanne, Faculty of Biology and Medicine Ethos - Interdisciplinary Platform of the University of Lausanne, University of Athens, Greece, Faculty of Law

T. Vidalis (✉) • G.-M. Gkotsi (✉)
National Bioethics Comission, Evelpidon 47, 11362 Athens, Greece
ETHOS, Interdisciplinary Ethics Platform, Quartier UNIL-Sorge, Batiment Amhipole, bureau 211, CH 1015, Lausanne, Switzerland
e-mail: t.vidalis@bioethics.gr; GeorgiaMartha.Gkotsi@unil.ch

T.M. Spranger (ed.), *International Neurolaw*,
DOI 10.1007/978-3-642-21541-4_10, © Springer-Verlag Berlin Heidelberg 2012

aforementioned judicial step towards the acceptance of these methods in criminal settings, as well as the innovative spirit that the Greek legislator shows in regulating biomedical matters during the last decades, lead to consider that a revision of the Greek legislation in the light of new neuroscience, should not be excluded for the future. Providing more information on neurolaw and its expected benefits could be, perhaps, the best motivation for taking action in this promising field.

1 Introduction

As neuroscience become an important source of information for understanding the formation of human self-determinism, they inevitably affect crucial topics both in ethics and in law.

Especially, after a decade of rapid developments in biomedical law, the thin line between the conscious, independent will and the unconscious acts of a person is considered as a central issue, not only in medical but also in legal terms.

In Europe, after adopting the first international instrument with binding force in the wider field of medicine, the Council of Europe's Convention on Human Rights and Biomedicine (Oviedo Convention), we have a general legal description of that thin line.[1] Although the Convention focuses on medical acts and relevant research, the established distinction between persons able and persons unable to consent in

---

[1] See Chapter II on consent:

Article 5 (General rule)

"An intervention in the health field may only be carried out after the person concerned has given free and informed consent to it.

This person shall beforehand be given appropriate information as to the purpose and nature of the intervention as well as on its consequences and risks.

The person concerned may freely withdraw consent at any time."

Article 6 (Protection of persons not able to consent)

"1. Subject to Articles 17 and 20 below, an intervention may only be carried out on a person who does not have the capacity to consent, for his or her direct benefit. . . .

3. Where, according to law, an adult does not have the capacity to consent to an intervention because of a mental disability, a disease or for similar reasons, the intervention may only be carried out with the authorisation of his or her representative or an authority or a person or body provided for by law.The individual concerned shall as far as possible take part in the authorisation procedure.

4. The representative, the authority, the person or the body mentioned in paragraphs 2 and 3 above shall be given, under the same conditions, the information referred to in Article 5.

5. The authorisation referred to in paragraphs 2 and 3 above may be withdrawn at any time in the best interests of the person concerned."

Article 7 (Protection of persons who have a mental disorder)

"Subject to protective conditions prescribed by law, including supervisory, control and appeal procedures, a person who has a mental disorder of a serious nature may be subjected, without his or her consent, to an intervention aimed at treating his or her mental disorder only where, without such treatment, serious harm is likely to result to his or her health".

such acts is critical, since self-determination in health matters lies in the core of personhood; other expressions of autonomy presuppose autonomy in health.

Bearing in mind this important role of the Oviedo Convention for neurolaw, in the following chapters we will present an overview of relevant national provisions of the Greek legal system, along with some comments that may clarify specific points in regulation.[2] In the first chapter, we describe the general framework that governs an adult's capacity of will, taking into consideration the constitutional aspect, the civil law's approach and special topics concerning medical law and research. The second chapter is dedicated exclusively to criminal law, since, there, the distinction between self-determinism (and imputation) and unconscious behaviour occurs not only as a personal but also as a societal problem.

1.1   The General Framework on Legal Capacity

1.1.1   The Constitutional Aspect

"Reading" the human cognitive and emotional functions by modern technology involves the interests of a number of constitutional provisions related to fundamental rights.

In the Greek Constitution, according to article 2 par.1 *respect and protection of the value of the human being constitute the primary obligations of the State.* This article establishes, in fact, the general principle of self-determinism, as a common characteristic of all persons.[3] Freedom of will and personal autonomy, in general, originate from that principle. Therefore, statutory provisions regarding capacity to consent to various legal acts or to form legally valid decisions are related directly to this constitutional article, in the meaning that human value is inevitably violated in case of unjustified loss of capacity. Generally speaking, this may be considered as an unacceptable treatment of the person concerned as a simple meaning rather than as a value per se.

Besides this programmatic principle, the most interesting articles of the Constitution for neurolaw, concerning fundamental rights, are:

– The article 5 par. 1 acknowledging the development of personality in social contexts as a general right. This covers not only conventional behaviors of a

---

[2] The Greek legal system belongs to the family of Roman systems. Its basic elements reflect the influence of both the German and the French law. Greece is also a member State of the EU (since 1981), therefore EU law covers a large part of internal law.

The Constitution protects a large catalogue of fundamental rights (civil liberties, political rights, social rights, and rights of "3d generation"). All courts have the power to control the constitutionality of laws (non-existence of a Constitutional Court). All ratified international conventions as well as all generally recognized rules of international law prevail over domestic laws.

[3] See *in extenso* Dagtoglou (1991), pp. 1133–1141.

person but even "strange" ones, including even behaviors tending to self-destruction,[4] on condition that the Constitution, the rights and legal interests of other persons and the common morality are respected. From the neurolaw's point of view, article 5 par. 1 stands as an institutional limit when assessing "deviating" behaviours as conscious or unconscious acts of a certain person. It is clear that, given the openness of this provision, unconscious acts (that may affect the legal validity of them) are considered as exceptional, and should be verified only if uncontestable objective medical, psychological or social indications exist.

– The article 5 par. 5 concerning the individual right to health and the protection from biomedical experimentation.[5] These relatively new constitutional provisions (existing from 2001) are of interest particularly when a person undergoes neurological, psychiatric or psychological treatment and, furthermore, when he/she participates in clinical trials related to experimental methods for such a treatment.

– The article 7 par. 2 prohibiting any sort of tortures, including those involving psychological violence.

– The (also new) article 9 A on the protection of personal data. Medical acts and, especially, experimental activities in medical or other research involving the collection and processing of sensitive neurological or psychological data are of interest, here, and should be addressed in conformity with the relevant Act 2472/1997.[6]

– The article 16 protecting the freedom of science and research, relevant to biomedical experimentation of neurosciences, especially when other individual rights of the persons undergoing research activities may be at a certain risk, and need to be balanced along with this freedom.

– The article 21 par. 1 protecting the family and, thus, prioritizing family members as caregivers and legal representatives of persons unable to form and express their will, due to their neurological or mental situation.

Other constitutional provisions have indirect effects in the specific context of neurolaw, since in any case they affect medical acts and health care in general.[7]

1.1.2 Civil Code: Legal Capacity in Regular Life

Neurolaw is related to the general capacity of concluding valid legal acts and especially contracts. Minors have limited such capacity, but our special interest here concerns the capacity of adults.

---

[4] See Manessis 1982, pp. 118–119.

[5] See Vidalis-Mitrou-Takis 2006, pp. 281–285 (Vidalis).

[6] See Vidalis-Mitrou-Takis 2006, pp. 308–312 (Mitrou).

[7] This is the example of articles 4 par. 1 pertaining to the equal treatment before the law, 5 par. 2, 3, 4 on the *habeas corpus* principle, 21 par. 3 concerning the social right to health etc.

The Greek Civil Code contains a general framework in this respect, according to which:

- adults living in whole and exclusive judicial assistance are deprived of legal capacity to conclude contracts (article 128);
- adults living in partial or concurring judicial assistance have limited such capacity (article 129);
- capable persons cannot express legally valid wills if, at the concrete moment, they acted unconsciously or they experienced a mental disorder restricting substantially the process of will (article 131).

The first two rules are enforced when judicial assistance of a person (whole and exclusive, partial or concurring, depending on the confirmed extent of the person's capacity) has been decided by the competent court, according to special provisions of the Civil Code (articles 1666–1668).

Judicial assistance may be decided, among others, also for medical reasons related to the situation of mental health. Crucial evidence for the court is, of course, medical indications deriving from neurological, psychiatric or psychological tests. Nonetheless, the court's decision can be modified if new evidence indicates differently,[8] therefore the involvement of scientific knowledge in decision-making regarding a person's capacity should be considered as continuous.

The practical outcome of the court's decision is the appointment of a custodian as legal representative of the person concerned, usually a close family member, which has similar duties to those of a minor's legal representative, needing to act in the best interest of the adult concerned.[9] The role of this custodian is supervised by a council of three to five members, composed by relatives or close friends.[10]

Finally, it is worth mentioning that the custodian or the supervising council cannot decide on its own for the adult's compulsory treatment in a mental health unit. This decision should be taken also by the competent court,[11] according to the specific legislation at stake.[12]

1.1.3 Medical Law and Research

Neurosciences have extensive applications directly related to medical acts and biomedical (or other) research. In this paragraph, we will address issues in these fields, as governed by the Greek legal system.

---

[8] See articles 1677, 1685, 1686 C.C.

[9] See article 1669, 1682 C.C.

[10] See article 1682 C.C.

[11] See article 1687 C.C.

[12] See below at (c).

Patients Unable to Consent

A crucial topic in relevance is that of "conventional" patient/physician relationships, when patients are adults unable to consent, and therefore unable to participate in the decision-making process for their own health.

As Greece has ratified the Oviedo Convention (Act 2619/1998), articles 6 and 7 of it, determining adults with mental disorders as persons unable to provide valid consent in medical settings, constitute the general framework in force. Similar provisions also exist in the national legislation (Act 3418/2005 on medical ethics). This Act explicitly states that close family members (parents, children, spouses) are the legal representatives of an adult unable to consent specifically for his/her health care.[13] Therefore, given the aforementioned arrangement on judicial assistance (where custodians may be other persons as well) it is possible to emerge differences (and maybe conflicts), when necessary acts in the best interest of the person concerned presuppose decisions on health care that need to be taken from another person, namely a non-custodian family member.

A difficult issue, in this respect, is also the handling of previously expressed wishes of the person concerned (that should be taken into account, according to both the Oviedo Convention and the Act 3418/2005[14]), when there are doubts about his/her mental health, at the crucial time of these wishes' expression. The lack of specific regulation on advanced directives and living wills in the Greek law makes this problem more serious, especially if relevant disagreements between physicians and/or family members occur.

Transplantations

Neurolaw is involved in the specific regulation concerning the collection of tissue and organs by deceased donors, for transplantation purposes.

The Greek Act 2737/1999 on transplantations explicitly refers to the medical confirmation of brain death, as the crucial criterium for proceeding to tissue and organ collection. Article 12 par. 6 determines brain death as the *necrosis of the encephalic stem, even if the function of other organs is maintained artificially.* Brain death needs to be confirmed by the attending physician, who should draw up a certification of death along with an anaesthesiologist and a neurologist or neurosurgeon.

Even if the wording of this provision may raise some doubts, it seems that the law does not hold the attending physician (who has to confirm the death) as the only responsible person, but rather it determines a shared responsibility of the three physicians who certify formally the event, given that they must have concrete medical expertise for that particular purpose.

---

[13] See articles 11 par. 4, 12 par. 2 bb of Act 3418/2005.

[14] See article 9 of the Oviedo Convention and article 29 par. 2 of Act 3418/2005.

The Status of Mentally Ill Patients

Compulsory hospitalization and treatment of mentally ill patients raise serious issues in neurolaw, directly associated with the protection of fundamental rights. A rich case-law originated from the European Court on Human Rights, regarding the application of the article 5 of the European Convention on Human Rights, is well known in that field. The Convention (ratified by the presidential decree 28/1974) and, indirectly, that case-law constitute the general framework to address such issues in the Greek legal system as well.

In addition, a detailed national legislation of a rather procedural nature exists, being part of the Act 2071/1992 on the National Health System. The procedural nature of that legislation intends to establish adequate guarantees (medical and judicial) at different levels, before and during the hospitalization of the patient concerned, in order to ensure that his/her fundamental rights are taken into consideration by family members, physicians and other health professionals involved.[15]

The basic points of this law[16] are:

- The necessary coexistence of four strict conditions, in order to justify compulsory hospitalization, namely, a confirmed mental disorder, a patient not able to judge for his/her own health, the medical adequacy of the treatment concerned, and a confirmed need for hospitalization in order to prevent potential violent actions. If any of these conditions does not stand permanently, the hospitalization should cease immediately. The scientific director of the hospital is responsible for deciding this.[17]
- A clear distinction between the situation of compulsory hospitalized patients and those detained in mental clinics for criminal reasons, being incapable for imputation. The law distinguishes also, explicitly, persons that cannot be considered as "patients" from the fact that they cannot be adapted to dominant social, moral or political values.[18]
- The need for the hospitalization to be decided only by judicial authority (prosecutor).[19]
- The priority of therapeutic purposes over any restrictive measures, in the meaning that therapy may contain even a controlled release from restrictions especially in movement and stay of the patient concerned.[20]

---

[15] In contrast, voluntary hospitalization does not raise special problems, since patients consent themselves to this kind of treatment, which has as an additional condition the confirmation of its necessity by the scientific director of the hospital concerned. Therefore, the law equates in this case mentally ill patients with other patients, regarding the protection of their fundamental rights (see article 94 par. 4 of Act 2071/1992).

[16] See, in detail, Vidalis (2007), pp. 56–63.

[17] Article 95 of Act 2071/1992.

[18] Article 95 of Act 2071/1992.

[19] Article 96 of Act 2071/1992.

[20] Article 98 par. 1 of Act 2071/1992.

– The maximum duration of 6 months for any compulsory hospitalization that may be extended only in "extremely exceptional cases" by the judicial authority (prosecutor). In general, the prosecutor stands as a legal supporter of the patient, and may decide his/her immediate release earlier, after relevant medical report from the attending physicians.[21]

Research in Neurosciences

Research is also a crucial topic to be discussed in the framework of neurolaw. Neurosciences develop experimental practices that have various aspects with particular problems. For instance, research in neurology and neuropsychiatry involves experiments with novel drugs and other interventional methods (implants, for example), and psychological studies collect information from empirical data related to behavioral investigation.

Issues on interventional experimentation on humans are governed by the relevant provisions of the Oviedo Convention and by Directive 2001/20. The latter has been transposed into the Greek law since 2003, and the National Organization of Medicines is the responsible authority to supervise its implementation.

A National Ethics Committee on Clinical Trials, functioning in fact as the only specific official body on human research ethics, ensures compliance of relevant protocols with the ethics requirements as mentioned in the Directive.[22]

This general framework is important, especially when specific problems in this field of research emerge, such as the participation of persons with limited consent capacity or the ethically acceptable use of placebo (that can lead not only to positive results, as such, in mental diseases, but also to produce confusion in assessing new drugs efficiency).

Apart from this kind of research, empirical research in psychological studies is not regulated by special provisions in the Greek legal system. Nevertheless, informed consent procedures are also applicable here, particularly when research may be proved stressful for vulnerable participants. In that case, research is in fact interventional; therefore, the Oviedo Convention needs to be taken into consideration.

Research, whether clinical or other, is directly related to the collection and processing of sensitive personal data, particularly of medical data. In this respect, compliance with the relevant provisions of Act 2472/1997 (which transposed into the Greek law the Directive 95/46 on data protection) should be ensured.

Yet, it is important to make a distinction between investigators officially submitted to rules of professional confidentiality (physicians, par excellence[23]) and

---

[21] Article 99 of Act 2071/1992.

[22] See, in general, Hellenic National Bioethics Commission (2005), III, 2.

[23] See article 13 of Act 3418/2005.

other investigators. According to the article 7 A of Act 2472/1997, the first can collect, have access and process sensitive data of persons undergoing research, without a license from the Data Protection Authority, whereas the latter need to obtain such a license.

In any case, data collection and processing presuppose specific and written informed consent by the subjects concerned, unless data are to be used in anonymous form or investigators are deprived of any access to their identity.[24] In addition, collected data may be used in future research on condition that the original consent covers explicitly the purposes of that research too, otherwise a fresh specific consent by the data subjects is needed.

1.2   Criminal Law and Neuroscience

1.2.1   Neuroscience and the Assessment of Criminal Responsibility Under the Greek Legal System

The Greek Penal Code, (PC), in article 14 defines the crime as *action which is unjust and imputable to the offender* The question is whether neuroscience, under the current legal framework, could provide evidence as to whether an action can be imputable to the offender or not. To answer it, we must consider the criteria that will determine the possibility of imputability or its absence, under the Greek criminal law.

These conditions are described in article 34 of the Penal Code: *The action is not imputable to the offender if, at the time he/she committed the crime, the existence of a morbid disturbance of mental function or disturbance of consciousness deprived him/her from the ability to understand the wrongfulness of his/her action or from the ability to act according to his/her perception of the wrongful character of the action.*

Respectively, the article 36 par. 1 PC describes the criteria for the diminished criminal responsibility: *If due to any of the mental states mentioned in article 34, the capacity for responsibility has not disappeared completely but has decreased to a significant degree, a reduced penalty should be imposed.*

There are two elements which compose the general notion of criminal irresponsibility under the Greek law: the inability to distinguish/understand the unjust nature of the action due to a rational defect (cognitive element) and/or the inability to act according to this understanding (volitional element).[25]

If recent neuroscientific discoveries and neuroimaging techniques manage to show some defect in one of these two elements, then they will have been able to prove or to play a significant role in proving the irresponsibility or diminished responsibility of the perpetrator.

---

[24] See article 7 of Act 2472/1997.

[25] Kotsalis (2002), p. 88.

However, two observations concerning the articles 34 and 36 PC previously mentioned make us sceptical over the role of neuroscience in the assessment of criminal responsibility under the Greek criminal law.

First, under the Greek legal system, an offender cannot be generally held irresponsible or partially responsible; according to the wording of the relative provisions, the assessment of capacity for criminal responsibility is always examined in relation with the *specific action* committed.[26,27] The Greek Penal Code, regarding the criteria of imputability, adopts a mixed system, composed of biological as well as psychological factors. According to this system, in order to assess a person's criminal responsibility, one has to look first for the existence of some biological factors (i.e existence of a morbid disturbance of mental functions or disturbance of consciousness) and second to examine a psychological question, i.e. what effect these biological factors had on the appreciation of the wrongful character of the specific action committed and on the ability to act according to this appreciation, at the time of crime.[28]

Therefore, even if neuroscientific techniques succeed in revealing a specific anomaly or malfunction in the brain of the accused, it does not follow that this is a reason per se to exculpate or diminish responsibility. What clearly needs to be established is a causal link between the brain dysfunction (or predispositions that this dysfunction entails for the individual) and the specific action committed by the perpetrator.[29]

Second, a crucial topic of relevance for the Greek penal law is to discover whether the *disturbance of consciousness* of the perpetrator of the crime (mentioned in articles 34 and 36) existed or not at the time of the action's commission.[30] The latter is defined, according to article 17 PC as the *time during which the perpetrator acted or ought to have acted. The time when the result occurred is indifferent.* However, in most cases, neuroscientific techniques illustrating the brain take place long after the crime was committed and, ultimately, not tell us much about the brain state of the accused at the moment of the crime, which is crucial for the law.

It is worth mentioning that the Penal Code in the articles concerning criminal responsibility does not explicitly refer to a "mental disorder" or "mental disease", but it makes a general reference to a "morbid disturbance of mental function" and "disturbance of consciousness". The specific criminal behaviour has to be the result of a disturbance of consciousness, which could have a morbid or nonmorbid

---

[26] Kotsalis (2002), p. 27.

[27] Kotsalis (2002), p. 28.

[28] Manoledakis (2001), p. 586: It is worth noting that according to the ruling opinion, the biological factors implied in the article 34 PC are used as legal concepts and they are not conceived as strictly medical terms; psychiatric science comes at this point to help legal science and not to replace it.

[29] Supreme Court 342/2010: "The reduced capacity of the offender should be evaluated always in relation to a specific crime and should be present at the time of the crime".

[30] Kotsalis (2002), p. 103, Anagnostopoulos and Magliveras (2000), p. 70.

character,[31] and could be of short or long duration.[32] The expression "morbid disturbance of consciousness" used by the law should be considered as a legal term covering a variety of psychological and psychopathological conditions.

With regards to the criminal procedure, according to article 80 of the Code on Criminal Procedure (CPP) entitled Mental Illness of the Accused, the court orders *a stay of proceedings when the accused is in a state of disturbance of his mental functions* and that *In order to evaluate the psychological condition of the accused, the court has to order an expertise.*

Therefore, at this point a neuroscientific expertise could contribute in establishing whether there is a mental illness or not and play a significant role in deciding whether the accused is capable of standing the trial.

1.2.2 Punishment: Security Measures

Under the Greek law, a sentence is imposed on the offender as a declaration of moral and social disapproval for the crime they committed. A general condition for the imposing of a sentence is the existence of criminal responsibility. If, during the criminal procedure the offender proves to be irresponsible due to the existence of a morbid disturbance of mental function or consciousness (article 34 PC) at the time of the crime, no sentence can be imposed. Instead, the offender may be subjected to a measure of security, the detention in a public sanitary institution, if they are found to be dangerous for the public safety[33] (article 69 PC).

If the offender is found to be partially responsible, then a proper sentence should be imposed, but this sentence should be reduced, due to his diminished responsibility (article 36 par. 1 PC). The notion of the dangerousness of the offender is also mentioned in the article 38 of the Penal Code according to which, in the case of criminals with diminished responsibility, the court should order their detention in a penitentiary or sanitary institution, if they have committed a serious felony and if, at the same time, they are dangerous for public safety.

As it can be seen, the notion of "dangerousness" of the offender is an essential prerequisite for the imposing of the security measure of detention in a sanitary institution (irresponsible criminals, article 69 PC) as well as for the particular

---

[31] Kotsalis (2002), p. 103.

[32] Androulakis (2000), p. 477; Supreme Court 2292/2003: "Morbid disturbance of mental functions" includes all forms of madness or insanity in the broadest sense. "Disturbance of consciousness" includes all the psychic disturbances, which do not arise from pathology of the brain but occur in mentally healthy people and are always transient".

See also Supreme Court 449/1996: "A morbid disturbance of mental function includes any form of disturbance of mental functions from pathological causes, any disease of the mind, such as specific forms of insanity, or insanity in the broad sense. Disturbance of consciousness includes any kind of mental disturbance which does not emerge from a general situation of the brain but can also occur in mentally healthy subjects and which is transient".

[33] Anagnostopoulos (1983), p. 779.

sentence of article 38 (detention in a penitentiary or sanitary institution for dangerous criminals with diminished responsibility).

In both cases, forensic psychiatry[34] and the opinion of an expert-psychiatrist are necessary for the diagnosis of the dangerousness of an offender[35]. In this context, neuroscientific techniques could offer a more precise diagnosis and contribute to the better evaluation of this vague and problematic concept.

The law makes an effort to clarify the notion of "dangerousness" in article 13 g PC, establishing some criteria: *The offender is characterized as particularly dangerous, when the severity of his/her act, the way and the circumstances under which it was committed, the causes that led him/her into committing the crime and his personality, reveal unsociability and clear tendency to commit new crimes in the future.*

As a result, under the current legal framework, as far as punishment and security measures are concerned, neuroscientific techniques theoretically could be used in court with the specific aim to evaluate the personality of the offender as one of the elements constituting their dangerousness. For example, an abnormality of the brain function detected by the neuroscientist that results in an impulsive personality or in a behaviour that makes the subject more susceptible to antisocial behaviour could be interpreted by the judge as a dangerous element of their personality or as an element of unsociability, and thus, could result in the imposing of the security measure of the detention in a therapeutic/psychiatric institution.

However, a question is if and how the "dangerousness" of an offender could be diagnosed. There are two main opinions with regards to this question. For some legal scholars, the notion of dangerousness is mainly characterized by the existence of a strong probability for a person to commit a new crime in the future[36]. In this case, diagnosis is identified with prognosis. According to others, despite the effort of the law to establish some criteria in order to clarify the notion of dangerousness in article 13.g, these criteria cannot be precisely defined. Because of the relativity and flexibility of the criteria mentioned, the diagnosis of "dangerousness" is impossible. Eventually, the concept of "dangerousness" is a legal construction used by the State, whose only purpose is to designate certain persons as "dangerous" and the relevant diagnosis based on objective criteria cannot be achieved[37]. For those who believe that "dangerousness" constitutes a clinical reality or a physical characteristic of the offender, the diagnosis is necessary in order to obtain an evaluation of the future behaviour of the same person,[38] and neuroscience could contribute significantly in this direction.

The "personality of the offender" is also a very important criterion of the decision on the length of a sentence, as described in article 79 PC. According to this article, a very important element of the personality of the perpetrator, is their character and its

---

[34] Kotsalis (2002), p. 73.

[35] Leivaditis (1994), p. 393.

[36] Panoussis (1978) p. 776.

[37] Alexiadis (1986), p.131.

[38] Dimopoulos (2008), pp. 322–326.

degree of development (par. 3, element b). The degree of development of a person's character constitutes an element which could be evaluated, to some extent, with neuroscientific techniques. Of course, the latter can not offer a complete picture of the personality of the person examined, but could reveal some elements which might be considered by the judge as indicative of the development of character and, thus, play an important role in the decision concerning the length of the sentence. It is worth noting that given the existing practice of Greek courts to provide insufficient explanation in their decisions concerning this particular issue (determination of the length of a sentence according to article 79 PC) – a practice which is often criticized by many legal scholars[39] – it would be perhaps desirable to give emphasis to the evaluation of the defendant's personality with more "objective" neuroscientific techniques, in order to make it a leading point of the criminal trial, as it should be.

The contribution of neuroscience seems to be important in the assertion of certain crimes related to drugs and alcohol (article 71 PC) and may help to establish whether these crimes are attributable to an abuse of alcohol or other substances, as well as to establish if the perpetrator is addicted to drugs or alcohol. If these two conditions are met, the court may order the detention into a sanitary institution, instead of imposing a sentence of imprisonment to the offender.

1.2.3   Juvenile Offenders

Concerning juvenile offenders in particular, neuroscience could play an important role in the evaluation of the conditions to be met in order for the juvenile offenders to undergo therapeutic measures.

These conditions are described in article 123 PC, according to which, therapeutic measures are ordered *especially if the juvenile offender suffers from a mental illness or any other morbid disturbance of mental functions or if they are . . . . . . epileptic or are addicted to alcohol or drugs and can not abort this habit with their own forces or if they show an abnormal delay in their mental and moral development.* These therapeutic measures can be ordered only after an opinion has been given by an expert physician. As a result, a an expertise including neuroscientific data related to the existence of a mental illness, an addiction, or a delay in the mental development of a juvenile offender, could influence to a considerable extent the decision on the necessity and the nature of the therapeutic measures which should be imposed to juvenile offenders.

1.2.4   Crime Prevention

According to the article 200A of the Code on Criminal Procedure, DNA analysis is allowed for the identification of the perpetrator of crimes described in article 187a PC, concerning the detection of organized crime.

---

[39] Kourakis (2007).

It could be argued that in analogy with this article, neuroscientific techniques should be admitted in criminal proceedings in order to ensure the struggle against terrorism, the protection of public safety and the detection of organised crime.

The Act 2472/1997 on the protection of sensitive personal data, in article 11, establishes the right of the subject whose data are processed, to be kept informed as to who has knowledge of these data. In paragraph 4, however, there is an exception to this general rule, *if the data are processed for reasons of national security or for the detection of particularly serious crimes.*

Therefore, according to this provision, neuroscientific data which are derived from neuroscientific research (and which constitute sensitive personal data since they are health related) could be used in criminal courts without the consent of the subject, for the purpose of the detection of serious crimes.

1.2.5 The Use of Neuroscientific Techniques in Criminal Trial: Admissibility in Court?

Legal Framework

According to the article 179 CCP, any kind of legal evidence can be allowed during the criminal trial. The article 177 CCP establishes the principle of Moral Evidence according to which, judges, during the criminal trial, are not obliged to follow legal rules of evidence, but can freely consider the evidence given during the trial and have to judge in accordance with their voice of consciousness.[40] General principles of law consider means of evidence to be unlimited. The article 178 CCP enumerates the main means of proof[41] but in an indicative and not in an exclusive way. The criminal judge may use any kind of evidence which is relevant and appropriate for the discovery of the substantial truth, as long as this evidence is not prohibited by law (exceptions are described in article 177 CPP).[42]

As a result, in principle, Greek criminal law does not seem to reject the use of neuroscientific methods and techniques in the courts, whose testing results are freely evaluated, according to the principle of Moral Evidence.

In addition, according to article 200A par. 1 of the CCP, *When there are strong indications that a person has committed a felony using violence or a crime against sexual freedom .... the competent judicial council may order an analysis of Deoxyribonucleic Acid - DNA) for the purpose of ascertaining the identity of the*

---

[40] Anagnostopoulos and Magliveras (2000), p. 160.

[41] Article 178 CCP: *The main means of proof in criminal proceedings are: a) indications. b) inspection. c) expertise. d) the confession of the accused. e) witnesses and f) documents.*

[42] Article 177 CCP: *Evidence that has been introduced into court with or through illegal acts must not be taken into consideration for the declaration of guilt and the imposing of a punishment or of coercive measures, except if it has to do with crimes for which the penalty is life sentence and if the court issues a reasoned decision for this subject.*

perpetrator of this crime .... The accused himself has the right to ask for the analysis of the DNA in order to defend himself.

It could be argued that, reasoning by analogy with this article, neuroscientific techniques could be admitted in criminal proceedings in similar cases, and particularly when the accused himself asks for their use, in the purpose of defending himself. In this way, the right to a fair trial is guaranteed. The right to a fair trial, is protected by the ECHR (article 6) and by the Greek Constitution (which establishes, with articles 2, 5 and 7 a primary obligation of the State to respect and protect human personality and dignity).

Furthermore, the Act 2472/1997 on data protection, in the article 7 par. 2 establishes an exception to the prohibition of the collection and processing of sensitive data if *processing of personal data is made by the subject itself or if the procession is necessary for the establishment, exercise or defense of a legal right before a court or disciplinary body.* This provision may be considered as a complementary argument for the use of neuroscientific techniques in criminal courts.

However, the use of neuroscientific techniques in the Greek criminal courts is restricted by some key provisions related to the protection of some rights and freedoms of citizens. One of these provisions is the article 7 par. 2 of the Greek Constitution, which generally prohibits any torture, bodily injury, damage to health or psychological violence, and any violation of human dignity, under form of punishment or under any other form (e.g. interrogation method). Psychological violence includes violation and investigation of the subconscious world of a person exercised by state institutions.

The article 137 PC on tortures considers any affront of human dignity as a torture strongly prohibited by the Greek Constitution. In the third paragraph of the article 137, the Penal Code explicitly refers to the use of a lie detector as an affront to human dignity. Thus, the combination of articles 2 par. 1 Const, 7 par. 2 Const. and 137 A par. 3 PC, can infer a general prohibition of the penetration and the forced investigation of a person's innate mind, especially with the use of techniques such as lie detectors.

The Issue of Consent

A difficult issue, in this respect, is whether the subject's consent can reverse the unacceptability of the use of neuroscientific techniques such as lie detectors in the criminal courts. The scientific community, with regards to this question, is divided, with some legal scholars arguing that even if the concerned person consents, their consent is invalid as contrary to morality and that in any case the application of such methods is contrary to the constitutional and ECHR provisions on the protection of personality and human rights. According to the opposite opinion, methods of lie detection could be applied in criminal proceedings only if the accused person consents, particularly for the purpose of strengthening the evidence for their innocence.

The Mixed Jury Court of Athens, however, shared the second opinion and gave permission to the accused to be submitted to a lie-detection method. This was the first time in the Greek criminal records that the use of a neuroscientific technique was admitted in a criminal trial.

Decision 93/2002 Mixed Jury Court of Athens

In one of the most sensational trials ever held in the Greek criminal courts, one of the defendants submitted the request to be subjected in the procedure of lie-detection with the method of Event-Related Potentials (ERPs) in order to prove his innocence. The defendant was facing – among others – the charge of man slaughter.

Brain waves have been used in neurophysiology for decades to uncover processes in the brain. One type of brain wave, called event-related potentials (ERPs), is of particular interest in lie detection research. ERPs are recorded by sensors placed on the scalp, and can reveal the timing and general location of electrical activity in the brain elicited by the presentation of sounds, words, text, and pictures. Using a particular type of test it has been shown that one kind of ERP, called the P300, can reveal whether an individual has guilty knowledge of crime-related information when it is placed in a list of other information that is unrelated to the crime. This could be useful to uncover whether a suspect knows something about a crime that only the guilty person would know.[43]

The prosecution claimed that the use of this scientific method should not be permitted; they argued that the use of a lie detector is generally prohibited because the Greek legal system explicitly considers it as a kind of torture (article 137 PC) and thus, it is unacceptable, even if the person who was submitted in this examination has consented. The prosecution also based their arguments on article 29 Sect. 1 of the Code of Prison Regulations which prohibits *the conduct of any medical or other relevant experiments which endanger the life, physical or mental health or offend the dignity and personality of the prisoner, even if the latter consented in carrying out these experiments.*

The defense, on the other hand, stated that this method was scientifically valid and reliable, already tested and regularly used in other countries and did not put at risk the defendant's physical and mental health. An additional argument was that the use of this method was absolutely necessary because it guaranteed the defendant's right to a fair trial and that it did not offend the defendant's dignity since the latter had already consented to its use. The defense claimed that the defendant had the right to undergo this examination as a prisoner, otherwise, he would be deprived of an important means of proving his innocence, which blatantly violated his rights to defend himself. With respect to article 137A PC on torture,

---

[43] Farwell and Donchin (1991), pp. 531–547.

which explicitly defines the use of a lie detector as an affront to human dignity, the defense argued that *a literal interpretation of the alleged provision leads to consider that the use of a lie detector is prohibited only when it is contrary to the will of the person to be submitted in the examination. The cases reported in the art 137A are considered as tortures only if they are imposed to a person, contrary to their will*[44].

For this reason, the defendant asked the court to give permission to two neuropsychiatrists to visit him in prison, in order to examine him with the method of ERPs. The Mixed Jury Court of Athens, with its Decision 93/2002, fully accepted the request of the accused, allowed him to be examined with this scientific method and allowed the use of the findings as evidence. The decision *upholds the claim of the accused and allows doctors of his choice to visit him in the prison so that he undergoes any medical examination he wishes.*[45]

Some months later, the court, with its decision 312/2002, acquitted the accused, without making special reference to the method of lie-detection.

2   Conclusion

Although the Greek legal system refers extensively to situations of interest for neurolaw, the acceptance of modern neuroscientific methods for determining the cognitive or mental status of persons involved in civil, medical, research and, especially, criminal relationships is rarely considered as important.

It is true that such methods are not mentioned explicitly in relevant legislation as appropriate evidence – means, and a question is whether that legislation should be revisited in this respect.

After the aforementioned important judicial step towards their acceptance in criminal settings, it seems reasonable to reflect on that question. Indeed, certainty in evidence might be promoted with an explicit introduction of certain techniques in the law, on condition that neuroscientists do not have objections on their reliability.

During the last decades, the Greek legislator shows a quite innovative spirit in regulating biomedical matters in various legal contexts. To provide more information on neurolaw and its expected benefits is, perhaps, the best motivation for taking action in this promising field as well.

---

[44] Praxis and Reasoning of Criminal Law (2003), 4th year, pp. 185–188.

[45] The neuropsychiatrists who conducted the examination testified in the court as "witnesses with special knowledge" in accordance with the article 203 CCP, and not as experts, since the criteria laid down in articles 183–186 CPR concerning the ordering of an expertise were not fulfilled, nor were the two neuropsychiatrists included in the tables of experts.

References

Alexiadis S (1986) The dangerousness of the criminals: a fake construction. Homage to Chorafas/ Gafos/Gardikas, vol. 2. (in Greek)

Anagnostopoulos I (1983) "Dangerous" criminals and preventive procedural measures. Poinika Hronika ΛΓ":769–784 (in Greek)

Anagnostopoulos I, Magliveras K (2000) Criminal law in Greece. Kluwer Law, The Hague-London-Boston/Sakkoulas, Athens

Androulakis N (2000) Penal law, general part. Sakkoulas PN ed, Athens (in Greek)

Dagtoglou P (1991) Constitutional law. Civil rights. Ant. Sakkoulas ed, Athens (in Greek)

Dimopoulos C (2008) Lectures on criminology. Nomiki Vivliothiki ed, Athens (in Greek)

Farwell LA, Donchin E (1991) The truth will out: interrogative polygraphy ("lie detection") with event-related brain potentials. Psychophysiology 28(5):531–547

Hellenic National Bioethics Commission (2005) Report on biomedical experimentations involving human subjects and clinical trials of medicinal products. In: Rap: Manolakou K, Vidalis T, http://www.bioethics.gr/media/pdf/reports/report_ct_en.pdf

Kotsalis L (2002) Introduction to legal psychiatry. Ant N. Sakkoulas, Athens (in Greek)

Leivaditis M (1994) Psychiatry and law. Papazisis, Athens (in Greek)

Manessis A (1982) Constitutional rights. a. Civil liberties. Sakkoulas ed, Thessaloniki (in Greek)

Manoledakis I (2001) Penal law - general part. Sakkoulas ed Athens (in Greek)

Panousis G (1978) The notion of dangerousness. Nomiko Vima: 776–782 (in Greek)

Praxis and reasoning of criminal law (2003) 4th year, pp 185–188 (in Greek)

Vidalis T (2007) Biolaw. The Person, T. 1. Sakkoulas ed, Athens (in Greek)

Vidalis T, Mitrou L, Takis A (2006) Constitutional reception of technological developments and "New" rights. In: Centre of European constitutional law, five years after the constitutional revision of 2001, T I., Ant. Sakkoulas ed Athens, pp 273–312 (in Greek)

Kourakis N (2007) The problem of inefficient explanation in the decisions on the length of the sentence. http://www.poinikos-logos.gr/arxeio/07t3e.html (in Greek)

Neuroscience and Converging Technologies in Italy: From Free Will Approach to Humans as *Not Disconnected Entities*

Amedeo Santosuosso

Abstract In recent years, a vast literature has developed on how neuroimaging may increase our understanding of deception, moral and legal responsibility, behaviour prediction, and much more.

Common approaches overlook the global reality of neuroscience and neurotechniques. This is the reason why (beyond controversial implications of neuroimaging techniques: i.e. lie detection, determination of mental impairment, or psychopathy) it is important to survey some technological applications of neuroscience on the human body (even beyond the field of criminal law), such as objective measurement of chronic pain, robots and artificial intelligence, brain–computer interfaces.

The review focuses on Italian case law on the concept of "moral damage" and the opportunities that neurotechniques offer in order to have a more objective evaluation. In addition, it is considered the responsibility for robot's actions (especially referring to learning robots) and the possible application of current Italian civil legislation (especially the responsibility of teachers).

Conclusive remarks are on the law and the way basic concepts as human individual are affected by neuroscience.

1 Neuroscience and Neurotechniques Within Converging Technologies

In recent years, a vast literature has developed on how neuroimaging may increase our understanding of deception, moral and legal responsibility, behaviour prediction, and much more.

A. Santosuosso (✉)
European Center for Law, Science and New Technologies, University of Pavia,
Corso Strada Nuova, 27100 Pavia, Italy
e-mail: Amedeo.santosuosso@unipv.it

T.M. Spranger (ed.), *International Neurolaw*,
DOI 10.1007/978-3-642-21541-4_11, © Springer-Verlag Berlin Heidelberg 2012

Indeed, cognitive neuroscience holds the promise of explaining operations of the mind in terms of the physical operations of the brain. It is claimed that brain imaging techniques now allow the discovery of neurophysiologic markers for almost any kind of behavioural phenotype, normal or pathological, both at explanatory and at predictive levels. Since early 2000, law has taken part in this debate within the perimeter of *neuroethics*, a new interdisciplinary field "at the intersection of the empirical brain sciences, normative ethics, the philosophy of mind, law and the social sciences of anthropology, economics, psychology and sociology".[1]

When issues like these are discussed in legal contexts, a normal first reaction is to wonder whether a more complete understanding of the neural mechanism for voluntary decision-making might undermine the legal notion of accountability (we can call it the "free will approach"). Would brain scans that reveal a brain feature correlated, even weakly, with a propensity for violence, influence a court's decision? Will brain imaging subvert the current nosography of mental diseases? Is *free will* still alive?

It seems to me that the *free will approach* is neither exhaustive, in theoretical terms, nor able to face the new reality and that a different one might be more productive. In my opinion, at least nowadays, the real overwhelming need is to know more on whether and how the presently available neurotechniques are actually used by experts before the courts in different countries. Indeed, it might be worthless going on wondering how and if neuroscientific findings change our idea of law and responsibility without having enough information about their real impact.[2] On the other side, we should look at neuroscience and neurotechniques as specific fields within the wider area of the new converging technologies. The *Nordmann Report* (2004) outlines the field in the following way:

> Information and communication technology helped produce the profound transformation of daily life in the 20th Century. Biotechnology is transforming agriculture, medical diagnosis and treatment, human and animal reproduction. Most recently, the transformative potential of nanotechnology has captured the imagination. Add to this that cognitive and neuroscience are challenging how we think of ourselves, or that the rise of the social sciences parallels that of bureaucracies and modern forms of governance.[3]

Once assumed such perspective, the issue immediately takes a shape, which is factually and conceptually wider and encompasses more practical applications. Under some respects, such (wider) an approach to neuroscience and law might seem to be less noble (almost everything is banal if compared with *free will*!), but,

---

[1] Glannon (2007).

[2] To survey the actual and current, and likely use of neuroscientific, behavioural genetic and neurogenomic techniques before the European Courts and in investigative activities and to create and implement an archive of cases and materials on NeuroLaw are among the aims of the newborn *European Association for Neuroscience and Law* (EANL). More information at http://www.unipv-lawtech.eu/.

[3] Nordmann A (Rapporteur) (2004). See also Roco and Bainbridge (2002).

in the end, the questions that arise might prove to be even more challenging (e.g. the social/mental/technical/biological boundaries of individuals).

In short, common approaches overlook the global reality of neuroscience and neurotechniques and, as a consequence, do not give attention to some extremely promising fields. This is the reason why I am not going to discuss here controversial implications of neuroimaging techniques (for example, lie detection, determination of mental impairment, or psychopathy). I am rather going to survey some techno-logical applications of neuroscience on the human body (even beyond the field of criminal law), such as objective measurement of chronic pain, robots and artificial intelligence, brain–computer interfaces. Conclusive remarks will focus on their impact on the law and the way basic concepts as *human individual* are affected.

2   Chronic Pain, Objective Measurement and the Law

Two cases clearly show the likely relevance of neurotechniques in a traditionally very subjective field such as chronic pain.[4]

On December 9, 2008, a patent on pain detection, entitled "Objective Determi-nation of Chronic Pain in Patients" (US Patent No. 7,462,155) was conferred to Dr. Robert England (an orthopaedic surgeon at Modesto, California). Dr England's method involves the use of functional magnetic resonance imaging (fMRI) to capture an image of the brain. It looks at neuron activity when the patient with chronic pain receives stimulation – such as excessive squeezing of a finger or mild electrical shock – and compares it against the neuron activity in the brains of pain-free people. According to the patent, the validation and measurement of chronic pain is accomplished without any input from the patient. Thus, the patent claims that the determination can be made *without any subjective input from the patient* (italics, mine).

In 2008, in California, the lawyers of a plaintiff who was claiming compensation for burns received on the workplace assembled evidence that chemical burns in the workplace had left their client with chronic pain. The evidence included fMRI scans of his brain that showed "heightened activity" in the brain regions, which have been involved in the neural basis of pain. The parties reached a settlement and the case was not brought before the court, but it is very plausible to think that other similar cases will make the headlines in the near future. In other words, the possibility of brain imaging techniques to be brought before the courts definitely is on the table.

Considering the possibility of future cases where fMRI scans are involved in order to assess chronic pain in a civil trial, the question whether brain imaging techniques could pass the tests for scientific admissibility in the courts (according to different national legal rules and standards) becomes therefore crucial.

---

[4] In this paragraph, I take advantage of Camporesi and Bottalico (2011). I thank the authors for kindly giving me the opportunity to use the results of their research.

2.1   Chronic Pain

Chronic pain represents an enormous problem to society. According to current statistics, approximately 20% of the adult population have chronic pain, and the financial cost to society is huge, in the range of more than € 200 billion per annum in Europe, and $150 billion per annum in the USA,[5] more than heart disease and cancer combined. Doctors and researchers have long sought to make assessments of pain more accurate and independent from numerical rating scales (NRS) or other kind of scales.[6] It is the lack of objective measurements for pain that originates the increasing attention towards brain imaging techniques (namely fMRI) before the courts.

In Italy, a huge legal and forensic medicine literature is dedicated to the issue and some recent case law try to deal with the issue in conclusive terms. It is worth noting the decision *Corte di Cassazione, Sezioni Unite*, 11 November 2008, n. 26972.[7] The Court stated that "moral damage" has to be conceived in his widest meaning as the violation of the individual's personal sphere, even if there are no immediate economic consequences. Damage is the result of the violation of constitutional rights such as psychological health and individual mental status. Such a holding is surely understandable and correct in theoretical terms, but leaves us with the entire huge problem of how to ascertain such a mental status and psychological dimension. The Italian law on *Private insurance*[8] regulates damages for psychophysical harm and leaves to judges the possibility to increase the level of damages when relational aspects are involved.

In general terms, we may say that, even if in Italy there is no specific provision on chronic pain and the way of measuring it, the problem is well known. A wide range of discretion is recognized to judges in order to evaluate not immediately measurable aspects.

Only recently chronic pain has been considered a disease of the brain. Non-invasive magnetic resonance imaging (MRI) methods, including morphological/anatomical imaging of grey matter (voxel-based morphometry, VBM), white matter tract connectivity (diffusion tensor imaging, DTI), fMRI, and magnetic resonance spectroscopy (MRS), have produced a shift in our understanding of chronic pain. From the original definition as an "unpleasant sensory and emotional condition," chronic pain is now understood to be a multidimensional "disease affecting

---

[5] Tracey and Bushnell (2009).

[6] Pain intensity scales used by researchers at the NIH Clinical Center to measure how intensely individuals are feeling pain and to monitor the effectiveness of treatments are at: http://painconsortium.nih.gov/pain_scales/index.html.

[7] When a case is very controversial or precedents are conflicting each other, the Italian *Corte di Cassazione* (Supreme Court) decides as a unique chamber (*Sezioni Unite*): several judges are added to a chamber and a bigger chamber is created *ad hoc* with the purpose to have a more authoritative and binding decision.

[8] Decreto Legislativo n. 209, 7 September 2005 (mainly art. 138).

the central nervous system," influenced by a variety of biological and psychosocial factors, such as genetics, hormones, emotions, memories, or social expectations.[9] While historically chronic pain has been understood as a syndrome, the recent evidence accumulated from neuroimaging studies points in the direction of redefining it as a neurodegenerative disorder. The most convincing data in support of the idea that chronic pain is a disease, rather than a syndrome, involve pathological modification both at the level of structure and neurochemistry of the brain of affected patients.

The claim made by the patent on pain accorded to Dr England goes against the traditional view, as it claims that fMRI as a tool to measure pain should have a crucial role in the assessment of pain, in the same way as the agent's utterance and subjective experience. It is an instance therefore of the realm of the claims belonging to the so-called "neurorealism", according to which only a subjective experience that shows to have a measurable physical correlate in the brain can be considered to be real. Neurorealism is a very controversial issue. Some authors maintain that neurorealistic claim, such as the one purported by the patent conferred to Dr England, is unrealistic and that reality of a subjective experience like pain cannot be trumped by a brain imaging technique. As provocatively put by Ben Goldacre: "For your own personal experience of pain, which is all that matters, if you say that your pain is relieved, then your pain is relieved (and I wish good luck to any doctor who tells his patient their pain has gone, when it hasn't, just because some magical scan says it has)".[10]

2.2   Case Law

The two above reported cases seem to tell us that neuroimaging is knocking on the courthouse door and offering a solution for a very controversial issue in medical, legal and insurance fields: how to measure patients' pain.

Subjective experiences such as feeling pain are private also in a legal sense. In some cases, we ought not to be forced to reveal information about what we are feeling, for example a person, when accused in the court, has a right to remain silent in all western countries.

This legal dimension of the subjective experience has various implications for the rights of the parties, which may be different in civil litigations. Chronic pain constitutes a huge issue not only for public health (as shown above), but also for the civil procedure. The relevance of chronic pain cases in civil cases is quantifiable in billions of dollars per year only in the United States, where in the courtroom pain always goes hand in hand with "suffering". In the legal context, the phrase "pain

---

[9] Borsook et al. (2007); Borsook et al. (2010).

[10] Goldacre (2010).

and suffering" is broadly construed to permit recovery "not only for physical pain but for fright, nervousness, grief, anxiety, worry, mortification, shock, humiliation, indignity, embarrassment, apprehension, terror or ordeal".[11]

Courts regularly admit imaging evidence (traditional radiography, CT scans and MRI) to provide insight into the extent of a person's pain and suffering. A large number of cases involve people unable to work due to serious and painful conditions that do not have objectively measurable symptoms or tests, and that may therefore face difficult problems when making and supporting a claim for disability insurance benefits. These hurdles are common also to people who suffer from conditions such as fibromyalgia, chronic fatigue syndrome, or chronic pain conditions such as complex regional pain syndrome (CRPS). Another frequent set of cases regards plaintiffs who sustain injuries in motor vehicle accidents and claim to have chronic pain well beyond the time that the objective injuries have healed.

The Supreme Court of Canada in *Martin and Laseur* v. *Nova Scotia* (Workers' Compensation Board, 2003) stated that Nova Scotia's restriction of workers' compensation benefits, for employees disabled by "chronic pain," to a limited 4-week "functional restoration program," violated equality rights guaranteed to the disabled in s.15(1) of the Charter, and that it was within the jurisdiction of the Workers' Compensation Appeals Tribunal to decide this issue because the Tribunal had authority, under the provincial Workers' Compensation Act, to decide questions of law. A very interesting point is the following one:

> There is no authoritative definition of chronic pain. It is, however, generally considered to be pain that persists beyond the normal healing time for the underlying injury or is disproportionate to such injury, and whose existence is not supported by objective findings at the site of the injury under current medical techniques. Despite this lack of objective findings, there is no doubt that chronic pain patients are suffering and in distress, and that the disability they experience is real. Despite this reality, since chronic pain sufferers are impaired by a condition that cannot be supported by objective findings, they have been subjected to persistent suspicions of malingering on the part of employers, compensation officials and even physicians.[12]

According to the Pennsylvania Supreme Court Justice Michael Musmanno, the subjectivity of pain is not even always a given, as there is no authoritative medical work which asserts that pain is wholly and always subjective: "The fact is that pain can be very objective and it can be detected by persons other than the one who states he feels it. There are symptoms of pain that write their story on one's countenance as clearly as lightning scribbles in the sky its fiery message of nature's discomfiture".[13]

---

[11] Kolber (2007).

[12] Nova Scotia (Workers' Compensation Board) v. Martin, Supreme Court of Canada, October 3, 2003. See also http://onlinedb.lancasterhouse.com/index.asp?navid=37&csid=2792&layid=87 &csid1=23&csid2=3467.

[13] City of Philadelphia v. Shapiro, 206 A.2d 308, 311 (Pa. 1965).

Lawyers depend on expert witnesses to demonstrate that their client is suffering from a serious condition or disability that impairs their ability to work and live. The plaintiff's attorney must still fight against subjectivity on two fronts: first in establishing that pain exists, or that pain as it is expressed is tied to a palpable and present pain as it is embodied, and second in precisely defining the contours, causation, and consequences of that pain. The law resolves the tension between subjectivity and objectivity by considering subjective clues of pain and suffering to the extent that they are not contradicted by objective evidence, and by crediting narrative constructions of pain to the extent that they do not conflict with medical evidence or other objective clues of disability.[14]

It would therefore have a great impact for the civil procedure if neuroimaging techniques could offer a tool for demonstrating chronic pain in the court, where malingering and feigning is very common, as litigants have strong financial interests to exaggerate their claims.

2.3   Limits and Opportunity of Neurotechniques

Neuroimaging techniques such as fMRI are not the first ones claiming to be an effective "pain-ometer", but only the most recent. In the 1960s, a number of researchers began touting an "objective" pain detection technique based on thermography and some still claim so today.[15]

Still, there are some reasons to be cautious, as other factors need to be kept in mind when discussing the potential of these techniques, among which (a) the variability in terms of baselines for pain of different individuals; (b) the heterogeneity of "chronic pain" (which subset of patients are we talking about?); (c) how beliefs (expectancy) can influence pain perception; (d) the unknown rate of false positive and false negatives. Indeed, negative and positive expectations are powerful modulating factors that influence behaviour, and many experiments have used simple verbal cues to manipulate expectations in both experimental and clinical pain studies.

Even taking into account the above reasons, it is beyond any doubt that brain-imaging techniques will play a pivotal role in ascertaining pain claim in the courtroom. In an adversarial system such as the UK or the US, where the burden of proof is upon the parties, such a scenario seems very plausible, and its ethical implications in terms of fairness do not escape us. In a civil-law system, such as the continental European ones, where the judge is entitled to appoint an expert, the scenario would be different, where an expert could be appointed with the task of demonstrating, as an "impartial" third-party, the actual medical situation and

---

[14] See also, Madeira (2006–2007).

[15] See for example the website http://www.thermographyclinic.ca/pain.html.

scientific underpinnings (assumed to be uncontroversial) for which the request of compensation is required.

In conclusion, new neuroscientific methods of assessment of chronic pain could have a great impact in tort litigation, where malingering and feigning is very common. Expert witnesses will have soon the possibility to use these methods to give a more scientific footprint to the expression of what has always been considered only a personal feeling. Avoiding neurorealist interpretation, it could be very useful having new tools in courts: the legal debate on their admissibility, and especially with regards to fMRI scans, is began and further indication from the scientific community will be crucial with this aim.

Although there are reasons to be cautious, according to Camporesi and Bottalico (footnote 4) it is reasonable to think that brain imaging techniques could pass the standards of scientific admissibility in the court. Brain imaging techniques are indeed reshaping the definition of chronic pain as a neurodegenerative disorder, highlighting functional, anatomical and chemical alterations that take place in the brain of affecting patients. Therefore, while the limits of the brain imaging techniques for the measurement and quantification of chronic pain need to be kept in mind, their application in the court as an aid to the pain claims seems to be a good opportunity, and an improvement over verbal and other kind of visual scales.

3   Artificial Intelligence and Robots

In October 1950 Alan Turing's article *Computing Machinery and Intelligence* is published on the journal *Mind* and gives a strong impulse to studies and research on mind–body relationship and on Artificial intelligence (AI).[16]

According to the original line of thought about AI (that now is called *Strong AI*), a machine that is able to reproduce and even surpass human intelligence can be created. This concept was based on the famous Turing test,[17] or "the imitation

---

[16] Turing (1950). The Stanford Encyclopedia of Philosophy outlines Alan Turing's very controversial biography and his contribution to contemporary philosophy as follows: *Alan Turing (1912–1954) never described himself as a philosopher, but his 1950 paper 'Computing Machinery and Intelligence' is one of the most frequently cited in modern philosophical literature. It gave a fresh approach to the traditional mind-body problem, by relating it to the mathematical concept of computability he himself had introduced in his 1936–7 paper 'On computable numbers, with an application to the* Entscheidungsproblem.' *His work can be regarded as the foundation of computer science and of the artificial intelligence program.. [...]Alan Turing's arrest in February 1952 for his sexual affair with a young Manchester man, and he was obliged, to escape imprisonment, to undergo the injection of oestrogen intended to negate his sexual drive. He was disqualified from continuing secret cryptological work. His general libertarian attitude was enhanced rather than suppressed by the criminal trial, and his intellectual individuality also remained as lively as ever.* (in: http://plato.stanford.edu/entries/turing/, accessed 7 Jan 2011).

[17] Turing (1950), pp. 433–434.

game". The basic version of the game involves a man, a woman and an interrogator, all of them in different rooms. The interrogator should be able to guess, through a series of questions, which of the two competitors is the man and which the woman. Alan Turing assumes a situation in which one of the two competitors is replaced by a machine. Does the interrogator's win percentage considerably change? If the results are similar, the thinking machine can be equated to humans. In Turing's opinion, the thinking skill defines a machine as intelligent. In general, *strong* AI assumes that the machine is acting as if it had a mind (mimetic ambiguity). Since the 1980s – also because of the failure of those projects, which were inspired to *strong* AI and the success of cognitive science as a discipline – new projects relating to AI have been focusing on individual defined problems, i.e. the execution of certain industrial activities. The machine can only simulate the cognitive processes of the human (*Light AI*), and therefore it can only *operate* in a similar manner to the behaviour of humans.[18]

The pragmatic attitude of Light AI gives up any mimetic ambition of human intelligence and adopts a substantially functionalist approach, which meets the old human dream of *automata*.[19] The convergence with new technologies (such as neuroscience and neurotechniques) is at the origin of contemporary *robotics*.[20]

3.1   About Robots

Continuing development in the field of robotics means that it is now an established fact that robots will gradually come to play an increasingly important role in our lives.

Robots will most probably develop to such a point that they will attain the level of human capability, at least in some specific activities, and it is not an unlikely prospect that robots' capabilities will even, sooner or later, surpass those of humans. There are currently several robotic projects underway whose aim is to create robots that are able to learn from interaction with the environment and, on the basis of their experience, are able to take autonomous decisions. These projects have already in part succeeded in their aim. The legal question is as follows: when a machine with such characteristics is actually created, should the law confer on it at least a minimum level of subjectivity as quasi-agent?

---

[18] Floridi (1999).

[19] Thomas Hobbes deals with *automata* in the first lines of the Introduction of his book Leviathan: T. Hobbes, *Leviathan*, 1651.

[20] In this paragraph I take advantage from a research in progress at the *European Centre for Law, Science and New Technologies*, University of Pavia (I). I am deeply in debt with Chiara Boscarato (Scholarship Fellow at the Centre) and thank her for giving me the opportunity to use her paper (draft) Boscarato (2011).

Robot's increasing levels of capacity requires distinguishing between robots, which act in a way that may be foreseen by the programmer, and those that act in an unplanned and/or unpredictable way. Starting from the lowest level of robot autonomy, the first kind of action corresponds to standard behaviour set up by the programmer in robots, which do not have the power of locomotion. This is the largest section and it contains the most elementary actions, which do not require any adaptive capability and are merely a reaction to user input.

Although it may not immediately come to mind, robots are already to be found in our homes. Appliances we use every day, such as washing machines and vacuum cleaners, are robots, i.e. machines that replace us in doing a job. In fact, the word "robot" comes from the Czech noun "*robota*" meaning "hard work", "labour".[21] Washing machines are now more technologically advanced than ever before. Their features make washing easier, less costly and more efficient. By operating various buttons and knobs, we can select a washing cycle appropriate for the type of load – temperature, duration of washing, release of detergent and spin speed. With some models we just need to put the washing in and leave the machine to it. It recognizes the weight of the load and calculates how much water, detergent and time are required.

This kind of robot is still regarded as a physical object and if it causes harm to others (for example, a user may get a slight electric shock from touching the display of a washing machine or be injured by the blade of a blender which becomes detached from the support on which it is mounted) the applicable legal framework is the traditional one relating to a manufacturer's liability for faulty products.[22]

3.2 Robots That Move and Animals (Roomba and Scooba)

The level of complexity of a robot increases when it is equipped with the means to move. Even should a robot act according to a set program, and therefore in a predictable manner, the fact that it can move, at various levels of autonomy, gives rise to the need for more caution (and, above all, the need for greater supervision). The robot could find itself in unpredictable situations due to its ability to move.

Roomba[23] is a first generation indoor cleaner robot, 5 million of which have already been sold. It consists of a disc that moves around the house, continually

---

[21] The word "robot" was used for the first time by Karel Čapek, a Czech writer, in his play *Rossum's Universal Robots*, published in 1920, on the advice of his brother Josef who had previously used the word "automa" in his short story "Opilec" published in 1917 (http://capek. misto.cz/english/robot.html).

[22] Consumer protection is regulated by EC Directive 1985/374, as amended by EC Directive 1999/34.

[23] http://www.irobot.it/.

turning on itself and sucking in dust (the second generation *Scooba* also washes floors). It has an internal mapping system that enables it to record the area to be cleaned and not to go over the same area more than three times, unless an area is particularly dirty. Thanks to its sensors, it can get between furniture and under tables, recognize corners and move along walls, and recognize and avoids stairs and other areas where there is the danger that it might fall. Its programming system allows a time to be set for the machine action and, when its battery is dead it comes back to its station to recharge.

Needless to say it is a harmless object to be used in all homes. However, let us imagine a situation in which the front door is inadvertently left open and the *Roomba* goes out and takes a stroll along the hallway of the building. It could trip up a neighbour loaded down with shopping or children while they are running up the stairs. *Roomba* has done nothing other than execute the function for which it was designed: to move across the available area for the purposes of cleaning it.

Its ability to move around and travel to other places without any human intervention means that this type of robot is similar to an animal. When a *Roomba* goes out of a door, it may be compared to an animal which is lost or escapes from the control of its owner.[24]

There are also robots which are designed with the specific aim of emulating real animals, in both appearance and behaviour. This interesting development is useful in order to break away from the level of "things" and move increasingly towards the world of robots with adaptive capabilities. Animals can be taken as a point of connection with and transition between the category of things, which lack any form of intelligence, even artificial intelligence, and that of humans, who possess not only intelligence but also consciousness.

The most famous animal-shaped robot in the world is the Artificial Intelligence roBOt (AIBO)[25] developed by Sony and available on the market between 1999 and 2006.[26] It is shaped like a dog and can reproduce all canine behaviour. It is equipped with cameras and sensors for the recognition of oral commands. Thanks to these tools, it is able to interact with its surroundings as if it were a real animal. Its face is a display showing lit-up LEDS. Any combination of an LED and a colour corresponds to an emotion or feeling. It works on AIBOware, a software developed by Sony and then put on open source for non-commercial purposes in response to numerous requests from customers. Thanks to this development kit, many people have been able to modify and customize the code of their AIBO and several universities have used it as a platform for artificial intelligence studies. In addition,

---

[24] According to Italian legislation Art. 2052 of the Civil Code should be applied: "The owner of an animal or anyone using it for a certain period is liable for damage caused by the animal, whether the animal was in his custody, had disappeared or had fled, unless he can prove that a fortuitous event occurred".

[25] http://www.aibosite.com/sp/gen/home.html.

[26] Production was discontinued on 2006 because of insufficient sales and high production costs (around 2.500 dollars).

through interaction with its owner, an AIBO evolves from a puppy to an adult dog. The robot puppy will thus go through many different stages of behaviour, up to full development with recognition of more than one hundred oral commands.

3.3 Robots That Learn and Grow Up (Icub and Nao)

The real interaction with neuroscience is at the higher level of robots which have so far been developed. They have a real form of learning and development of problem-solving skills. This is the last frontier for research projects which study the development of cognitive skills in robots. The neural processes of the brain are reproduced through artificial neural networks and learning algorithms. Such robots are able to have *new* reactions and to learn new skills through their own direct experience. Such actions/reactions were not originally intended by the programmer. He/she simply added the algorithm for their learning. What follows, a greater or lesser ability to take action, or the behavioural "choices" of the robot, cannot be entirely predicted at the outset. We are still most certainly in the field of unpredictable action which depends on the type of programming carried out and, thus, indirectly, on the programmer. As if such robots were children that have been guided by their parents and who react on the basis of the education received.

The best example of such robots is *iCub*. This is a humanoid robot developed by the IIT Centre of Genoa (I).[27] It is about the size of a 3-year-old child and simulates the movements and learning abilities of a child of that age. This is an extremely challenging project, in terms of robotics. The robot's humanoid body has 53 degrees of freedom; its hands have complete powers of manipulation; its head and eyes are fully articulated. Thanks to its cameras and sensors, it has visual, auditory and sensory (tactile sensing with objects) skills and also has a sense of balance. It can crawl and sit and make several *facial* expressions.

The aim of the project is to construct a robot with cognitive skills, which is able to rework data acquired through its own experience and which will become an useful tool in a two-way study (from man to machine and vice versa) of cognitive systems. The key aspect of the project is its aim to develop a learning machine, based on knowledge of human behaviour and the human mind. The approach is thus multidisciplinary involving a team of experts in robotics, bioengineering and neuroscience. At the current stage of the project, the robot can feel and pick up objects such as small balls. This action, taken for granted with regard to humans (such a movement is directed by the brain in humans), may seem banal but the movement requires a precise amount of force and pressure. The challenge is to create a robot capable of learning from its mistakes and learning from experiences, step by step so that it eventually makes the right move. Just like a child. The final

---

[27] http://www.robotcub.org/. This is an open source project funded by the European Commission and used by more than 20 laboratories worldwide.

result will be a machine, which can simulate human mental processes by means of complex algorithms installed in its software. For example, after being instructed how to hold a bow and release an arrow, it learns by itself how to shoot an arrow and hits the centre of the target after only eight tries.

In short, the aim of the project is that the *iCub* will be able to learn new skills, behaviour and concepts. Moreover, it is worth noting that *iCub* project does not try to revive the old concept of *Strong AI* but responds to *Light AI* assumptions.

If we move into the legal field, it is clear that law cannot escape dealing with such a new generation of robots. For example, once *iCub* (or similar projects) will fully develop its *new* unforeseeable experience and actions, who will be liable for damage it will cause as a result of its new behaviour and on what legal basis?

Just considering Italian law, the most appropriate legal rule would seem to be that of Art. 2048[28] of Civil Code, which concerns the liability of parents, guardians, tutors and teachers of art. The first paragraph concerns parental liability in the case of damage caused by the unlawful acts of a minor living with his parents. It is the second paragraph which is pertinent. This stipulates that tutors and those who teach a craft or art are liable for damage caused by the unlawful acts of their students and trainees when under their supervision.

In this case, liability rests with the guardian on the grounds that he/she (supposedly) neglected the child, in terms of both *culpa in educando* and *culpa in vigilando*. No longer does the mere fact of having a *de facto* relationship with the perpetrator of damage gives rise to liability (as would be the case with a keeper). On the contrary, there is no presumption of guilt, but only a presumption of liability. This approach greatly benefits the victim, who is not required to prove the guilt of the parties involved. A tutor may be exempt from liability only if he/she can prove that he/she could not prevent the incident occurring (thus we cannot speak of objective liability).

The very interesting aspect is that this kind of liability presupposes the freedom to move and act and seems best suited for regulating the harmful consequences of harmful events caused by robots such as *iCub* (once the project is completed). This set of rules relating to liability assumes a certain level of material ability in the agent (minor), and, also, of legal subjectivity. Anyone who thinks that legal reasoning is rushing too far ahead must consider another example of cutting-edge robots.

---

[28] Article 2048 Italian Civil Code: *Il padre e la madre, o il tutore, sono responsabili del danno cagionato dal fatto illecito dei figli minori non emancipati o delle persone soggette alla tutela, che abitano con essi. La stessa disposizione si applica all'affiliante. I precettori e coloro che insegnano un mestiere o un'arte sono responsabili del danno cagionato dal fatto illecito dei loro allievi e apprendisti nel tempo in cui sono sotto la loro vigilanza. Le persone indicate dai commi precedenti sono liberate dalla responsabilità soltanto se provano di non avere potuto impedire il fatto.*

Nao[29] is a humanoid robot created for the purpose of carrying out functions of assistance. Apart from its skill of being able to communicate with its owner, who may thus teach it new behaviour, and its participation in the RoboCup,[30] *Nao* is a very special robot since it is the first one into which an ethical code has been inserted. Its designers have inserted, in an automatic learning algorithm, a series of situations which present ethical problems and their correct solution. Actions are classified on the basis of three principles: beneficence, non-maleficence and fairness. On the basis of ethical choices preloaded by the programmer, which show how it must act in a certain standard situation, the robot obtains a general rule of ethical conduct. The robot is thus able to independently assess the situation based on this new scale of values and to therefore make the right decision. The robot's scope of use is hospital care. The robot will deal with patients, starting from the standard three situations: reminding a patient to take his/her drugs but not interfering with his/her refusal to do so unless such a refusal could lead to serious consequences for the patient's health; deciding who will use the TV remote control and delivering food to a patient. If these three situations all occur at the same time, *Nao* can also make an independent assessment of priority, always based on the three ethical principles above.[31]

The existence of *Nao* opens up new scenarios of interplay between robotics, ethics and the law. If a code of ethics can be installed in robots, then it should be a specific responsibility of manufacturers to install that code of ethics in machines which are capable of taking independent decisions. If a manufacturer does not do so, it cannot prove that it could not have avoided the harmful event occurring. A user may also be required to install a code of ethics and conduct, if a robot is equipped with a program which may be managed by a user. Failure to install such codes may be equated with failure to supervise. In any case, these kinds of liability being understood, it must be recognized that producers, owners and users are increasingly taking on the role of external controller of an entity that has ever more autonomy.

3.4 Humans and Machines

Completely equating a robot to a human being would mean continuing to focus on the old concept of artificial intelligence, a concept which was abandoned in the

---

[29] This was developed by the French company Aldebaran Robotics http://www.aldebaran-robotics.com/eng/Nao.php.

[30] *Robocup* is an international robotics competition founded in 1997. The aim is to develop autonomous soccer robots with the intention of promoting research and education in the field of artificial intelligence. The name *RoboCup* is a contraction of the competition's full name, "Robot Soccer World Cup", but there are many other stages of the competition such as "Search and Rescue" and "Robot Dancing". http://en.wikipedia.org/wiki/RoboCup.

[31] Anderson and Anderson (2010).

1980s. Robots, however, are no longer simple mechanical objects. In the not too distant future everyone will have a "personal robot", just as almost everyone now has a personal computer. Today it is virtually unthinkable to leave the house without our mobile phone, or travel without GPS. Such devices are simple but they will continue to develop exponentially. Unbeknown to us, robots are entering all areas of our lives. It is not inconceivable that, sooner or later, they will be given a minimum of subjectivity and *ad hoc* legal status. The degree of legal liability to be attributed to robots directly depends on the level of legal subjectivity they may be given.

Although this vision may now seem rather futuristic and virtually unrealizable, it is highly likely that it will actually come to pass. The relationship between man and machine is becoming increasingly closer, above all in the medical, rehabilitation and care sectors. Knowledge about the human brain and cognitive development are being used to create robots with ever more sophisticated and responsive artificial intelligence. At the same time, studies on the development of cognitive robots could be useful in better understanding the functioning of the human brain.

4   Still Self-Determination?

The above-reported cases and legislation and the picture they give of the legal boundaries of human beings are not and do not have any pretence to be exhaustive. In addition, we can remind brain-machine-web connections and the creation of cyborgs that are now less a futuristic issue, given that new organic/synthetic interfaces allow computers to read, interpret and interact with human nerve fibres. This represents new opportunities for people affected by diseases such as muscular dystrophy. Moreover, unforeseen chances have opened up for healthy people wanting to improve their level of communication and mental performance.

As a matter of fact, the list of technological applications to human body lengthens every day more and, in the end, we should accept that *we are not disconnected entities, but informational organisms (inforgs), who share with other kinds of agents a global environment, ultimately made of information, the infosphere (Turing revolution).*[32]

From the legal point of view, the issue could be framed in terms of sovereignty, a question which is often overlooked: the boundary litigation on shaping human individuals. Of course, the issue is extremely broad and encompasses the more traditional challenges as well as the greater and more complex challenges arising from biotechnologies, informatics, cognitive sciences, nanotechnologies and their convergence. In any case, both old and new technological applications on the human body challenge old naturalistic assumptions of bodily boundaries. Coming

---

[32] Floridi (2008), p. 95.

back to neuroscience, it is confirmed that the impact of neuroscience on law is factually and conceptually wider than usually considered under *the free will approach*.

It seems to me that the neuro-induced redefinition of the biological and mental boundaries of any individual is the most critical point. Of course this is not new and neuroscience is only the latest reason for reopening the issue. In recent decades, human biological limits look like a field where battles are fought, peace treaties signed and boundary lines drawn. All these, of course, depend on the extensive application of human genetics, new biotechnologies and medical techniques in health services, biological research and society. All human individuals, as biological entities, are deeply affected by these developments. Advances such as artificial ventilation, new resuscitation techniques, artificial nutrition and hydration have prolonged people's lives. Assisted reproduction techniques have widened the opportunities of bearing children. Furthermore, individuals are given the possibility to radically change their physical state even when there is no disease (at least in traditional terms), such as in the case of transsexuals. As a result, the possibility of re-determining human biological limits, and selecting options, is further increased.[33]

The contribution of neuroscience to the battlefield might be summarized in the following terms. If the question of individual boundaries is given priority, the age-old question of free will may no longer be at the forefront. Furthermore, we may discover that an individual's will is intertwined with those of other people (or even machines). If so, should we move from the concept of an *individual (supposed free) will* towards a *social group's will*? Are associations like this merely temporary? Is immediate opting-out guaranteed? Who is the individual that will opt out? Will it be the "who" that freely decided to join the association? Or will it be the "new who" that is the result of the experience of association? Are they the same person? In other words, does the individual identity survive the association? Is the free, informed decision to associate a sufficient guarantee? Or should a guardian be appointed in order to assure both the freedom of (initial) decision and the respect for the conditions necessary for opting out? Who is the sovereign in such decisions?

Questions like these, even though not new in philosophical debate,[34] are now on the legal agenda and require a social response. In other words, once *nature* is out of play and is no longer able to tell us what to do or not to do, the question *Who has the power and is entitled to draw the boundary line for each individual?* becomes crucial. At the moment, there is no better idea than to entitle and empower each individual.[35]

---

[33] Amedeo et al. (2007).

[34] Parfit (1984).

[35] This is the conclusion we reached in Santosuosso and Bottalico (2009), Art. 46.

References

Amedeo S, Valentina S, Pavone IR (2007) Drawing the boundary lines of humans: in whose Bailiwick? In: Derecho y Religiòn, vol. II, pp 11–36

Anderson M, Anderson SL (2010) Robot be good: a call for ethical autonomous machines. Scientific American, October 2010

Borsook D, Moulton EA, Schmidt KF et al (2007) Neuroimaging revolutionizes therapeutic approaches to chronic pain. Mol Pain 3:25

Borsook D, Sava S, Becerra L (2010) The pain imaging revolution: advancing pain into the 21st century. Neuroscientist 16(2):171–85

Boscarato (2011) Who is responsible for robot's actions? An initial examination of Italian law within a European perspective, paper submitted to "*TILT*ing Perspectives 2011 – Technologies on the stand: Legal and ethical questions in neuroscience and robotics" Tilburg University, The Netherlands, April 2011

Camporesi and Bottalico (2011) Can we finally 'see' pain? Brain imaging techniques and implication for the law. J Conscious Stud (Forthcoming)

Floridi L (1999) Philosophy and computing. An introduction. Routledge, London, pp 132–216

Floridi L (ed) (2008) Philosophy of computing and information. 5 Questions. Automatic Press/VIP, Copenhagen, p 95

Glannon W (2007) Defining right and wrong in brain science. Essential readings in neuroethics, Dana Foundation Series on Neuroethics. Dana Press, New York

Goldacre B (2010) Lost your libido? Let's try a little neuro-realism, madam. The Guardian, Saturday 30 October, 2010. http://www.guardian.co.uk/commentisfree/2010/oct/30/ben-goldacre-bad-science-neuroscience. Accessed October 30 2010

Kolber AJ (2007) Pain detection and the privacy of subjective experience. American Journal of Law & Medicine 33:433–456

Madeira JL (2006–2007) Regarding pained sympathy and sympathy pains: reason, morality, and empathy in the civil adjudication of pain. S C L Rev 58:415

Nordmann A (Rapporteur) (2004) Converging technologies – shaping the future of European societies, Report 2004, reperibile in: http://ec.europa.eu/research/conferences/2004/ntw/pdf/final_report_en.pdf

Parfit D (1984) Reasons and persons. Clarendon, Oxford

Roco MC, Bainbridge WS (eds) (2002) Converging technologies for improving human performance. Nanotechnology, biotechnology, information technology and cognitive science (National Science Foundation/DOC-sponsored report), June 2002, Arlington, Virginia

Santosuosso A, Bottalico B (2009) Neuroscience, accountability and individual boundaries. In: Front Hum Neurosci 3:45

Tracey I, Bushnell C (2009) *How neuroimaging studies have challenged us to rethink: is chronic pain a disease?* J Pain 10(11):1113–20

Turing AM (1950) Computing machinery and intelligence. Mind 59(236):433–460

Neurolaw in Japan

Katsunori Kai

Abstract In Japan, we are now discussing neuroethics [We can know the detailed contents of neuroethics and the various problems by Illes (Neuroethics-defining the issues in theory, practice and policy, 2006). And concerning the situations of neuroethics in Japan, see Fukushi et al. (Neuroscience Research 57:10–16, 2007)], but have not yet argued on neurolaw in earnest. Right from the beginning, neuroethics in itself is a very new field, which has only begun within the last few years in the world [See Chiaki Kagawa (Gendaishiso (Modernthought), 34(11): 188ff, 2006), Chiaki Kagawa (Gendaishiso (Modernthought), 36(7):69ff, 2008)]. Also neurolaw is a newer field and concept of law, so we are now discussing on the problem of free will, the criminal responsibility, and the problem of the limit of intervention into human brain in the field of human experimentation or enhancement as much as possible. In the field of Bioethics, however, we have accumulations of arguments on neuroethics in bioethics in Japan.

Therefore, in this paper I must start to follow the situations of arguments on neuroethics in Japan, and then advance toward legal issues in the field of neuroscience in Japan, and finally consider the way to legal regulation. The decisive question is whether it is possible to shift from neuroethics to neurolaw in Japan.

1 Discussions on Neuroethics in Japan

1.1 Short History of Arguments on Neuroethics in Japan

First, I begin with the short history of arguments on neuroethics in Japan.

K. Kai (✉)
Center for Professional Legal Education and Research (CPLER), Waseda Law School,
1-6-1 Nishi-Waseda, Shinjuku-Ku, Tokyo, Japan
e-mail: kai@waseda.jp

T.M. Spranger (ed.), *International Neurolaw*,
DOI 10.1007/978-3-642-21541-4_12, © Springer-Verlag Berlin Heidelberg 2012

In Japan, the starting point age of neuroethics was in 2005. The workshop of "Neuroscience and Ethics" was held as a project of "Brainscience and Society" by the Japan Science and Technology Agency (JST) in 2005. Originally, this project was derived from the project of "Brain-Science and Ethics" in the middle of 1990s.[1] Especially, it was very impressive to me that Dr. Hideaki Koizumi, who was a Fellow of Hitachi Co., had a lecture on neuroethics in our invitation lecture in the 17th Annual Conference of Japan Association for Bioethics. He presented the scientific and ethical meaning of brainscience and neuroscience from the viewpoint of bird's-eye-view integrationism and trans-disciplinarity,[2] and he pointed out the importance of the role of Brain Machine Interface (BMI) or Brain Machine Optical Interface (BMOI) for ALS- patient's communication by reading his or her mind.[3] At the same time, he emphasized the importance of trans-disciplinarity between brainscience and cultural-social science.[4]

Since 2006, the issue of neuroethics has often been taken up as a new topic in our Annual Conference of Japan Association for Bioethics by Ph.D. Tamami Fukushi's playing a key role. The main discussions were concerning the use of Deep Brain Stimulation (DBS).

In 2007, the report of "Brainscience Renaissance" was published by the Ministry of Education, Culture, Sports, Science and Technology, and emphasized harmonization with society. Furthermore, at Waseda University, the symposium of "Neuroscience and ethics" was held by the ASMeW (Advanced Science and Medical Care, Waseda University) project team in 2008. The contents of this symposium were the following. Opening Speech "Ethics of Neuroscience and Policy" by Prof. Yutaka Hishiyama (Director, Life Science Division, Research Promotion Bureau, Ministry of Education, Culture, Sports, Science and Technology), Invited Lecture "The Diversity and Possibility of Neuroethics" by Tamami Fukushi, Ph.D. (Research Institute of Science and Technology for Society (RISTEX), JST), Lecture "Functional Analysis of the Candidate Gene Product for Schizophreinia: an Approach to the Molecular mechanism of Schizophreinia" by Naoya Sawamura Ph.D. (Domain of Molecule-Based Medical Treatment, ASMeW), Lecture "The Role of Psychotherapy in the Treatment of Panic Disorder" by Shuhei Izawa, (Domain of Medical Care, ASMeW), Lecture "Neuroscience and Robotics; Ethical and Social Implications" by Massimiliano Zecca, Ph.D. (Domain of Robotics for Medical Care, ASMeW), Panel Discussion (Coordinator: Assoc. Prof. Naoto Kawahara, Domain of Ethics for Science and Bioengineering, ASMeW), and General Lecture by Prof. Shigetaka Asano, M.D., Ph.D. (Domain of Ethics for Science and Bioengineering, ASMeW) with Opening Remarks and

---

[1] See Koizumi (2006), p. 12ff.; Kagawa (2006), p. 188ff.; Kagawa (2008), p. 69ff.

[2] See Koizumi (2006), p. 12ff.

[3] See Koizumi (2006), pp. 23–25.

[4] See Koizumi (2006), pp. 25–28.

Closing Remarks by Prof. Toru Asahi, Ph.D. (Institute for Biomedical Engineering. Executive Office Manager, ASMeW).[5]

From this point the discussions on neuroethics in Japan have become more intense; still, the intensity is not yet that high, due to the fact that there are few actors in the discussion.

1.2 Outline of Current Situations of Arguments on Neuroethics in Japan

Second, what are the current situations of the ethical issues concerning neuroethics then in Japan? There are some aspects in this field. I will show the outline of them.

Ethical status of neuroethics is not clear in relation to bioethics. Is neuroethics independent from bioethics or a part of bioethics? Some specialists say it is independent from bioethics and therefore a new ethics, and others place it as a part of bioethics.[6] If neuroethics is independent from bioethics, what is the difference from bioethics?

Prof. Tatsuya Mima, who is a specialist of neuroscience in Japan, points out that neuroethics is in connection with enhancement or cyborg beyond traditional bioethics, and has no aspect of "movement" with social impact.[7] And Prof. Yoko Matsubara points out that we cannot yet find out "victim" in the field of neuroethics differently from bioethics, and it is very difficult in principle for to draw a line up to which there can be "permissible human experimentation".[8] Furthermore, Mr. Mizuki Matoiba (political thought) points out that bioethics tries to set ethics to regulate scientific technology from outside of it, or to the contrary, neuroethics tries to deduce ethics from scientific technology in itself.[9] Furthermore, there is a large problem e.g. whether a change of personality is to be considered as a benefit for the human subject in BMI.[10]

These opinions seem very important to me when thinking about neurolaw. However, we must add to them further two points which should be taken into consideration to connect them with the discussion of neurolaw. One is whether it is invasive or not. And the other is a fundamental problem, like Mr. Jiro Nudeshima (scientific policy) emphasizes, whether we can originally clarify the function of brain by means of present brain image processing.[11] Concerning the latter, of course, I do not know the scientific prospect of it.

---

[5] See Ethics for Bioscience and Bioengineering, vol. 5, (2009).

[6] See Matsubara and Mima (2008), p. 50ff. This talk is very stimulating.

[7] Mima (Matsubara and Mima 2008), pp. 54–55, and p. 60.

[8] Matsubara (Matsubara and Mima 2008), p. 60 and p. 65.

[9] Matoiba (2008), p. 138.

[10] Mima (Matsubara and Mima 2008), p. 66.

[11] Nudeshima (2008), p. 157.

2 Legal Issues in the Field of Neuroscience in Japan

Now in Japan, we do not have enough legal approaches to neurolaw. Therefore, we should consider neurolaw from the viewpoint of Japanese law based on the above ethical discussions.

2.1 Informed Consent and Protection of Patients or Human Subjects

First, we should start from taking legal rules of human experimentation or clinical research into consideration.[12] Especially informed consent and protection of patients or human subjects are very important.[13] However, informed consent is very complicated in this field.

In Japan, we had a relevant civil case called as "Lobotomy Case" in 1978.[14] In this case, X was rough and made acts of violence to his wife, therefore he was hospitalized in Y hospital due to his mental disorder in 1973. And then he was compelled to have an operation of lobotomy, which was a kind of operation by cutting a part of the frontal lobe, without his own consent. As a result, he had an aftereffect of frontal lobe syndrome, and claimed damages to Y including the surgeon. On 29th September 1978, Sapporo District Court decided that this operation had been unlawful medical treatment without the patient's consent. Although the court did not recognize it as unlawful human experimentation, it was decisive for the Japanese legal system that it declared the operation without the patient's consent as unlawful also in such a case. In my opinion, this case is a kind of unlawful human experimentation. However, this was a civil case, but generally speaking, arbitrary medical treatment can consist of a charge of inflicting bodily injury in Japan.

Taking this case into consideration, we should think about legal issues on neuroscience. When the patient or the human subject is competent, the doctor should give him or her accurate information and then get consent to this treatment from him or her. Even so, there is a problem whether the technique has an accurate predictability to the result or not. In the described case, as a rule the treatment is legally justifiable due to his or her consent. However even in such a case, if the identification of his or her personality is remarkably changed, such treatment does not automatically seem justifiable. The reason is that this acceptance leads to permission of enhancement which can contravene human dignity. Although

---

[12] See Kai (2005).

[13] See Miller and Fins (2006), p. 210ff.; Illes (2006), everywhere.

[14] Hanreijiho 914, p. 85; Hanreitaimuzu 368, p. 132. These journals are Japanese documents of judicial precedents.

of course, as Prof. Tsuyoshi Awaya (medical law, bioethics) points out,[15] it is not appropriate for us to decide a priori alternatively the problem by "Yes" or "No" we should continue thinking about human dignity in this field; in this context, the identification of personality seems to be very important to me. In the case of remarkably changing the identification, his or her consent can be legally invalid. However, even if the technique had any harmful effects connected to an arbitrary medical treatment, we cannot always impose civil or criminal responsibility on the doctor.

On the other hand, when the patient or the human subject is incompetent, things are complicated because his or her mental illness may be serious. Can the doctor intervene in his or her brain by force without consent or by proxy consent? It is not so easy to justify such a treatment, because intervention in his or her brain is such a serious infringement of the legal interest concerning personality that we cannot justify it very easily. However, there can be permissible exceptional cases under certain strict conditions, where the treatment might be "ultima ratio" and in best the interest for the patient or the human subject.[16]

2.2 Criminal Responsibility and Problem of Free Will

Second, in the field of criminal law, we can consider neuroscience in relation to the problem of free will.[17] The problem of free will had been discussed as an important argument of "Modern School" and "Classical School" in the field of criminal law in Japan modeled Italy and Germany till the middle of 1990s. The former insisted that free will did not exist from the view point of determinism. To the contrary, the latter insisted that free will existed from the viewpoint of indeterminism and principle of culpability (nulla poena sine culpa = no punishment without culpability).[18] As a rule at the present time, the latter is rather predominant, but also the third position of soft determinism (theory of relative free will) is influential in Japan.

I think that these arguments are in relation to neuroscience. Generally speaking, determinism trends to accept to intervene in brain as a social treatment or a measure for preserving public health. However, I think that we should not approach neuroscience, and therefore neurolaw from this viewpoint, because it can lead to fatalism, which enables psychiatric patients to be a simple means to defend a society. Thus, we should approach neuroscience in harmonization with human right of these patients by keeping free will.[19]

---

[15] Awaya (2004), p. 180.

[16] To this point, see Greely (2006), p. 399ff.

[17] See Shimada (2009), p. 225ff; Masuda (2007), p. 9ff. Incidentally, see Hilenkamp (2006).

[18] To details of a principle of culpability, see Kaufmann (1976).

[19] Akiba (2010), p. 242.

2.3 Protection of Personal Data

Third, we should consider protection of personal data concerning the patient or the human subject. Especially, personal neurodata which are gained from reading mind, etc., are very sensitive information as well as genetic data. They are different from general medical information in the protection and use of them.

Recently, in Japan, we are discussing the legal protection and the use of genetic information.[20] Genetic information is often used not only in medical areas, but also in commercial areas. In the field of medicine, the number of genetic tests just keeps on increasing, and at the same time, the legal and ethical issues concerning uses of genetic test attract attention. However, in Japan, we have no specific legal system of protection and use of genetic information and medical information. And genetic information is not legally and clearly ranked.

Furthermore, also in the area of criminal investigation, we do not have the legal system of the use of DNA. We have only some guidelines concerning genetic information in Japan; e.g. Guideline of the Japan Society of Human Genetics (non-official guideline), Ethics-Guideline for Human-Genome/Gene Analysis Research (2001, revised 2004 by Ministry of Education, Culture, Sports, Science and Technology; Ministry of Health, Labor and Welfare; Ministry of Economy, Trade and Industry: official guideline; Guideline for the Protect of Personal Information for Business Operations Handling Personal Genetic Information (2004; Ministry of Economy, Trade and Industry: official guideline).

Thus, we can get some principles from these guidelines; (1) informed consent by documents, (2) genetic counseling, (3) setting up committee, (4) specifying strictly the aim of use, (5) prohibition of getting sensitive information, (6) safe risk management including anonymity of materials, (7) general prohibition of providing it to the third party, (8) withdrawal of consent, (9) setting up the window for consultations. But I think that we should make the legislation for the legal protection and the use of genetic informations. In the legislation, we should consider avoiding genetic discriminations in the areas of employment, insurance and marriage, etc. And we should protect genetic information from abuses of them, because it belongs not only to the person him or herself, but also to their family members.

It is true that genetic information is not the same as neurodata, but I think that these factors are applicable also in the framework or area of neurolaw in Japan. We can appropriately manage neuroscience in the neurolaw by using this framework.

---

[20] See Kai (2007).

3 The Way to Legal Regulation: From Neuroethics to Neurolaw – Is It Possible in Japan ?

3.1 Fundamental Stance

The area of life science is very dynamic and flexible. Also neuroscience is one of them. Aldous Huxley had already such a symbolic novel "Brave New World" in connection with in-vitro-fertilization in 1932, and recently, we have just known a new symbolic invention of "induced pluripotent stem cell (iPS Cell)" by Prof. Shinya Yamanaka of Kyoto University in Japan in 2007, by which we have had just possibilities to use "regenerative medicine" or "tissue engineering" without breaking human embryos like in case of using "embryonic stem cell (ES Cell)". Then many efforts are made to overcome the risk of cancer which will derive from the technique of iPS Cell.

In the post-genome era, it may cause disadvantages for mankind that the law tries to regulate scientific and medical activities of these fields extensively because it can obstruct the progress of life science or medicine. It is true that the freedom of study and research is guaranteed by the Art. 23 of the Constitution in Japan on the one hand. But on the other hand, we must examine carefully whether this freedom is unlimited or not. Prof. Koichi Bai, who is the founder of medical law in Japan, had already pointed out some important fundamental perspectives on this aspect in 1974, among those:

1. Awareness of the margin of legal intervention into natural facts and progresses of natural science,
2. Role of law in adjusting conflict between one interest and the other interest, and
3. Awareness of positive meaning of legal approach, or guarantee and establishment of fundamental rights.[21]

These perspectives seem to be very useful even today. We must consider the balance between promotion of life science or medical science and protection of human right in this field. Thus, we must rethink how we should regulate illegal misconducts in this field.

3.2 Objects of Legal Regulation

We can classify objects of regulation into three categories in this field. The first is objects to regulate clearly; e.g. crimes, social harmful conducts (trafficking), and abuse of eugenics, discrimination. We should legally prohibit these conducts due to

---

[21] Bai (1974), p. 197ff. Especially pp. 200–201.

being so harmful to our society, and therefore impose criminal sanction on these conducts in certain cases.

The second is objects to promote; e.g. genome research. Naturally, it needs due process in going on the study plan, but it is not necessary to regulate it legally.

The third is objects to permit with conditions; e.g. therapeutic cloning, use of ES-cell, stem cell, and iPS-cell. As we cannot predict or foresee concretely any risks, we should watch these researches with certain conditions. Neuroscience belongs to this category. We can hope that they may bring about possibilities to cure some curable diseases in near future. Naturally, also it needs guarantee of due process in going on the study plan, but it is not necessary to regulate legally.

3.3 Grounds of Regulation

What can we think about the ground of the regulation? In my opinion, first it should be based on "human dignity", which derives from German philosopher Immanuel Kant. "Human dignity" is in "Sein mit Menschen-Dasein" and should be behind each human being, human tissues, corpse and human embryo.

Then what should we think about criminal regulation? Criminal regulation is the last means (ultima ratio). There are some fundamental principles in applying criminal law. Incidentally, Japanese criminal law has been strongly influenced from German criminal law.

The first principle is the so-called "Tatprinzip" (conduct-principle in English). According to this principle, we cannot punish a conduct without certifying an external harmful conduct. It includes causation. In Anglo-American jurisdiction, it is concerned with actus reus.

The second principle is "Nulla poena sine lege, nullum crimen sine lege" (No penalty without law, no crime without law). According to this principle, we cannot punish a conduct without a clear provision of law.

The third principle is the so-called"Schuldprinzip "(principle of culpability: Nulla poena sine culpa; No penalty without culpability). According to this principle, we cannot punish a conduct without intention or negligence, and criminal responsibility). In Anglo-American jurisdiction, it is concerned with mens rea.

These three principles should be considered into also in the field of medical science, life science, and neuroscience. At least, we should not use criminal sanction to such cases in which people feel or have merely vague and slight misgivings in this field.

3.4   Model of Regulation: Proposal of Legal Doctrine of Medical Due Process

Then what should we think about a future model of regulation? We can classify it into three categories. The first is the hard law style like in Germany. For example, the German Embryo Protection Act of 1990 is typical of it, because it is a special criminal law. I think, however, that the German legal system is not suitable for regulation to these fields because it is too hard to keep up flexibly with the trend of life science or neuroscience. Indeed in Germany, the Stem Cell Act has been enacted in 2002 (revised in 2008), and by this law, they have been able to use human stem cell for research in Germany. However, it seems strange that they can use only stem cell which is imported from foreign countries.

The second is the soft law style like in Japan. We have many official guidelines in this field in Japan; for example, Ethics-Guideline for Human-Genome/Gene Analysis Research (2001, revised 2004 by Ministry of Education, Culture, Sports, Science and Technology; Ministry of Health, Labour and Welfare; Ministry of Economy, Trade and Industry), the Guideline for the Protect of Personal Information for Business Operations Handling Personal Genetic Information (2004; Ministry of Economy, Trade and Industry). The last Guideline is to enterprises (except use for research), which includes (1) informed consent by documents, (2) genetic counseling, (3) setting up committee, (4) specifying strictly the aim of use, (5) prohibition of getting sensitive information, (6) safe risk management including anonymity of materials, (7) general prohibition of providing it to the third party, (8) withdrawal of consent, (9) setting up the window for consultations.

However, these guidelines have no legal sanctions; therefore, they cannot ensure more effectiveness to exclude remarkable abuses. And as they are so-called a kind of patch work, we cannot understand the fundamental viewpoint. Thus, this model is not enough suitable in this field although they are flexible.

The third is the mixed style of hard and soft law like in UK and Australia, etc. The Human Fertilization and Embryology Act 1990 (HFEA 1990) and the Human Tissue Act 2004 are typical of it, and furthermore they are supplemented by some guidelines. According to this model, we can normally correspond with various new medical and scientific technologies and problems.

Thus in Japan, we should aim at this mixed type between hard law and soft law. And yet, we should consider " the Doctrine of Medical Due Process".[22] This is the legal theory which I have insisted for a long time. According to this theory, as a rule, medical innovation/medical research without due process is unlawful. And Medical Due Process contains (1) informed consent, (2) balancing between risks and benefits, (3) due review by appropriate ethical committee, and (4) compensation system to human subjects because we cannot foresee concrete risks. Furthermore, (5) it contains some exceptional legal sanctions to extreme abuses. Due to

---

[22] See Kai (2006), p. 7f. and p. 30ff.

this doctrine, we can build a bridge between law, bioethics, neuroethics and medical and neuroscientific research and practice. I think that we can realize better it by enacting the Fundamental Law of Bioethics including neuroethics in Japan.

4 Conclusion

Nowadays we should trans-nationally consider the problems of this field, because for example the biobank system has become more and more important in the world. The first thing we should have to do is to make the Fundamental Law of Bioethics in Japan in harmonization with foreign countries. We are now preparing this draft with Prof. Ryuichi Ida (Kyoto University).[23] Concerning to important points in bioethics or neuroethics, we should make a fundamental legal system. The Fundamental Law of Bioethics will be in the center of bioethics or neuroethics. Thus, I think it better that the model of regulation on medical innovation/medical research including neuroscience should be the mixed type of hard law and soft law, that is to say, four steps which consist of public guideline(soft law), civil regulation, administrative regulation, and finally criminal regulation(hard law).[24] Thus, we can establish the neurolaw system in Japan.

References

Akiba E (2010) Psychiatric medicine (in Japanese). In: Kai K (ed) Lecture: bioethics and law (in Japanese). Horitsubunnkasha, Kyoto, p 242
Awaya T (2004) Does mankind begin to have his own wing? In: Nishinihon Seimeirinri Kenkyuukai (West Japan Bioethics Study Group) (ed) Toward remaking bioethics: perspectives and subjects (in Japanese). Seikyuusha, Tokyo, p 180
Bai K (1974) Science, law and life (in Japanese). In: Matsuo T (ed) Life science note (in Japanese). Tokyo University Press, Tokyo, p 197ff
Fukushi T, Sakura O, Koizumi H (2007) Ethical considerations of neuroscience research: the perspectives on neuroethics in Japan. Neurosci Res 57:10–6
Greely HT (2006) The social effects of advances in neuroscience: legal problems, legal perspectives (by translation in Japanese). In: Illes (ed) Neuroethics-defining the issues in theory, practice and policy. Oxford University Press, New York, p 399ff
Hilenkamp T (ed) (2006) Neue Hirnforschung-Neues Strafrecht? (in Germany). Nomos, Baden-Baden
Ida R (2009) Bioethics and law in the post-genome society (in Japanese). In: Kai K (ed) Post-genome society and medical law (in Japanese), A series of medical law, vol 1. Shinzansha, Tokyo, p 211ff

---

[23] See Ida (2009), p. 211ff.

[24] See Kai (2009), p. 191ff.

Illes J (ed) (2006) Neuroethics-defining the issues in theory, practice and policy. Oxford University Press, New York (Translation into Japanese by Takahashi T, Kume K (ed), 2008), Shinohara-Shuppannshinsha, Tokyo)

Kagawa C (2006) Newness of neuroethics (in Japanese). Gendaishiso (Modernthought) 34(11): 188ff

Kagawa C (2008) Critisism to Balkanierung Bioethics and Neuroethics (in Japanese). Gendaishiso (Modernthought) 36(7):69ff

Kai K (2005) Protection of human subjects and criminal law (in Japanese). Seibundo, Tokyo

Kai K (2006) Importance of bio-ethics in brain-science research (in Japanese). J Jpn Assoc Bioethics. 16(1):12ff

Kai K (ed) (2007) Genetic information and legal policy (in Japanese). Seibundo, Tokyo

Kai K (2009) Model of regulation on medical innovation/medical research from the perspective of comparative law (in Japanese). In: Kai (ed) Post-genome society and medical law (in Japanese), A series of medical law, vol 1. Shinzansha, Tokyo, p 191ff

Kaufmann A (1976) Das Schuldprinzip-Eine strafrechtlich-rechtsphilosophische Untersuchung, 2. Aufl., Carl Winter Universität, Heidelberg (translation into Japanese by Katsunori Kai, 2000, Kyushu University Press, Fukuoka)

Koizumi H (2006) Importance of bio-ethics in brain-science research (in Japanese). J Jpn Assoc Bioethics 16(1):12ff

Masuda Y (2007) Arguments concerning results of brain science and criminal responsibility (sequel) (in Japanese). Horitsuronso 79(6):9ff

Matoiba M (2008) Neuroethics as political theory (in japanese). Gendaishiso (Modernthought) 36(7):138

Matsubara Y, Mima T (2008) A talk: creation of neuroethics (in Japanese). Gendaishiso (Modernthought) 36(7):50ff

Miller FG, Fins JJ (2006) Protecting human subjects in brain research: a pragmatic perspective (by translation in Japanese). In: Illes (ed) Neuroethics-defining the issues in theory, practice and policy. Oxford University Press, New York, p 210ff

Nudeshima J (2008) Can be brainscience "non-invasive"? (in Japanese). Gendaishiso (Modernthought) 36(7):157

Shimada M (2009) Theory of free will and neuroscience (in Japanese). Report of graduate school of law in Chuo University, No. 38, p 225ff

Neuroscientific Evidence and Criminal Responsibility in the Netherlands

Laura Klaming and Bert-Jaap Koops

Abstract Insights from neuroscientific research are increasingly advancing our understanding of the neural correlates of human behaviour, cognition and emotion and can therefore be of significant practical use in a legal context. One of the most fundamental legal applications of neuroscience refers to the assessment of criminal responsibility. Recent empirical studies have established links between certain brain structures and antisocial or criminal behaviour. Three areas of brain abnormalities that are relevant for assessments of criminal responsibility can be differentiated: (1) impairments in the frontal lobes and associated problems with impulse control, aggressiveness and the processing of information that is evocative of moral emotions, (2) abnormalities in the limbic system and associated problems in affective processing and (3) the potential side effects of neurotechnologies and associated problems with impulse control, aggressiveness and disinhibited behaviour. This chapter addresses recent research findings in these three areas and how these could affect responsibility assessments. In addition, eight cases are discussed in which insights from neuroscientific research have been used by Dutch courts in responsibility assessments. By illustrating how neuroscientific evidence has already entered the courtroom in the Netherlands, the possible conditions and implications of such practice are addressed.

1 Introduction

Neuroscience is rapidly increasing our knowledge of the functioning of the brain. Recent studies have, for instance, shown that reduced prefrontal volume and abnormal activity in the prefrontal cortex – which is an area in the brain that is

L. Klaming (✉) • B.-J. Koops
Tilburg Institute for Law, Technology, and Society, Tilburg University, PO Box 90153, 5000 LE Tilburg, The Netherlands
e-mail: L.Klaming@uvt.nl

T.M. Spranger (ed.), *International Neurolaw*,
DOI 10.1007/978-3-642-21541-4_13, © Springer-Verlag Berlin Heidelberg 2012

involved in various cognitive processes and executive functions including attention, working memory, planning and decision making – are associated with antisocial behaviour (Barkataki et al. 2006; Brower and Price 2001; Laakso et al. 2002; Raine et al. 1997, 1998; Sapolsky 2004; Volkow et al. 1995). Additionally, research has revealed that the frontal cortex plays an important role in moral reasoning and response inhibition (Greene et al. 2004; Horn et al. 2003; Liddle et al. 2001; Moll et al. 2002) and is associated with intentional behaviour (Haynes et al. 2007). For instance, Moll and his colleagues (2002) demonstrated that certain areas in the prefrontal cortex and superior temporal sulcus – which is an area in the brain that is involved in tasks such as the interpretation of other people's actions and intentions, the comprehension of speech and the perception of biological motion – are engaged when individuals view scenes that are evocative of moral emotions, which suggests that these brain regions are involved in moral processing. Damage to these areas may therefore compromise an individual's ability to know right from wrong or to act upon knowledge of the morality of behaviour. Neuroscience research has furthermore indicated that certain brain areas that are associated with high-level executive functions including areas in the frontal cortex are more active during deception, which has been argued to be due to the increased cognitive effort involved in lying (Langleben et al. 2002; Spence et al. 2004).

These and other insights from neuroscientific studies can be of relevance to the law. As outlined by Goodenough and Tucker (2010), the interactions between neuroscience and the law can be grouped into the following three categories: (1) *the law of neuroscience*, (2) *neuroscience of the law* and (3) *neuroscience in law*. The first category refers to the regulation of neuroscientific research and applications and focuses on issues such as informed consent, privacy and dealing with incidental findings in experimental research involving human participants. Additionally, advances in neuroscience are believed to have implications for intellectual property issues such as the patenting of mental processes (Goodenough and Tucker 2010; Greely 2004; Tovino 2007). The second category of neurolaw refers to the neuroscience of normative judgement and decision-making. Research on neuroscience of the law has studied the neural correlates of normative and legal judgement (Casebeer and Churchland 2003; Goodenough and Prehn 2004; Schleim et al. 2010) and decisions about punishment (Buckholtz et al. 2008; Knoch et al. 2010; Seymour et al 2007). Research in this area has, for instance, demonstrated that legal judgements involve other brain areas as compared to moral judgements, namely areas associated with reflecting explicit rules (Schleim et al. 2010). The third category of neurolaw refers to the study of cognition and behaviour relevant to the law, for instance truth-telling and memory, impulsive behaviour (Barkataki et al. 2008; Raine et al. 1998), moral reasoning (Greene et al. 2004; Moll et al. 2002), psychopathy (Kiehl et al. 2001), and drug addiction (Hyman et al. 2006). These findings can be of great practical use in the legal context, for instance to detect deception (Kozel et al. 2005; Langleben et al. 2002; Spence et al. 2001: Wolpe et al. 2005), to develop treatment options for criminal behaviour (Greely 2008), to enhance eyewitness memory (Klaming and Vedder 2009; Vedder and Klaming 2010) and to detect physical and emotional pain and suffering (Grey 2007; Jones et al. 2009; Kolber 2007; Ochsner et al. 2006; Peyron et al. 2000; Tovino

2007). One of the most fundamental legal applications of neuroscience within this third category refers to the possibility of determining criminal or civil responsibility,[1] i.e. the degree to which a person can be held legally accountable for her actions. Since up to now the use of neuroscience in a legal context is confined to assessments of responsibility, we have chosen this as the topic of focus in this chapter to illustrate the potential implications of neuroscience for the law.

There are two ways in which neuroscience can impact on criminal responsibility. The first is on the theoretical and most general level: do or should the insights appearing from neuroscientific research alter our conception of criminal responsibility? As research seems to suggest that decisions are taken in the brain before the subject is conscious of them, can we then still hold the subject accountable for her actions, in the most general and fundamental sense? A significant body of research is discussing notions of responsibility and accountability, free will, and consciousness in light of neuroscientific research (e.g. Aharoni et al. 2008; Greene and Cohen 2004; Morse 2004, 2007; Roskies 2006; Sapolsky 2004; Vincent 2010a). While some researchers believe that advances in neuroscience will show that neuroscience challenges our ideas of free will and will therefore have to lead to changes in the law (Greene and Cohen 2004; Sapolsky 2004), others have argued that existing legal principles and practice can accommodate whatever new information neuroscience will provide, mainly because there is no brain correlate of responsibility (Gazzaniga 2005; Morse 2004, 2007). Instead of being a property of an individual, responsibility is a normative concept (Gazzaniga 2005). Still others have argued that neither neuroscience nor criminal responsibility are unified concept, which complicates answering the question whether neuroscience is relevant to the law (Vincent 2010a).

The second way in which neuroscience impacts on criminal responsibility is on a more applied and specific level: if neuroscientific research shows abnormalities in brain functioning of a specific subject, for example in areas associated with morality or impulse control, the question arises whether the subject can be held accountable for specific acts she committed. Research on the underlying neural mechanisms of moral judgement and intentional behaviour is still in its infancy, and the responsible use of neuroscientific evidence to answer questions of liability therefore seems questionable at this point in time. Nevertheless, the fact that recent empirical studies have established links between brain structures and antisocial or criminal behaviour warrants the assumption that neuroscience could play a more important role in assessments of criminal responsibility in the future. In fact, neuroscientific evidence has already been used in order to argue for diminished responsibility in various cases both in the Netherlands and internationally. Additionally, new technologies such as Deep Brain Stimulation that are intended to treat neurological and psychiatric disorders raise new concerns for questions of criminal responsibility.

Contrary to the general and fundamental discussion of criminal responsibility, the use of neuroscientific evidence of brain abnormalities in concrete criminal cases has been less a topic of research so far, even though it is more immediately relevant

---

[1] The terms *responsibility*, *accountability* and *liability* will be used interchangeably in this chapter.

to actual practice of criminal law. For these reasons, we focus our discussion in this chapter on the second type of question: how does neuroscientific evidence impact the attribution of criminal responsibility in concrete cases? Since this question depends to a significant extent on the theory and practice of the specific legal system at issue, we base our discussion on Dutch law.

Neuroscientific research of brain abnormalities that are relevant for assessments of criminal responsibility can be broadly separated into three main areas. The first line of research focuses on impairments in the frontal parts of the brain and associated problems with impulse control and aggressiveness as well as with the processing of information that is evocative of moral emotions. The most frequently cited case of an individual with frontal lobe damage is that of Phineas Gage who, after having suffered severe damage to his left ventromedial frontal cortex caused by an accident, changed as a person. While his intelligence remained unaffected and he was not impaired in movement, speech and memory, he had lost any social inhibitions, became capricious and indulged in profanity which often offended those around him (Damasio et al. 1994). This case demonstrates the importance of the frontal parts of the brain in self-control and the ability to act appropriately in social situations. The second line of research studies limbic abnormalities and affective processing. The limbic system is a network of brain regions that includes the hippocampus, hypothalamus and amygdala and is involved in various functions such as memory and regulating emotion. In the case of Brian Dugan, who was found guilty for murdering a 10-year-old girl, evidence pertaining to his malfunctioning limbic system was presented as a mitigating factor in the sentencing phase of the trial. The expert witness assigned by Dugan's defence lawyers argued that Dugan, like other psychopaths, had reduced levels of activity in his limbic system (Hughes 2010). Neuroscientific research has indicated that psychopaths show less activity in limbic structures when they are presented with affective materials (Kiehl 2006; Kiehl et al. 2001). Apparently, anomalies in affective processing such as a lack of empathy which are often found in criminal psychopaths may be linked with inadequate activation of the limbic system.

In addition to the link between antisocial behaviour and frontal lobe impairments as well as reduced activation of limbic structures, we believe that there is a third link between activities in certain brain structures and criminal behaviour that is relevant in discussions about neuroscience and responsibility. The application of neurotechnologies including psychopharmaca and Deep Brain Stimulation (DBS) can have unwanted side effects. These technologies may in rare cases induce changes in a patient's personality which may conceivably result in criminal behaviour (Breggin 2003/2004; Healy et al. 2007; Klaming and Haselager 2010; Okado and Okajima 2001). It is, for instance, known that certain antidepressants can have effects such as mania, agitation and akathisia, which is an inner agitation that typically manifests itself in the inability to stop moving and that has been associated with increased aggressiveness (Breggin 2003/2004; Healy et al. 2007; Okado and Okajima 2001). These behavioural reactions can result in violence and other forms of abnormal behaviour, including abnormal sexual behaviour in some patients, especially in the initial period of taking the medication. In addition to psychopharmaca, DBS can have unwanted side effects that may in rare cases lead to criminal behaviour. DBS is a is a well-accepted treatment for movement disorders, including Parkinson's disease (PD), Dystonia and Essential Tremor, if symptoms are medically intractable and/or medical treatment has serious side effects (Houeto et al. 2002; Limousin et al. 1998; Weaver et al. 2009). It is also currently explored as a treatment option for a variety of neurological and psychiatric disorders (Gabriëls et al. 2003; Mayberg et al. 2005; Sturm et al. 2003). Although to our knowledge no case in which a DBS patient became criminal as a result of the treatment has yet been described, side effects have been reported in the literature, such as increased impulsivity and aggressiveness (Hälbig et al. 2009; Houeto et al. 2002; Sensi et al. 2004; Frank et al. 2007) or inappropriate sexual behaviour (Houeto et al. 2002; Leentjens et al. 2004), both of which could result in wrongful or even criminal behaviour. We therefore believe that certain treatments that affect brain functions can cause changes in an individual's behaviour patterns in such a way that he acts in morally or legally questionable ways. Consequently, the use of neurotechnologies such as certain psychopharmaca and DBS and their impact on an individual's brain pose important questions for criminal responsibility.

In light of these three distinct areas of neuroscientific research, we will describe how recent research findings in these areas are being or could be used to determine a suspect's responsibility. We start this discussion with further examining the potential impact of neuroscientific research into brain abnormalities in relation to criminal responsibility in general (Sect. 2). We will then discuss the same issue, applied to the Dutch legal system and practice. We will briefly explain the concept of criminal responsibility in Dutch criminal trial and forensic practice, and, based on the consecutive stages in which cognitive and other brain functions could be used in the court's decision-making on attributing criminal liability, we will discuss whether and how relevant findings of neuroscientific research on frontal lobe impairments, reduced activation in the limbic system and the potential impact of neurotechnological interventions are or could be used by the Dutch courts. We will illustrate the discussion with case descriptions (Sect. 3). We will conclude by discussing the actual and potential relevance of neuroscientific findings for the assessment of responsibility in Dutch criminal law and possible conditions and implications of such practice (Sect. 4).

2 Neuroscience and Criminal Responsibility

Since insights from neuroscience are greatly advancing our understandings of human behaviour and cognition, it seems only logical that neuroscience can contribute to our understanding of what it means to act intentionally and knowingly. As described above, neuroscientific research on responsibility has focused on the link between frontal lobe dysfunction and criminal behaviour on the one hand and limbic system dysfunction and criminal behaviour on the other hand. Damage to either of these brain regions could be seen as indicative of poorly developed or pathologically disturbed mental capacities and hence of diminished responsibility or complete absence of responsibility. In addition to these two types of dysfunction and their implications for criminal responsibility, we will discuss the role of potential side effects of neurotechnological interventions and their implications for criminal responsibility.

2.1 Frontal Lobe Dysfunction and Criminal Behaviour

The frontal lobes, which are located in the front part of the brain and which are involved in various cognitive processes and executive functions including attention, working memory, planning and decision-making, seem to play an important role in moral judgement and intentional behaviour and therefore might be relevant to responsibility assessments. A link between frontal lobe dysfunction and violent behaviour has been reported in the literature (Brower and Price 2001; Raine et al. 1997, 1998; Sapolsky 2004, Volkow et al. 1995). For instance, research has found that murderers have reduced glucose metabolism in the prefrontal cortex (Raine et al. 1997). This finding suggests prefrontal deficits in antisocial individuals which may predispose them to higher impulsivity and lower self-control. In line with this finding, another study has shown that affective murderers in contrast to predatory murderers and a non-criminal control group had decreased activity in their prefrontal cortex, which suggests that affective murderers are less able to control their aggressive impulses due to insufficient prefrontal regulation (Raine et al. 1998). Hence, dysfunctions in certain areas in the frontal cortex can interfere with an individual's ability to control his impulses and foresee the consequences of his behaviour.

Additionally, studies have revealed that the orbitofrontal regions, including the orbitofrontal cortex and the ventromedial prefrontal cortex, are activated during moral reasoning. For example, when individuals were confronted with pictures that were morally charged, for instance pictures that depicted a physical assault or a war scene, they showed increased activation in the orbitofrontal cortex and superior temporal sulcus (Moll et al. 2002). The involvement of orbitofrontal regions in moral processing and decision-making was replicated in several studies (Casebeer and Churchland 2003; Greene et al. 2004), indicating that these brain regions play an important role in moral processing. Damage to these areas may therefore compromise an individual's ability to know right from wrong.

Based on the findings of empirical research, frontal lobe dysfunction can implicate several deficits in planning and foresight, in moral and social judgement and in the regulation of behaviour; it therefore seems to play a substantial role in explaining some types of criminal behaviour. Damage to the frontal lobes can, for instance, be caused by mild or traumatic brain injuries (for instance due to a motor vehicle accident, a sport accident or a fall) or by a brain tumour. An interesting case in this regard is that of a 40-year-old man with so-called *acquired paedophilia*. The man experienced sudden and uncontrollable paedophilia for which he was ultimately ordered by a judge either to undergo inpatient rehabilitation in a program for sexual addiction or to go to jail. Although he knew that his behaviour was morally wrong and he furthermore had a strong desire to complete the treatment program and to avoid prison, the man was unable to control his sexual urges and was eventually expelled from the rehabilitation centre. Right before his prison sentence, he received a neurological examination because of sudden severe headaches. During this examination, an egg-sized tumour was found in the orbitofrontal cortex. After the tumour had been removed, the inappropriate urges disappeared and the man successfully completed the rehabilitation program. When about a year later the man started secretly collecting pornography again, re-growth of the tumour was detected. Again, after removing the tumour, the urges disappeared (Burns and Swerdlow 2003). This case demonstrates that damage to the frontal lobes, in this case caused by a tumour, can significantly change an individual's behaviour.

Based on this case and the findings of several studies, it seems that frontal lobe damage can cause a number of deficits that might excuse criminal behaviour or lead to treatment rather than imprisonment of perpetrators. However, it is also important to note that most people with some kind of damage to their frontal lobes never show any antisocial or criminal tendencies (Brower and Price 2001). For instance, the literature describes the case of a woman who was found to lack more than 75% of her cerebral cortex and nevertheless did not have any problems controlling her behaviour (Glannon 2005). In addition to this case, the decision of the United States Supreme Court in the case Roper v. Simmons is relevant for the argument that brain abnormalities imply that an individual cannot be held responsible for his actions. In 2004, the Supreme Court reviewed the case of Christopher Simmons, who at the age of 17 had robbed and murdered a woman (Roper v. Simmons). Simmons was initially convicted of first-degree murder and sentenced to death. However, the Supreme Court later ruled that it is unconstitutional to impose the death penalty on individuals under the age of 18 based on the argument that the frontal cortex is not yet fully developed in adolescents and they are therefore incapable of acting rationally and controlling their impulses (Beckman 2004). However, as outlined by Glannon (2005) and others, there are problems with this argument. An abnormality in the frontal cortex does not necessarily imply a lack of impulse control. After all, if Simmons had an immature brain at the age of 17 due to the normal development of the brain, all adolescents have immature frontal lobes and yet only an extremely small number of adolescents ever display any antisocial or criminal behaviour. Hence, abnormalities in the frontal cortex do not automatically imply a lack of impulse control.

2.2 Limbic System Dysfunction and Criminal Behaviour

While the frontal lobes seem to be particularly relevant to responsibility assessments, dysfunction in other brain areas may of course also be associated with violent or criminal behaviour. Studies have, for instance, shown that reduced activity in the thalamus (Barkataki et al. 2008) and a dysfunctional amygdala–hippocampal complex (Kiehl et al. 2001) are linked with antisocial behaviour. Research on psychopathy has shown that psychopathic individuals have decreased activation in the limbic system, a structure in the middle of the brain including the hippocampus, hypothalamus and amygdala, when they view negative affective pictures (Kiehl et al. 2001). The limbic system is involved in various functions, most notably memory and the regulation of emotion. In contrast to normal individuals, psychopaths fail to differentiate between negative and neutral material. They lack empathy, are unable to experience guilt or remorse, are highly manipulative and are typically unconcerned about the consequences of their actions. The finding that psychopaths have decreased limbic activation has been taken as an indication for a biological basis for the affective abnormalities observed in psychopaths. In addition to decreased activation in the limbic system when being confronted with information that is morally charged, psychopaths were found to have increased activity in prefrontal regions (Kiehl et al. 2001; Müller et al. 2003). This finding has been interpreted as supporting the notion that psychopathic individuals use cognitive strategies to process affective information instead (Kiehl et al. 2001). This finding is supported by a study exploring the neural correlates of conscious self-regulation of emotion in healthy individuals (Beauregard et al 2001). The participants in this study watched erotic film excerpts and were either instructed to respond normally or to suppress their emotional responses. In the first group of participants, limbic and paralimbic structures were activated, while in the latter group, prefrontal regions were activated instead, leading the researchers to conclude that humans have the capacity to self-regulate their emotional responses. Additionally, they concluded that "a defect of this neural circuitry [...] may have disastrous psychological and social consequences" (Beauregard et al. 2001, p. 5).

Based on the findings of research on the neural correlates of psychopathy, some researchers believe that psychopathy is as much an illness as, for instance, schizophrenia. However, there is a debate about whether the fact that psychopathy has a biological basis is an argument to excuse or condemn criminal behaviour (Maibom 2008; Miller 2008; Reimer 2008; Vincent 2010b). A diagnosis of psychopathy is therefore currently in some cases (mainly and probably exclusively in the United States) used to argue for mitigation of the sentence rather than for diminished responsibility. The above described case of Brian Dugan is the most well-known case in this regard.

2.3   Neurotechnologies and Criminal Behaviour

Most medical treatments or interventions can have unwanted side effects, some of which may even lead to morally or legally questionable behaviour. For instance, it is well known that antidepressants can have effects such as mania, agitation and akathisia, which is an inner agitation that typically manifests itself in the inability to stop moving and that has been associated with increased aggressiveness (Breggin 2003/2004; Healy et al. 2007; Okado and Okajima 2001). These behavioural reactions can result in violence and other forms of abnormal behaviour, including abnormal sexual behaviour in some patients, especially in the initial period of taking the medication. In the literature, several cases in which individuals without any prior history of aggressive or violent behaviour suddenly behaved violently after taking antidepressants are reported (Breggin 2003/2004; Healy et al. 2007; Okado and Okajima 2001). Additionally, several legal cases linking violent behaviour with the use of antidepressants have been reported (Breggin 2003/2004; Healy et al. 2007; Merckelbach et al. 2009). For instance, Merckelbach and his colleagues (2009) describe the case of a 56-year-old man who had murdered his girlfriend. When confronted with what he had done, the suspect stated that he had problems remembering what exactly had happened, but that he believed that he had probably murdered her. According to the authors, it is possible that the violent behaviour of the suspect was caused by the antidepressant that he had started taking only a few days before the murder. The suspect committed suicide before the court could make a decision in this case (Merckelbach et al. 2009).

In addition to psychopharmaca, invasive neurotechnologies such as deep brain stimulation (DBS) can have unwanted side effects that may in rare cases lead to criminal behaviour. DBS is a well-accepted treatment for movement disorders, including Parkinson's disease, Dystonia and Essential Tremor if symptoms are medically intractable and/or medical treatment has serious side effects (Houeto et al. 2002; Limousin et al. 1998; Weaver et al. 2009); it is currently also explored as a treatment option for a variety of neurological and psychiatric disorders (Gabriëls et al. 2003; Mayberg et al. 2005; Sturm et al. 2003). Although to our knowledge no case in which a DBS patient became criminal as a result of the treatment has yet been described, side effects have been reported in the literature, such as increased impulsivity and aggressiveness (Hälbig et al. 2009; Houeto et al. 2002; Sensi et al. 2004; Frank et al. 2007) or inappropriate sexual behaviour (Houeto et al. 2002; Leentjens et al. 2004), both of which could result in wrongful or even criminal behaviour. For instance, Sensi and his colleagues (2004) describe the case of a 64-year-old patient who received DBS to treat the symptoms of Parkinson's disease and who a few days after the implantation of the electrodes displayed spontaneous, unprovoked aggressive outbursts. In the days following the operation, he physically attacked other patients, medical staff and members of his family. "When asked about his excessive and unusual conduct the patient denied being aggressive; he was not able to control himself if asked" (Sensi et al. 2004, p. 248). Another case that is discussed in the literature concerns a 62-year-old Dutch man suffering from a severe case of Parkinson's disease. He showed remarkable improvement of his physical condition during DBS treatment, but at the same time manic and megalomanic symptoms as well as boundary-crossing sexual behaviour were observed (Leentjens et al. 2004). On the basis of these and other cases discussed in the literature, it seems that interventions that aim at changing specific brain activity can have unwanted side effects, some of which may lead to morally or legally questionable behaviour and therefore raise the question whether an individual who acted under the influence of such an intervention can or should be held (completely) responsible for his actions.

3 Neuroscience and Criminal Responsibility in Dutch Law

3.1 Criminal Responsibility in Dutch Law

A prerequisite for culpability in most jurisdictions is that the accused must have acted intentionally or knowingly. However, even if someone acted intentionally or knowingly, there may be circumstances that prevent attribution of criminal liability, for example if circumstances point to a case of *force majeure*, self-defence, or that someone acted in the exercise of government authority. For the purpose of this chapter, the exception of an unsound mind is relevant. In Dutch law, an individual cannot be held responsible if his mental capacities were seriously deficient at the time he committed a crime: "Not punishable is he who commits an act which cannot be attributed to him because of poorly developed or pathologically disturbed mental capacities" (Article 39 of the Dutch Criminal Code [DCC]). This provision determines that an individual cannot be held accountable in criminal law for his actions if he lacks particular mental capacities that enable him to act (sufficiently) intentionally and knowingly. The emphasis is on whether the act can (in a more objective sense) be attributed to the defendant (*toerekenbaarheid*) rather than on whether (in a purely subjective perspective) the defendant was able to determine his will during the act – the train driver who fell asleep could not subjectively do much about his negligence, but the fact can be attributed to him nonetheless (De Hullu 2003, p. 343). An important change in this provision was the replacement in 1928 of "powers of reason" (*verstandelijke vermogens*) by the broader "mental capacities" (literally: "mind capacities": *geestvermogens*), so that not only cognitive but also emotional deficiencies could be included in the scope of the exception (De Hullu 2003, p. 345).

The defect or disorder of the mental capacities must have existed at the time of the crime and must have contributed to the perpetration of the crime, i.e. there must be a relation between the crime and the defect or disorder of the suspect's relevant mental capacities (Mooij 2005).

The background of excluding culpability, also in relation to mental capacity, was briefly explained in the Explanatory Memorandum when a new Criminal Code was introduced in 1886: "No criminal responsibility without accountability of the act to the perpetrator, and no accountability in case either the freedom of acting – choosing between doing or not doing what the law prohibits or requires – is excluded or the perpetrator is in such a state that he cannot realise the unlawfulness of his act and cannot calculate its consequences" (quoted in De Hullu 2003, p. 284, our translation). This shows that two capacities are considered key to criminal responsibility: the freedom of choosing to act or not to act (which may be related to impulse control) and the capacity to distinguish right from wrong.

In the Netherlands, an assessment of the suspect's accountability can be ordered by the court if it believes that the suspect may be suffering from a defect or disorder of his mental capacities. Typically, the court orders a mono-disciplinary assessment, which involves a psychiatrist or a psychologist, or a bi-disciplinary assessment, which involves both a psychiatrist and a psychologist. In such cases, the expert typically only has one or two interviews with the suspect. In severe cases, i.e. if the crime and the assumed psychopathology are severe, the court can order a more intensive multidisciplinary assessment, which means that the suspect is assessed during a 7-week observation period in a forensic institution (Barendregt et al. 2008).

A 5-point scale (complete responsibility – slightly diminished responsibility – diminished responsibility – severely diminished responsibility – complete absence of responsibility) is used in forensic practice in the Netherlands in order to determine the degree of responsibility. Responsibility is typically measured by a number of clinical variables, which are determined by means of an anamnesis and standardized behavioural and/or (neuro)psychological tests, as well as by demographic and crime-related variables (Barendregt et al. 2008). Complete absence of responsibility is a reason for discharge, i.e. the suspect committed the crime but is not liable to punishment or the criminal act cannot be attributed to the suspect, while diminished responsibility is typically a reason for mitigated sentencing and typically leads to an order of detention during Her Majesty's pleasure (*terbeschikkingstelling*, TBS).

3.2 Neuroscientific Evidence in the Courtroom

Although the number of cases is unknown, neuroscientific evidence aimed at the assessment of responsibility has already entered the courtroom in the Netherlands.[2] In some cases, the expert witness – a behavioural neurologist – did not find any brain damage or not sufficient connections between the brain damage and the behaviour that constituted the criminal act, and hence the suspect was considered fully responsible (cases LJN BA9671, BO0306, and BK3854). In other cases, however, brain damage and a link between the damage and the behaviour were found that did influence the decision about the degree of responsibility.

We briefly describe eight court cases to illustrate the use of neuroscientific evidence in Dutch criminal cases. Not all of them directly deal with criminal

---

[2] In all cases discussed in this chapter in which a neurologist was involved as one of the expert witnesses, he explained the suspect's behaviour in terms of deficits caused by specific brain damage (with an exception of case LJN BM1948, in which the brain damage was not believed to have affected the suspect's behaviour). On the basis of the summaries of the court decisions, it is unclear how exactly the expert assessed the brain damage and associated deficits. In all cases however, the expert referred to some kind of brain damage and explained the suspect's behaviour in terms of this damage.

238 L. Klaming and B.-J. Koops

liability, but they all involve an assessment of the suspect's capacities conducted by inter alia a neurologist. The first case refers to the use of neuroscientific evidence to determine the defendant's capacity to understand prosecution (LJN BM8774). The second case refers to the use of neuroscientific evidence to determine intentionality (LJN BC9296) and the third case refers to the use of neuroscientific evidence to determine premeditation (LJN BB2861). We then briefly describe five cases in which neuroscientific evidence was used to determine culpability. In three of the five cases, the suspect had frontal lobe damage which according to the expert witness impaired his ability to control his impulses (LJN AV1864, LJN BA3923, LJN BK5962). In the fourth of the five cases, a defect at the pituitary gland was considered during the responsibility assessment (LJN BM1948). Cases in which reduced activity in the limbic system plays a role in the determination of culpability are, however, extremely rare. In the last case, the evidence pertains to the potential side effects of the antidepressant taken by the suspect (LJN BK4178). To our knowledge, there have not yet been any cases in the Netherlands in which the expert charged with the responsibility assessment has argued for diminished responsibility due to the unwanted side effects of DBS. So far, frontal lobe dysfunction seems to play the most important role in neuroscientific responsibility assessments in the Netherlands.

3.2.1 Use of Neuroscience to Determine Defendant's Capacity to Understand Prosecution

Before the court assesses whether the charge can be proven and whether the defendant can be held responsible for her act, the court has to answer preliminary questions: whether the summons is valid, whether the court has jurisdiction, whether the prosecutor is allowed to prosecute, and whether there are reasons to suspend the prosecution (art. 348 Dutch Code of Criminal Procedure, DCCP). A possible reason to suspend the prosecution is that the defendant is incapable of understanding the prosecution: "If the suspect suffers from poorly developed or pathologically disturbed mental capacities that are such that he is not capable of understanding the gist of the prosecution brought against him, the court will suspend the prosecution, regardless of its stage" (art. 16 para 1 DCCP).

District Court Amsterdam 21 June 2010/Court of Appeals Amsterdam 27 August 2010

A man was charged with murder in early 2010 (LJN BM8774). The defence called for suspension of the prosecution, arguing that the defendant was incapable of understanding the gist of the prosecution brought against him. A report from, *inter alia*, a psychologist, psychiatrist and behavioural neurologist showed that the defendant suffered from brain damage in the form of a lesion in the nucleus caudatus, which constituted a frontal syndrome of permanent damage. Together with the personal observation of the court in its questioning of the defendant, the court concluded that the defendant did not sufficiently understand the charge, why he was in preventative custody, nor the relationship between the preventative custody and the court hearings. The attorney was not able to have a meaningful conversation with the defendant and could not prepare a defence accordingly. Consequently, the defendant was incapable of understanding the gist of the prosecution, and hence, the court determined the prosecution to be suspended on the basis of article 16 DCCP.

Part of this case hinged on the interpretation of article 16 DCCP, a very rarely used provision with not very clear case-law. Historically, article 16 was meant for situations in which the defendant became mentally ill *after* the crime, but when the provision was changed in 1988, the formulation of the provision allowed an interpretation that the mental illness already existed at the time of the crime. And although the provision suggests that the legislator had temporary mental disturbances in mind (para 2 of article 16 indicates that the suspension be lifted as soon as the defendant has recovered), temporariness was not a necessary condition either. The court based both these interpretations on the phrase "poor development" of the mental capacities, arguing that a poor development is likely to have existed for a long time and to continue to exist.

However, the district court's verdict was quashed in appeal (LJN BN5666), since the court of appeal interpreted article 16 as, in principle, still applying only to cases where a mental illness occurred after the crime. For situations in which the defendant already suffered from a mental defect at the time of the crime, the rules on absence of or diminished culpability (art. 37 et seq. DCC) and on how to prosecute defendants with poorly developed or pathologically disturbed mental capacities (art. 509a et seq. DCCP) will be sufficient for securing a fair trial, according to the appeals court. It also found that the medical experts had not been asked the correct questions: instead of the question "Can the defendant be considered capable of standing trial?", they should have been asked the specific legal questions whether the defendant could be held accountable for his act (art. 37 DCC) and whether he was capable of sufficiently securing his interests (art. 509a DCCP), as well as whether he was capable of understanding the gist of the prosecution (art. 16 DCCP). The appeals court therefore referred the case back to the district court for further behavioural examination.

The follow-up of the case is still pending at the time of writing.

The case is interesting in that it shows that impairments of brain functions can have different effects on the criminal proceedings, depending on the type of impairment in relation to the stage of the procedure. A distinction is made between the (mental) capacity to defend one's interests (art. 509a DCCP) and the capacity to understand the gist of the prosecution (art. 16 DCCP). If the former is impaired, the defendant can be prosecuted, but he is likely not to be held culpable if the same impairment is found to have influenced his behaviour at the time of the crime; he can then, for example, be sentenced to detention during Her Majesty's pleasure.

If the latter capacity is impaired, the defendant will no longer be prosecuted, as the very idea of the prosecution has become meaningless. (Although this does not necessarily imply that the defendant is released, he can still be kept in preventative custody, as provided by article 17 para 2 DCCP, awaiting future examination in case recovery is considered possible.) If the impairment of the latter capacity – to understand the gist of the prosecution – already existed at the time of the crime, however, the impairment is likely to be related to the criminal behaviour, and most courts then find article 16 DCCP inapplicable (in line with the decision of the Court of Appeal in the above case), and instead opt for continuing the prosecution while taking into account art. 509a et seq. DCCP. Only if the brain impairment occurred after the crime does the understanding of the gist of the prosecution come into play. An example of this is a defendant who was charged with having illegally disposed of a dead body in 2001, and who subsequently, in 2004, had various cerebral haemorrhages and infarcts leading to vascular dementia. A neuropsychologist and neurologist reported that the man suffered from multiple cognitive functional impairments directly caused by cerebrovascular accidents, as a result of which no recovery was expected. The court therefore suspended the prosecution on the basis of article 16 DCCP, and – because recovery was extremely unlikely – determined the case closed on the basis of article 36 DCCP (LJN AY8840).

3.2.2 Use of Neuroscience to Determine Intentionality and Premeditation

If the court has answered the preliminary questions of article 348 DCCP in the affirmative – and hence the prosecution can be continued – the court has to answer two subsequent questions: first, whether it can be proven that the charged fact has been committed by the suspect and if so, which crime this charged fact constitutes, and, second, if the fact is indeed a crime, whether the defendant can be held culpable and which punishment or measure is appropriate (art. 350 DCCP). If the first part of the first question is answered negatively (i.e. if the suspect is not proven to have committed the charged fact), the suspect will be acquitted (art. 352 para 1 DCCP); if he has committed the fact but it is not a crime, or if the defendant cannot be held culpable, he will be discharged. If the suspect is not held culpable because of article 39 DCC (mental defects), the court can order him to be admitted to a mental hospital or be detained during Her Majesty's pleasure (art. 352 para 2 DCCP).

Neuroscientific evidence can, in principle, play a role in both stages of the court decision. In this subsection, we discuss the first issue: whether it can be proven that the defendant committed the charged facts and which crime this charged facts constitutes. Many criminal provisions contain an explicit element of criminal intent that must be proven, and if a mental impairment is found to be of such a nature that the person cannot be considered to have acted intentionally or with premeditation, he will be acquitted.

District Court Amsterdam 28 March 2008

In 2007, a 63-year-old man stabbed a friend nine times as a result of which she deceased. The suspect declared that he was annoyed by the victim's behaviour. He furthermore declared that he saw that the victim lost a lot of blood as a result of the stabbing and furthermore lost her consciousness several times. When she regained consciousness and tried to get up he stabbed her again. At the time of the incident, the suspect was intoxicated with alcohol and cocaine. According to the expert witness, a behavioural neurologist, the suspect's behaviour during the incident was affected by damage to his frontal lobes. More specifically, the brain damage had rendered the suspect unable to control his impulses and reflect on his actions in difficult situations. The alcohol and cocaine were believed to have aggravated his impulsive behaviour. Additionally, the expert witness stated that the suspect's brain damage interfered with his free will. The presiding judge decided that the suspect had acted intentionally. More specifically, he stated that although the suspect's behaviour was affected by the frontal lobe damage, he did not lack complete insight into the consequences of his actions. According to the judge, the suspect was aware of the possibility that the victim would die as a result of the harm that he was inflicting on her. Consequently, the court decided that the suspect had committed the act of intentionally killing someone (manslaughter). On the basis of the expert witness' report, however, the judge decided that the suspect had severely diminished responsibility for his actions as a result of the frontal lobe damage, which eventually resulted in reduced sentencing of 18 months' imprisonment, plus detention during Her Majesty's pleasure (LJN BC9296).

This case demonstrates the use of neuroscientific evidence to determine intentionality, which is important in light of deciding whether the suspect committed the charged facts. The case is interesting, because it demonstrates that a diagnosis of serious impairments of the suspect's cognitive functions apparently has little impact on the court's decision concerning the intentionality of the suspect's actions. The brain damage that was diagnosed by the behavioural neurologist was considered serious, after all it resulted in the decision of severely diminished responsibility. Despite its severity however, the damage was not deemed to have affected the suspect's intentionality. The court believed that although the suspect's actions were influenced by the brain damage, he knew what he was doing and therefore acted intentionally. As discussed by Stevens and Prinsen (2009), who analyzed manslaughter cases in the Netherlands in which the presiding judge decided that the suspect lacked intentionality because of a mental disorder, the concept of intentionality is purely legal with little psychological content. The suspect's mental disorder – apparently regardless of its severity – typically has little influence on the court's decision concerning the intentionality of the suspect's actions. It seems that the only diagnosis that convinces the court that the suspect has not acted intentionally is that of dissociation, which is characterized by a disruption of the normal integration of an individual's conscious functioning (Stevens and Prinsen 2009). A case in which the court decided that the suspect did not act intentionally and was therefore acquitted (art. 352 para 1 DCCP) concerns that of a man who caused an accident by knocking over a biker with his car. The suspect himself indicated that he probably had a light epileptic seizure, which the three expert witnesses (a psychiatrist, a psychologist and a neurologist) could not rule out as cause of the suspect's behaviour (LJN BN0983). The above described case is furthermore interesting, because the same evidence that was used to determine that the suspect had acted intentionally, was subsequently used to determine the degree of responsibility. Although the court claimed that the brain damage did not affect the suspect's intentionality, it believed that the damage was severe enough to deem the suspect severely diminished responsible for his actions.

District Court's-Hertogenbosch 5 September 2007

A 50-year-old man had murdered his wife during a fight about money by smashing her head against the wall, choking and repeatedly stabbing her in the presence of their 9- and 7-year-old children. He gave several accounts of what had happened when he was interrogated by the police and after being confronted with the statements of his children, he said that they were not lying but that he was unable to remember what had happened. On the basis of expert testimonies – from a psychologist, a psychiatrist and a behavioural neurologist – the court judged that the suspect had acted without premeditation and that he had diminished responsibility for his actions. According to the psychiatrist, the suspect was suffering from an autistic spectrum disorder, mental retardation, depression and damage to his frontal lobes at the time he committed the crime, as a result of which he is diminished or severely diminished accountable for his actions. The psychologist's report stated that, at the time he committed the crime, the suspect suffered from a pervasive developmental disorder not otherwise specified. In addition, he might suffer from a mental impairment, both of which led the psychologist to conclude that the suspect should be held diminished responsible for his actions. According to the behavioural neurologist, the suspect possibly suffers from damage to his frontal lobes, which is accompanied by disturbances in his executive functions and a loss of impulse control. The behavioural neurologist stated in his report that if the report of the psychiatrist supports the diagnosis of frontal lobe damage (which it does), the suspect's actions were influenced by his frontal lobe damage. According to the behavioural neurologist, as a result of the frontal lobe damage, the suspect had less control over his impulses, was unable to adequately evaluate the situation, and could no longer disrupt his actions once he had started them, which is why he repeatedly stabbed his wife. The expert witnesses' reports were considered during the determination of which crime the charged fact constitutes; the court interpreted the statements of the experts in favour of the suspect and determined that he had not acted premeditatedly, because his mental disorder had probably caused an unexpected and extreme reaction towards his wife during the fight. Additionally, the statements were considered in the deliberation of the verdict and the suspect, being held diminished responsible, was sentenced to 9 years in prison for manslaughter (LJN BB2861).

This case is interesting for several reasons. First of all, the fact that the suspect claimed memory loss could have been taken as an indication for some kind of dissociative disorder, which as described above can be taken as a reason for lack of intent (although of course the possibility that he falsely claimed memory loss to avoid prosecution or at least receive a reduced sentence cannot be excluded). The suspect's mental disorder did, however, not affect the decision concerning intentionality. It was, however, taken as evidence for a lack of premeditation. The court interpreted the expert witnesses' testimonies as indicating that the suspect's mental disorder led to an unexpected and extreme reaction as a result of which he acted the way he did. As a consequence of this decision, the suspect was charged with manslaughter. The case is furthermore interesting because although they come to the same conclusion of diminished responsibility, the psychiatrist's and the psychologist's diagnoses are fairly different. The psychiatrist diagnosed an autistic spectrum disorder, mental retardation, depression and damage to the suspect's frontal lobes at the time he committed the crime, whereas the psychologist diagnosed a pervasive developmental disorder not otherwise specified. The psychiatrist's and the neurologist's testimonies are clearly aligned to each other in so far as they both include a diagnosis of frontal lobe damage. From the neurologist's testimony, it seems that he based his diagnosis on the report of the psychiatrist: "If the other assessments ([...], psychiatric report) confirm a frontal syndrome, the suspect's behaviour was influenced by this damage at the time of the crime. Because of the frontal syndrome, the suspect had less control over his impulses, was unable to adequately evaluate the situation and could no longer disrupt his actions once he had started them" (LJN BB2861, our translation). Since the consequences of frontal lobe damage are rather severe and so are the conclusions and decisions that the court bases on these assessments, it is somewhat surprising that the neurologist's diagnosis depended on the psychiatrist's diagnosis.

As in the previous case, the experts' testimonies were not only used to determine intentionality and premeditation, respectively, but were furthermore used to determine the suspect's responsibility. On the basis of the expert testimonies, the court decided that the suspect was diminished responsible.

3.2.3 Use of Neuroscience to Determine Culpability

As the previous subsection showed that defendants, despite possible mental impairments, are usually considered to have acted with intent, most emphasis comes to lie with the second part of article 350 DCCP: whether the defendant can be held culpable for his act. Culpability in the sense of attributability is a basic element of criminal liability; Dutch criminal law is often characterised in that respect as a "guilt-based criminal law system" (*schuldstrafrecht*) (De Hullu 2003, p. 284). This, in principle subjective, approach is contrasted with a more objective approach of an "act-based criminal law system" (*daadstrafrecht*). In common-law terms, Dutch law could be said to place more emphasis on *mens rea* than on *actus reus* when it comes to attributing criminal responsibility. (Although it is often questioned whether Dutch criminal law still retains such a subjective approach – an issue we will briefly discuss in Sect. 4.)

In determining culpability, the exception of article 39 DCC plays a central role. Whether a defendant can be considered to suffer from impairments in his mental capacities, will depend crucially on expert statements: as soon as the possibility of the exception is raised, behavioural experts enter the criminal procedure. Article 37 para 2 DCC determines that – if a court wants to order forced hospitalisation on the basis of poorly developed or pathologically disturbed mental capacities, it can only do so after being advised by two or more behavioural experts from different disciplines, including a psychiatrist, who have examined the defendant; article 37a para 3 DCC provides the same for the measure of detention during Her Majesty's pleasure. As a result, case-law very frequently discusses reports from behavioural experts when the culpability of the defendant is being questioned. Most often, this concerns psychologists and psychiatrists, but sometimes, also behavioural neurologists have entered the picture, who provide evidence based on their examination of the suspect's brain functioning.

District Court Utrecht 14 February 2006

In 2005, a 74-year-old man was charged with sexual and physical abuse of four children. The suspect was working as an alternative healer and had gotten into contact with the four children through their mother who was one of his patients. The children's mother was convinced of the suspect's healing qualities and decided that her four daughters should also see him. The suspect sexually and physically abused the four girls over a period of several years. During the criminal trial he was examined by a psychiatrist, a psychologist and a behavioural neurologist. The psychiatric/psychological report stated that the suspect was suffering from a personality disorder with narcissistic, antisocial and schizotypical characteristics. In addition, he had suffered two cerebrovascular accidents (CVA) which contributed to cognitive impairments. According to the psychiatrist and the psychologist, the suspect's actions were primarily affected by his personality disorder, while the cognitive impairments only had a marginal influence. They furthermore reported that although the suspect was able to realize the wrongness of his actions, he had less capacity to freely determine his will as compared to most people. The psychiatrist and the psychologist believed that as a consequence of this, the suspect has diminished responsibility for his actions. According to the behavioural neurologist, the CVAs caused serious brain damage, which affected his motoric and cognitive functions. The behavioural neurologist furthermore reported that it is therefore very likely that at the time he committed the crimes, the suspect's behaviour was influenced by an impaired judgment, which made it impossible for him to foresee the consequences of his actions. The behavioural neurologist concluded that on the basis of this, the suspect can be considered slightly diminished responsible. The court accepted diminished responsibility and sentenced the suspect to 4 years in prison, which is the sentence that was demanded by the prosecutor (LJN AV1864).

This case demonstrates the use of neuroscientific evidence to determine the degree of responsibility. All experts came to the conclusion that the CVAs had caused cognitive impairments and that the suspect's capacity to freely determine his will was diminished. However, according to the neurologist, this was the result of the cognitive impairments, whereas according to the psychiatrist and the psychologist, this was the result of the suspect's personality disorder and the cognitive impairments only had a marginal influence. Regardless of this incongruence, all expert witnesses declared diminished responsibility. Although the court adopted the experts' opinion of diminished responsibility, due to the nature and the severity of the charges, it nevertheless passed the sentence that was demanded by the prosecution. Hence, diminished responsibility did lead neither to mitigated sentencing nor to an order of detention during Her Majesty's pleasure in this case.

District Court Alkmaar 24 June 2008

A man was charged with sexual abuse of a girl under the age of 12. The victim was the daughter of one of the suspect's neighbours who used to visit him to play with his pets. During the criminal trial, the suspect was examined by a behavioural neurologist and a psychologist. The behavioural neurologist detected symptoms of a beginning fronto-subcortical dementia as a result of which the suspect suffers from cognitive limitations and increased impulsivity. According to the expert, the suspect's behaviour was influenced by the brain damage that is related to the dementia. More specifically, due to the cognitive limitations and his impulsivity, the suspect was unable to control his impulses and to foresee the consequences of his behaviour and could no longer disrupt his actions once he had started them. In addition, the behavioural neurologist believed that as a consequence of his brain damage related to the beginning dementia, the suspect lacked the ability to reflect on his own actions which prevented him from checking whether his behaviour was adequate. The psychologist referred to the neurologist's testimony and stated that the suspect was not sufficiently able to control his impulses. He furthermore stated that the dementia was in an initial phase at the time he committed the crimes and that the suspect did not show any other types of disinhibited behaviour which is why he is at least partially accountable for his actions. On the basis of the expert testimonies, the court decided that the suspect is diminished responsible for his actions and imposed a prison sentence of 279 days of which 60 days are provisional and a probation of 2 years (LJN BK5962).

In contrast to the previous case, the suspect's brain damage in this case was believed to have had a greater impact on his behaviour; the suspect was unable to control his impulses and to disrupt his actions once he had started them. Since at the time of the assessment and therefore also at the time of the crime, the dementia was in an initial stage, the suspect was hold at least partially accountable for his actions. The responsibility assessment did not result in an order of detention during Her Majesty's pleasure in this case.

District Court Amsterdam 26 April 2007

In June 2006, a man visited the Rijksmuseum in Amsterdam, randomly chose a painting, poured petrol over the painting and lit it on fire. He was charged with arson and damage to goods. During the criminal trial, the suspect was examined by a psychiatrist, a psychologist and a behavioural neurologist. After consultation with each other, the experts testified that the suspect suffered from frontal lobe damage which resulted from a previous leucotomy. As a result of the brain damage, the suspect had developed a personality disorder with obsessive, neurotic and narcissistic features. He was characterized as living in his own world and being primarily driven by his obsessive beliefs. The experts contemplated that the suspect was aware of the wrongness of his actions, but nevertheless was unable to act upon his free will. The court accepted the experts' judgement of severely diminished responsibility and sentenced him to 1 year in prison and additionally imposed an order of detention during Her Majesty's pleasure and a compulsory admission to a psychiatric institution (LJN BA3923).

In this case, the suspect was determined severely diminished responsible for his actions due to severe frontal lobe damage which had resulted from a leucotomy, which is a neurosurgical procedure that consists of cutting the connections to and from the prefrontal cortex. Apparently, according to the expert witnesses, this damage had interfered with the suspect's free will. It seems that in this case the order of detention during Her Majesty's pleasure and the compulsory admission to a psychiatric institution were based on the rather severe mental disorder as well as the suspect's poor physical condition and the high risk for recidivism.

District Court's-Gravenhage 22 April 2010

A man was charged with the murder of his wife. The suspect had shot at his wife several times during the night in their bedroom and had subsequently reported himself to the police. According to the psychiatrist and the psychologist who assessed the suspect, he was somewhat mentally retarded which however according to them does not imply diminished responsibility. The neurologist detected a tumour in the suspect's pituitary gland, but concluded that there is not sufficient indication for a link between this defect and the suspect's behaviour during the crime. The suspect was considered completely responsible, was convicted for murder and received a sentence of 16 years' imprisonment (LJN BM1948).

Although extremely rare, there seem to be cases in which a defect in the limbic system is considered during the responsibility assessment. In this case, the suspect was suffering from a tumour in his pituitary gland. The pituitary gland is located at the base of the brain underneath the hypothalamus and produces hormones which regulate the other glands in the body. It is part of the limbic system and is functionally connected to the hypothalamus. The neurologist who detected the tumour did not find any evidence for a link between the defect in the limbic system and the suspect's behaviour. He therefore concluded that the suspect was fully responsible for his actions. Whether the defect did actually not play any role in the suspect's behaviour or whether a defect in the limbic system is typically not used as evidence for diminished responsibility or accepted as excuse for criminal behaviour is unclear from this case. The fact that there seem to be very few cases in which the limbic system plays a role in the assessment of responsibility suggests that it is not a common reason to assume diminished responsibility.

District Court Haarlem 24 November 2009

The last case has received a considerable amount of media attention in the Netherlands. It concerns a 63-year-old woman who had murdered her husband and daughter with an axe while they were sleeping before trying to commit suicide. The woman had suffered from depressive episodes since several years and had taken antidepressants to relieve the symptoms. However, a few weeks before the incident she had stopped taking the medication upon the advice of her family, who wanted her to get psychological help instead. Two days before the murder, the suspect had an appointment with her general practitioner who subscribed a higher dose of the antidepressants she had already taken earlier, since there was no improvement of the symptoms. The expert testimonies concerning the impact of her depression and the medication on her behaviour were inconsistent, ranging from slightly diminished responsibility to complete absence of responsibility. According to a court-ordered assessment by a psychiatrist and a psychologist, several factors including the depressive episodes and the woman's personality had contributed to the murder. The experts furthermore stated that they could not exclude the possibility that the effects of the antidepressants played a role in the causation of her behaviour. The defence lawyer subsequently ordered a counter opinion. According to these experts (a psychiatrist and a psychologist), there is a direct causal link between the effects of the antidepressants and the murder, which according to the experts implies that the suspect could not exert her free will during the night she murdered her family. The woman was ultimately convicted for murder, but received a reduced sentence of 8 years' imprisonment due to her diminished responsibility (LJN BK4178).

As this case demonstrates, medical interventions are sometimes considered during the responsibility assessment. The court did not accept the opinion of one of the experts who stated that there was probably a causal link between the medication and the suspect's actions. The court did, however, assume diminished responsibility on the basis of personality problems including the depressive episodes. In addition, the court acknowledged that the medication might have played some role in the suspect's behaviour, probably by contributing to a certain disinhibition. The decision of diminished responsibility resulted in a mitigated sentence in this case. There are some other cases in which a causal link between the psychopharmaca taken by the suspect and the subsequent violent actions was not accepted by the court and therefore did not lead to a decision of diminished or complete absence of responsibility (LJN BI6332, LJN BL5774). Apparently, the possible side effects of medical interventions that directly affect specific brain activity is typically not considered as being a reason for diminished responsibility. The reverse could, however, apply, in that people suffering from a brain disorder could be held responsible for *not* having taking medication. In the case of a woman who caused a lethal traffic accident due to an epileptic attack, her not having taken medication was not considered sufficiently negligent to convict her for criminally negligent homicide, because her neurologist had not explicitly prescribed medication. However, because she was a professional taxi driver, the court found she should have asked the neurologist more clearly about the consequences if she was indeed suffering from epilepsy, and she was convicted to 60 hours' community service for causing danger on the road (LJN BN7251).

Overall, it seems that in cases in which psychopharmaca might have played some role in the suspect's behaviour, it is typically the mental disorder for which the psychopharmaca were prescribed rather than the possible side effects of the medication that influenced the responsibility assessment.

4 Discussion

The above briefly described cases demonstrate that neuroscientific evidence has already entered the courtroom in the Netherlands. We found a number of cases in which neuroscientific evidence played a role at some stage of the criminal trial – to determine the defendant's capacity to understand prosecution or to determine intentionality, premeditation or the degree of responsibility. However, it seems that at present neuroscientific evidence is not used frequently in criminal procedure. In cases in which the court based its decision inter alia on neuroscientific evidence, this evidence was always used in connection with a behavioural assessment by a psychiatrist and in some cases also a psychologist. Apparently, neuroscientific evidence is at least currently only an addition to evidence based on behavioural assessments.

Interestingly, in all cases briefly described above, the same neurologist acted as an expert. It is not entirely clear why this is the case, but it is possible that the neurologist was approached in these cases by one of the other experts ordered with the responsibility assessment, because he expected additional insights from the neurologist's assessment. Nevertheless, without questioning the eligibility and qualifications of this neurologist, including other neurologists during the responsibility assessment might increase the objectivity and validity of neuroscientific evidence in criminal procedure. At least it is important that there are other neurological experts available, if only for counter-appraisal purposes.

On the basis of the cases briefly discussed above, it seems that the frontal lobes play a predominant role in responsibility assessments if some kind of brain damage is believed to underlie the suspect's behaviour. In five of the eight cases, some kind of damage to the frontal lobes was diagnosed, which interfered with the suspect's capacity to act upon his free will and to foresee the consequences of his actions.

The frontal lobes are involved in various cognitive processes and executive functions including attention, working memory, planning, and decision-making and seem to play an important role in moral judgement and intentional behaviour. It is therefore not surprising that they are deemed relevant to the assessment of responsibility in some cases. As described above, research on frontal lobe dysfunction supports this assumption as it has demonstrated that damage in this area can implicate several deficits in planning and foresight, in moral and social judgement and in the regulation of behaviour. The other three cases briefly described above concern vascular brain damage (in which the affected brain region was not specified), damage to the limbic system in the form of a tumour in the pituitary gland and the influence of psychopharmaca on the suspect's behaviour. From the literature review in Sect. 2, it is clear that other brain damage or impairment of brain functions besides damage to the frontal lobes might as well play a role in explaining some types of criminal behaviour. Defects in the limbic system were found to be linked with antisocial behaviour, in particular to a lack of empathy, remorse and guilt. It seems that cases in which the defence argues for reduced responsibility due to a defect at the limbic system are extremely rare. This may at least be partially due to a reluctance to accept a defect in the limbic system and associated cognitive, affective and behavioural problems as an excuse for criminal behaviour. It seems that intuitively, lacking the capacity to empathise is seen as indicating that someone is bad rather than mad (Maibom 2008; Vincent 2010b). Similarly, the possible side effects of psychopharmaca are typically not accepted as explaining (some of) the suspect's behaviour. A causal link between the medication and the behaviour is typically not considered plausible by the court, which is why cases in which the defence argues for diminished responsibility due to the side effects of psychopharmaca typically do not result in a decision of diminished or complete absence of responsibility. If the court assumed diminished responsibility in these cases, it is typically the mental disorder for which the psychopharmaca were prescribed rather than the possible side effects of the medication that influenced the responsibility assessment. To our knowledge, no case in which a DBS patient became criminal as a result of the treatment has yet occurred in the Netherlands. However, given the possible side effects of this intervention, it seems merely a matter of time before a court is confronted with the first case in which a DBS patient argues for diminished or complete absence of responsibility due to the neurotechnological intervention and its effect on his behaviour.

Moreover, neuroscientific evidence seems to play a predominant role in determining culpability and to a much lesser degree in determining intentionality or premeditation. Apparently, brain damage associated with diminished impulse control or cognitive deficits may imply that a perpetrator is less to blame for his actions, but they do not easily lead to the judgement that he did not want those actions to happen. This makes sense from a legal perspective, as the threshold for assuming intent is quite low in Dutch law; intent can be assumed, from a criminal-law perspective, when a person knowingly accepts the substantial chance that a consequence of his act will happen (in other words: he should be conscious of the considerable probability of the effect occurring) (De Hullu 2003, p. 236). It seems that from a legal perspective, the capacity to foresee the consequences of certain behaviour is not easily believed to be impaired even if the suspect suffers from cognitive or emotional deficits. From a psychological perspective, however, it is not evident how diminished impulse control can lead to diminished culpability while not leading to absence of intent – can someone really be said to "knowingly accept the chance" that his behaviour will have a certain effect if he cannot control his impulses? The same may be true for other cognitive or emotional impairments. If insights from a behavioural and/or neurological assessment can be used to answer questions of responsibility, they might as well inform to what degree an individual had certain capacities that make it plausible to assume he acted intentionally or premeditatedly. In this respect, we think that the relationship between current psychological and neurological insights into brain-behaviour correlates and the legal assessment of intent and premeditation merits further research.

A somewhat related issue is the notion, mentioned in Sect. 3.2.3, that Dutch criminal law is traditionally characterised as a "guilt-based" rather than an "act-based" criminal-law system. If this is true, then the importance of establishing guilt in a context- and case-specific sense would seem to suggest that insight into the mental state of a defendant would play a substantial role in criminal procedure. The criminal law system seems to be reticent in this respect: over the past decades, "guilt" in Dutch criminal law has been "objectivised" or "normalised", as evidence of someone's intent and culpability is assessed on the basis of an external perspective, focusing on the act and all its circumstances as much as, or more than, on the person. Partly this has to do with respecting privacy of the mind, but partly it has also a pragmatic reason, in that objectivised intent and guilt are easier to prove than subjective elements (De Hullu 1998, p. 181). Since also other tendencies – such as the rise of the risk society – reinforce the focus on guilt that is determined from an external perspective, the literature is questioning whether the objectivised approach to guilt should not be shifted back to a more subjectivised approach (Buruma 1998). Here, we think, the potential of neuroscientific evidence is important to factor into the equation, as neuroscience could, in principle, provide more insight into the brain-cognition behaviour interrelationships of individual defendants. Also in this respect, we recommend that the legal, doctrinal interpretation of intent and guilt are studied further in relation to current insights from neuroscientific research.

Despite the potential value of neuroscientific evidence in the assessment of criminal responsibility, however, it is important to mention that there are several significant legal implications of using neuroscientific evidence in legal proceedings. Probably one of the most important challenges of applying neurotechnologies for legal purposes refers to the possibility that neuroscientific evidence is inappropriately persuasive and may therefore unduly affect legal decision-making. This assumption is supported by recent research demonstrating that people view explanations of psychological phenomena as more believable if these explanations contain a neuropsychological component (Weisberg et al. 2008). Additionally, including visual information, i.e. brain images, with explanations of cognitive neuroscience data was found to increase judgements of scientific reasoning (McCabe and Castel 2007). Besides these two studies, that analyzed the influence of neuroscientific explanations on the public's perception of scientific research and not within a legal context, a more recent study empirically supports the concern that neuroimaging evidence unduly affects legal decision-making by showing that students were more likely to find a hypothetical offender not guilty by reason of insanity if he had some kind of brain damage as presented in a brain image (Gurley and Marcus 2008). These preliminary empirical data support the concerns of many researchers who believe that judges and juries may perceive evidence derived by means of insights from neuroscience without sufficient critical appraisal (Aharoni et al. 2008; Garland and Glimcher 2006; Gazzaniga 2005; Jelicic and Merckelbach 2007; Morse 2006; Reeves et al. 2003; Sinnott-Armstrong et al. 2008). It is therefore important to further empirically explore the effect of neuroscientific evidence on legal decision-making in order to ensure the responsible use of this type of evidence.

5 Conclusion and Outlook

Insights from neuroscience seem to have the potential to contribute to our understanding of what it means to act intentionally and knowingly. At present, these insights only play a marginal role, which is mainly due to the fact that research on the underlying neural mechanisms of moral judgement and intentional behaviour is still in its infancy, and hence the responsible use of neuroscientific evidence to answer questions of liability is questionable at this point in time. Nevertheless, the fact that recent empirical studies have established links between brain structures and antisocial or criminal behaviour as well as the fact that neuroscientific evidence has already entered the courtroom in the Netherlands (and other countries) warrants the assumption that neuroscience could play a more important role in assessments of criminal responsibility – and potentially also questions related to other legal concepts such as intentionality and premeditation – in the future. We believe that more empirical research into the neural correlates of impulse control, moral judgement, intentional behaviour, and other mental capacities related to the legal concept of responsibility is necessary in order to better understand some of the causes of criminal and antisocial behaviour on the one hand and to assure the responsible use of neuroscientific evidence in the courtroom on the other hand. In this respect, we furthermore think that the relationship between current psychological and neurological insights into brain-behaviour correlates and the legal assessment of intent and premeditation merits further research. Despite the fact that intentionality and premeditation are legal concepts, they are nevertheless strongly linked with mental functioning, which is why insights from behavioural and/or neurological assessments might be useful to determine whether a suspect likely acted with intent or premeditation. Additionally, from a legal perspective, there is a difference in case-law between intent and culpability, and between intent and premeditation. It therefore seems interesting to further explore how neuroscience explains or could explain correlations between brain, cognition and behaviour in terms of these differences; does neuroscience provide insight into cognitive and volitional functions involved in (what legal doctrine interprets as) intent and premeditation? More multidisciplinary research on whether and how insights from neuroscience might be useful in answering questions related to these normative, legal issues seems beneficial. With regard to the responsible use of neuroscientific evidence in the courtroom, we furthermore believe that more research on the (potentially overly persuasive) effect of this type of evidence on legal decision-making is essential.

With regard to the three distinct areas of brain abnormalities that are relevant for assessments of criminal responsibility, it seems conducive to further discuss why certain types of brain damage are accepted as excusing some types of criminal behaviour, whereas others are not. Why is it that intuitively we find psychopaths, who according to recent neuroscientific research suffer from limbic abnormalities, bad rather than mad and are therefore inclined to refuse brain deficiencies as an excuse for their behaviour? With regard to the third area – the potential side effects of neurotechnologies such as psychopharmaca and DBS – it seems that these at present only play a minor role in questions of criminal responsibility. Since empirical research has demonstrated that some psychopharmaceuticals can have severe side effects that are associated with increased aggressiveness, this raises the question whether the fact that the side effects of these medications are only rarely seen as interfering with the suspect's mental capacities is warranted. With regard to DBS, it seems that the growing use of this technology promises for crimes being committed by a DBS patient who subsequently argues for diminished or complete absence of responsibility due to the side effects of the intervention. Since neurotechnologies raise new concerns for questions of criminal responsibility, there is a great need for further research into ways to deal with these cases once they arise.

References

Aharoni E, Funk C, Sinnott-Armstrong W, Gazzaniga M (2008) Can neurological evidence help courts assess criminal responsibility? Lessons from law and neuroscience. Ann N Y Acad Sci 1124:145–160

Barendregt M, Muller E, Nijman H, de Beurs E (2008) Factors associated with experts' opinions regarding criminal responsibility in the Netherlands. Behav Sci Law 26:619–631

Barkataki I, Kumari V, Das M, Taylor P, Sharma T (2006) Volumetric structural brain abnormalities in men with schizophrenia or antisocial personality disorder. Behav Brain Res 169(2):239–247

Barkataki I, Kumari V, Das M, Sumich A, Taylor P, Sharma T (2008) Neural correlates of deficient response inhibition in mentally disordered violent individuals. Behav Sci Law 26:51–64

Beauregard M, Levesque J, Bourgouin P (2001) Neural correlates of conscious self-regulation of emotion. The Journal of Neuroscience 21 RC165: 1–6

Beckman M (2004) Crime, culpability, and the adolescent brain. Science 305:596–599

Breggin PR (2003/2004) Suicidality, violence and mania caused by selective serotonin reuptake inhibitors (SSRIs): a review and analysis. Int J Risk Saf Med 16:31–49

Brower MC, Price BH (2001) Neuropsychiatry of frontal lobe dysfunction in violent and criminal behaviour: a critical review. J Neurol Neurosurg Psychiatry 71:720–726

Buckholtz JW, Asplund CL, Dux PE, Zald DH, Gore JC, Jones OD, Marois R (2008) The neural correlates of third-party punishment. Neuron 60:930–940

Burns JM, Swerdlow RH (2003) Right orbitofrontal tumor with pedophilia symptom and constructional apraxia sign. Arch Neurol 60:437–440

Buruma Y (1998) Het schuldig subject. In: Borgers MJ, Koopmans IM, Kristen FGH (eds) Verwijtbare uitholling van schuld? Ars Aequi Libri, Nijmegen, pp 1–9

Casebeer WD, Churchland PS (2003) The neural mechanisms of moral cognition: a multiple-aspect approach to moral judgment and decision-making. Biol Philos 18:169–194

Damasio H, Grabowski T, Frank R, Galaburda AM, Damasio AR (1994) The return of Phineas Gage: clues about the brain from the skull of a famous patient. Science 264(5162):1102–1106

De Hullu J (1998) Bedreigingen van het schuldbeginsel? In: Borgers MJ, Koopmans IM, Kristen FGH (eds) Verwijtbare uitholling van schuld? Ars Aequi Libri, Nijmegen, pp 179–187

De Hullu J (2003) Materieel Strafrecht. Kluwer, Deventer

Frank MJ, Samanta J, Moustafa AA, Sherman SJ (2007) Hold your horses: impulsivity, deep brain stimulation, and medication in Parkinsonism. Science 318:1309–1312

Gabriëls L, Cosyns P, Nuttin B, Demeulemeester H, Gybels J (2003) Deep brain stimulation for treatment refractory obsessive-compulsive disorder: psychopathological and neuropsychological outcome in three cases. Acta Psychiatr Scand 107(4):275–282

Garland B, Glimcher PW (2006) Cognitive neuroscience and the law. Curr Opin Neurobiol 16:130–134

Gazzaniga MS (2005) The ethical brain. Dana Press, New York

Glannon W (2005) Neurobiology, neuroimaging, and free will. Midwest Stud Philos 29:68–82

Goodenough OR, Prehn K (2004) A neuroscientific approach to normative judgment in law and justice. Philos Trans R Soc B 359:1709–1726

Goodenough OR, Tucker M (2010) Law and cognitive neuroscience. Annu Rev Law Soc Sci 6:61–92

Greely HT (2004) Prediction, litigation, privacy, and property: some possible legal and social implications of advances in neuroscience. In: Garland B (ed) Neuroscience and the law: brain, mind, and the scales of justice. Dana Press, New York, pp 114–156

Greely HT (2008) Neuroscience and criminal justice: not responsibility but treatment. Kansas Law Rev 56:1103–1138

Greene J, Cohen J (2004) For the law, neuroscience changes nothing and everything. Philos Trans R Soc B 359:1775–1785

Greene JD, Nystrom LE, Engell AD, Darley JM, Cohen JD (2004) The neural bases of cognitive conflict and control in moral judgment. Neuron 44:389–400

Grey BJ (2007) Neuroscience, emotional harm, and emotional distress tort claims. Am J Bioeth 7(9):65–67

Gurley JR, Marcus DK (2008) The effects of neuroimaging and brain injury on insanity defenses. Behav Sci Law 26:85–97

Hälbig TD, Tse W, Frisina PG, Baker BR, Hollander E, Shapiro H, Tagliati M, Koller WC, Olanow CW (2009) Subthalamic deep brain stimulation and impulse control in Parkinson's disease. Eur J Neurol 16:493–497

Haynes J-D, Sakai K, Rees G, Gilbert S, Frith C, Passingham RE (2007) Reading hidden intentions in the human brain. Curr Biol 17(4):323–328

Healy D, Herxheimer A, Menkes DB (2007) Antidepressants and violence: problems at the interface of medicine and law. Int J Risk Saf Med 19:17–33

Horn NR, Dolan M, Elliott R, Deakin JFW, Woodruff PWR (2003) Response inhibition and impulsivity: an fMRI study. Neuropsychologia 41:1959–1966

Houeto JL, Mesnage V, Mallet L, Pillon B, Gargiulo M, Tezenas du Moncel S, Bonnet AM, Pidoux B, Dormont D, Cornu P, Agid Y (2002) Behavioural disorders, Parkinson's disease and subthalamic stimulation. J Neurol Neurosurg Psychiatry 72:701–707

Hughes V (2010) Science in court: head case. Nature 464:340–342

Hyman SE, Malenka RC, Nestler EJ (2006) Neural mechanisms of addiction: the role of reward-related learning and memory. Annu Rev Neurosci 29:565–598

Jelicic M, Merckelbach H (2007) Hersenscans in de rechtzaal: oppassen geblazen! Nederlands Juristenblad 44:2794–2800

Jones OD, Buckholtz JW, Schall JD, Marois R (2009) Brain imaging for legal thinkers: a guide for the perplexed. Stanford Technology Law Review 5. http://stlr.stanford.edu/pdf/jones-brain-imaging.pdf

Kiehl K (2006) A cognitive neuroscience perspective on psychopathy: evidence for paralimbic system dysfunction. Psychiatry Res 142:107–128

Kiehl K, Smith AM, Hare RD, Mendrek A, Forster BB, Brink J, Liddle PF (2001) Limbic abnormalities in affective processing by criminal psychopaths as revealed by functional magnetic resonance imaging. Biol Psychiatry 50:677–684

Klaming L, Haselager P (2010) Did my brain implant make me do it? Questions raised by DBS regarding psychological continuity, responsibility for action and mental competence. Neuroethics. doi:10.1007/s12152-010-9093-1

Klaming L, Vedder A (2009) Brushing up our memories: can we use neurotechnologies to improve eyewitness memory? Law Innov Technol 2:203–221

Knoch D, Gianotti LRR, Baumgartner T, Fehr E (2010) A neural marker of costly punishment behavior. Psychol Sci 21:337–342

Kolber AJ (2007) Pain detection and the privacy of subjective experience. Am J Law Med 33:433–456

Kozel FA, Johnson KA, Mu Q, Grenesko EL, Laken SJ, George MS (2005) Detecting deception using functional magnetic resonance imaging. Biol Psychiatry 58:605–613

Laakso MP, Gunning-Dixon F, Vaurio O, Repo E, Soininen H, Tiihonen J (2002) Prefrontal volume in habitually violent subjects with antisocial personality disorder and type 2 alcoholism. Psychiatry Res Neuroimaging 114:95–102

Langleben DD, Schroeder L, Maldjian JA, Gur RC, McDonald S, Ragland JD, O'Brien CP, Childress AR (2002) Brain activity during simulated deception: an event-related functional magnetic resonance study. Neuroimage 15:727–732

Leentjens AFG, Visser-Vandewalle V, Temel Y, Verhey FRJ (2004) Manipuleerbare wilsbekwaamheid: een ethisch probleem bij elektrostimulatie van de nucleus subthalamicus voor ernstige ziekte van Parkinson. Nederlands Tijdschrift voor Geneeskunde 148:1394–1397

Liddle PF, Kiehl KA, Smith AM (2001) Event-related fMRI study of response inhibition. Human Brain Mapping 12:100–109

Limousin P, Krack P, Pollak P, Benazzouz A, Ardouin C, Hoffmann D, Benabid A-L (1998) Electrical stimulation of the subthalamic nucleus in advanced Parkinson's disease. N Engl J Med 339(16):1105–1111

LJN AV1864, District Court Utrecht, 14 February 2006

LJN AY8840, District Court Breda, 26 September 2006

LJN BA3923, District Court Amsterdam, 26 April 2007

LJN BA9671, District Court Utrecht, 16 July 2007

LJN BB2861, District Court 's-Hertogenbosch, 5 September 2007

LJN BC9296, District Court Amsterdam, 28 March 2008

LJN BI6332, District Court Leeuwarden, 4 June 2009

LJN BK3854, Court of Appeals Amsterdam, 19 November 2009

LJN BK4178, District Court Haarlem, 24 November 2009

LJN BK5962, District Court Alkmaar, 24 June 2008

LJN BL5774, District Court 's-Gravenhage, 26 February 2010

LJN BM1948, District Court 's-Gravenhage, 22 April 2010

LJN BM8774, District Court Amsterdam, 21 June 2010

LJN BN0983, District Court Maastricht, 13 July 2010

LJN BN5666, Court of Appeals Amsterdam, 27 August 2010

LJN BN7251, District Court Alkmaar, 16 September 2010

LJN BO0306, District Court Utrecht, 6 October 2010

Maibom HL (2008) The mad, the bad, and the psychopath. Neuroethics 1:167–184

Mayberg HS, Lozano AM, Voon V, McNeely HE, Seminowicz D, Hamani C, Schwalb JM, Kennedy SH (2005) Deep brain stimulation for treatment-resistant depression. Neuron 45(5):651–660

McCabe DP, Castel AD (2007) Seeing is believing: the effect of brain images on judgments of scientific reasoning. Cognition 107:343–352

Merckelbach H, M Jelicic C, de Ruijter (2009) De B. heeft een persoonlijkheidsstoornis en doodt zijn vriendin. Maandblad Geestelijke Volksgezondheid 9:747–759

Miller G (2008) Investigating the psychopathic mind. Science 321:1284–1286

Moll J, de Oliveira-Souza R, Eslinger PJ, Bramati IE, Mourao-Miranda J, Andreiuolo PA, Pessoa L (2002) The neural correlates of moral sensitivity: a functional magnetic resonance imaging investigation of basic and moral emotions. J Neurosci 22:2730–2736

Mooij AWM (2005) De vraag naar de toerekeningsvatbaarheid. Voordrachtenreeks van het Lutje Psychiatrisch-Juridisch Gezelschap 11:7–20

Morse SJ (2004) New neuroscience, old problems. In: Garland B (ed) Neuroscience and the law: brain, mind, and the scales of justice. Dana Press, New York, pp 157–198

Morse SJ (2006) Brain overclaim syndrome and criminal responsibility: a diagnostic note. Ohio State J Criminal Law 3(2):397–412

Morse SJ (2007) The non-problem of free will in forensic psychiatry and psychology. Behav Sci Law 25:203–220

Müller JL, Sommer M, Wagner V, Lange K, Taschler H, Roder CH, Schuierer G, Klein HE, Hajak G (2003) Abnormalities in emotion processing within cortical and subcortical regions in criminal psychopaths: evidence from a functional magnetic resonance imaging study using picture with emotional content. Psychiatry Res Neuroimaging 54:152–162

Ochsner KN, Ludlow DH, Knierim K, Hanelin J, Ramachandran T, Glover GC, Mackey SC (2006) Neural correlates of individual differences in pain-related fear and anxiety. Pain 120:69–77

Okado F, Okajima K (2001) Violent acts associated with fluvoxamine treatment. J Psychiatry Neurosci 26:339–340

Peyron R, Laurent B, Garcia-Larrea L (2000) Functional imaging of brain responses to pain: a review and meta-analysis. J Clin Neurophysiol 30(5):263–288

Raine A, Buchsbaum M, LaCasse L (1997) Brain abnormalities in murderers indicated by positron emission tomography. Biol Psychiatry 42:495–508

Raine A, Meloy JR, Bihrle S, Stoddard J, LaCasse L, Buchsbaum MS (1998) Reduced prefrontal and increased subcortical brain functioning assessed using positron emission tomography in predatory and affective murderers. Behav Sci Law 16:319–332

Reeves D, Mills MJ, Billick SB, Brodie JD (2003) Limitations of brain imaging in forensic psychiatry. J Am Acad Psychiatry Law 31(1):89–96

Reimer M (2008) Psychopathy without (the language of) disorder. Neuroethics 1:185–198

Roper v. Simmons, United States Supreme Court, 1 March 2005

Roskies AL (2006) Neuroscientific challenges to free will and responsibility. Trends Cogn Sci 10(9):419–423

Sapolsky RM (2004) The frontal cortex and the criminal justice system. Philos Trans R Soc B 359:1787–1796

Schleim S, Spranger TM, Erk S, Walter H (2010) From moral to legal judgment: the influence of normative context in lawyers and other academics. Soc Cogn Affect Neurosci. doi:10.1093/scan/nsq010

Sensi M, Eleopra R, Cavallo MA, Sette E, Milani P, Quatrale R, Capone JG, Tugnoli V, Tola MR, Granieri E, Data PG (2004) Explosive-aggressive behavior related to bilateral subthalamic stimulation. Parkinsonism Relat Disord 10:247–251

Seymour B, Singer T, Dolan R (2007) The neurobiology of punishment. Nat Rev Neurosci 8:300–311

Sinnott-Armstrong W, Roskies A, Brown T, Murphy E (2008) Brain images as legal evidence. Episteme J Soc Epistemol 5(3):359–373

Spence SA, Farrow TFD, Herford AE, Wilkinson ID, Zheng Y, Woodruff PWR (2001) Behavioural and functional anatomical correlates of deception in humans. Neuroreport 12(13):2849–2853

Spence SA, Hunter MD, Farrow TFD, Green RD, Leung DH, Hughes CJ, Ganesan V (2004) A cognitive neurobiological account of deception: evidence from functional neuroimaging. Philos Trans R Soc B 359:1755–1762

Stevens L, Prinsen M (2009) Afwezigheid van opzet bij de geestelijk gestoorde verdachte. Expertise en Recht 5(6):113–118

Sturm V, Lenartz D, Koulousakis A, Treuer H, Herholz K, Klein JC, Klosterkötter J (2003) The nucleus accumbens: a target for deep brain stimulation in obsessive-compulsive- and anxiety-disorders. J Chem Neuroanat 26(4):293–299

Tovino S (2007) Functional neuroimaging and the law: trends and directions for future scholarship. Am J Bioeth 7(9):44–56

Vedder A, Klaming L (2010) Human enhancement for the common good: using neurotechnologies to improve eyewitness memory. Am J Bioeth Neurosci 1(3):22–33

Vincent NA (2010a) On the relevance of neuroscience to criminal law. Criminal Law Philos 4:77–98

Vincent NA (2010b) Madness, badness and neuroimaging-based responsibility assessments. In: Freeman M (ed) Law and neuroscience, current legal issues. Oxford University Press, Oxford

Volkow ND, Tancredi LR, Grant C, Gillespie H, Valentine A, Mullani N, Wang GL, Hollister L (1995) Brain glucose metabolism in violent psychiatric patients: a preliminary study. Psychiatry Res 61:243–253

Weaver FW, Follett K, Stern M, Hur K, Harris C, Marks WJ Jr, Rothlind J, Sagher O, Reda D, Moy CS, Pahwa R, Burchiel K, Hogarth P, Lai EC, Duda JE, Holloway K, Samii A, Horn S, Bronstein J, Stoner G, Heemskerk J, Huang GD (2009) Bilateral deep brain stimulation vs bestmedical therapy for patients with advanced Parkinson disease: a randomized controlled trial. J Am Med Assoc 301(1):63–73

Weisberg DS, Keil FC, Goodstein J, Rawson E, Gray JR (2008) The seductive allure of neuroscience explanations. J Cogn Neurosci 20(3):470–477

Wolpe PR, Foster KR, Langleben DD (2005) Emerging neurotechnologies for lie-detection: promises and perils. Am J Bioeth 5(2):39–49

Neuroscience and the Law in New Zealand

Mark Henaghan and Kate Rouch

Abstract The New Zealand Court of Appeal has rejected evidence of neuroimaging to help juries assess the capacity of the accused in an insanity plea. This chapter says the Court of Appeal was right to do so because neuroimaging should not replace the role of the jury. The chapter explains; that neuroscience will help us better understand how the brain functions and what relationship there is between that functioning and how we make decisions. The chapter concludes that neuroscience will be helpful for insight into the human condition but cannot replace the moral choices of what we think is right or wrong or whether we should be culpable or should not be.

1 Introduction

In *R v Dixon*[1] the New Zealand Court of Appeal has emphatically rejected evidence of neuroimaging. Antonie Dixon pleaded not guilty by way of insanity under s 23 of the New Zealand Crimes Act 1961 which states that:

23 Insanity

(1) Every one shall be presumed to be sane at the time of doing or omitting any act until the contrary is proved.

Professor Mark Henaghan is the Dean of the Faculty of Law at the University of Otago. Kate Rouch is a summer research assistant studying for the LLB degree at the University of Otago Faculty of Law.

[1] *R v Dixon* (2008) (CA).

M. Henaghan (✉) • K. Rouch (✉)
Faculty of Law, University of Otago, PO Box 56, Dunedin, New Zealand
e-mail: mark.henaghan@otago.ac.nz; kate.rouch@hotmail.co.uk

T.M. Spranger (ed.), *International Neurolaw*,
DOI 10.1007/978-3-642-21541-4_14, © Springer-Verlag Berlin Heidelberg 2012

(2) No person shall be convicted of an offence by reason of an act done or omitted by him when labouring under natural imbecility or disease of the mind to such an extent as to render him incapable

   (a) of understanding the nature and quality of the act or omission; or
   (b) of knowing that the act or omission was morally wrong, having regard to the commonly accepted standards of right and wrong.

Dixon argued that when he killed one person and badly mutilated another, he did not know that his actions were morally wrong. As the Court of Appeal said, the question for the jury was "whether the defence had established, on the balance of probabilities that Mr Dixon, because of his disease of the mind, did not know that what he was doing was morally wrong".[2] The Court of Appeal said that this was a "comparatively simple inquiry", and that "simplicity in this area is highly desirable".[3] According to Dixon J in *R v Porter*[4] the issue is not one of right or wrong in the abstract, but "whether [the accused] was able to appreciate the wrongfulness of the particular act he was doing at the particular time".[5]

The Court of Appeal went out of its way to avoid any suggestion that "under this limb of the insanity defence, the jurors' task is to perform a neurological or psychiatric assessment of the accused's brain or its workings, with a view to establishing its capacity".[6]

This chapter will explain the neuroscience of neuroimaging, assess its potential fields of use in law and analyse the potential problems with using neuroscience findings in law. Ultimately, this chapter concludes that the New Zealand Court of Appeal was correct to avoid complicating and confusing what are ultimately moral issues.

Neuroscience is likely to show us how brains function differently depending on genetic and environmental factors.[7] We then need to decide as a society how we should respond to the different ways brains function. If some people's brains give them less self control, or lessen their ability to make appropriate decisions in certain circumstances, we need to decide whether the law should accommodate this by having different levels of responsibility, or by determining a particular threshold of control necessary to be deemed responsible. These are moral choices that a society has to make about where lines are to be drawn for criminal responsibility to be established. What neuroscience is never going to be able to determine is culpability once that line has been drawn. Also, any brain scan is likely to occur long after the

---

[2] *R v Dixon* (2008), at p. 32.

[3] *R v Dixon* (2008), at p. 32.

[4] *R v Porter* (1933) 55 CLR 182.

[5] *R v Porter* (1933) 55 CLR 182, at p. 33.

[6] *R v Dixon* (2008) (CA) at p. 32.

[7] Merriman and Cameron (2007), pp. 62–67.

behaviour in question, which makes it difficult to draw firm conclusions as to the state of the brain at the actual time of the behaviour.[8]

2   What Is Cognitive Neuroscience?

Cognitive neuroscience is a relatively new sub-field of neuroscience, described as "an investigational field that seeks to understand how human sensory systems, motor systems, attention, memory, language higher cognitive functions, emotions and even consciousness arise from the structure and function of the brain".[9] It can also be described as a "bridging discipline" between biology and neuroscience on one hand, and cognitive science and psychology on the other.

The most common use of neuroscience in law is neuroimaging technology. This allows researchers to see structures and functions in the brain, and "search" for brain patterns that match the known patterns of capacity, sanity, innocence and guilt.[10]

The current status of neuroscience in law is still at that stage where it is not possible to try someone simply on the basis that a particular brain pattern means a particular thought. Thought is elusive and is not capable of being described merely by a scientific formula.

2.1   Types of Neuroimaging

The most advanced neuroimaging technology is functional magnetic resonance imaging (fMRI), which uses an MRI scanner to measure blood flow to potentially show causal links between brain function and behavioural patterns. However, this technology is only in a developing stage. There are only two fully operational fMRI businesses performing brain scans in the United States.[11] Other commonly used neuroimaging technologies are electroencephalography (EEG), and positron emission tomography (PET). There have been 130 reported cases in the United States involving PET and EEG brain imaging evidence, and two involving fMRI evidence.[12]

---

[8] Jones et al. (2009).

[9] Gazzaniga (2005), p. 88.

[10] Compton (2010), p. 333.

[11] Moreno (2009), p. 725.

[12] Yang et al. (2008), pp. 77–78.

2.2   Fields of Uses for Neuroimaging

Neuroimaging in one form or another is currently used in intellectual property law, tort law, consumer law, health law, employment law, constitutional and criminal law.[13] Case law involving neuroscience evidence is still relatively scarce as "lie detector pattern matching" technologies such as fMRI scans are still in their infancy and what role neuroscience ought to play is yet unclear.[14]

Foundations such as the Law and Neuroscience Project funded by the John D and Catherine T MacArthur Foundation are currently bringing together top scientists and legal scholars to attempt to determine the proper role of neuroscience in the legal system. There is also concern that this evidence will influence the role of the jury, sentencing practices and bring an unbalanced focus on offender propensities.

3   Use of Neuroscience in Civil Cases

3.1   Proving Harm

Neuroscience evidence can be used to prove actual harm in tort law. In America at least, the prevailing customary view of physical and emotional harm is that the former brings a stronger claim to compensation, as it is objectively verifiable and arguably more important.[15] Neuroimaging may provide empirical evidence of emotional distress, as well as readdressing approaches towards emotional and physical harms.

3.2   Uses in Health Law

In terms of health law and for proving injuries to the brain or spine for accident compensation purposes, neuroimaging can provide an objective assessment of the state of these body parts in the same way that medical certificates certify that an individual was seen and found ill.[16] Additionally, it can be used in decisions regarding the removal of a vegetative state person from life supports.

---

[13] Shafi (2009), p. 27.

[14] Morse (2008).

[15] Grey (2011).

[16] Pettit (2007).

3.3   Uses in Contract Law

It can also be used to show that a party lacked the sufficient cognitive capacity required to form a valid contract,[17] or lacked testamentary capacity.

4   Use of Neuroscience in Criminal Cases

Mostly, defendants in criminal cases provide neuroimaging evidence to show diminished capacity or insanity during trial, and as supporting mitigation during sentencing.

The use of neuroscience in criminal law has largely come from the research done by Californian neurophysicist Benjamin Libet, who in the 1980s had a profound impact on the scientific understanding of the association between the brain and behaviour, while challenging the notion of free will. Through a series of experiments in which Libet had subjects make voluntary hand movements while measuring their brain activity, he concluded the brain was active even before the subject was aware of having made the conscious decision to move their hand. This inferred that choice may be determined in the brain before the mind acts, making free will illusory.[18] According to Libet, this has major implications for individual responsibility and the concept of guilt.[19]

In terms of New Zealand case law, recent observations in the New Zealand Court of Appeal in *R v Dixon*[20] excludes the possibility of admitting neuroimaging evidence to provide an assessment of the accused's brain, or its working order, in establishing its capacity to know right from wrong.

4.1   Negating Mens Rea

Neuroscience evidence may be used to negate mens rea. In *State v Anderson*,[21] the defendant provided an expert testimony supported by neuroscience evidence that his brain damage induced depression and paranoia to such an extent that he was incapable of the degree of premeditation required for first-degree murder. The jury was not convinced and Anderson was convicted on all counts. Conversely, on

---

[17] Compton (2010).

[18] Choong (2007).

[19] Libet (1999), p. 339.

[20] *R v Dixon* (2008) (CA).

[21] *State v Anderson* (2002).

appeal in *United States v Erskine*[22] the defendant was permitted to introduce evidence of a brain scan which his counsel claimed showed his lack of specific intent. The conviction was reversed.

According to Joshua Greene, the impact of Benjamin Libet's research to the concept of mens rea is that since human beings are their individual brains, this requires a redefinition of a "guilty mind" and also of punishment.[23] If an individual personhood is co-extensive with their brain and nothing more, then predestination by forces hard-wired into our brains is determinative and the idea of free and rational choice is illusion. The integrities of moral cognition and demands of justice so far as this goes, collapse.[24]

4.2 Supporting Insanity

Alternatively, the neuroscience evidence may be used to support a claim of not guilty by reason of insanity. The defendant in *United States v Hinckley*,[25] who had attempted to assassinate President Ronald Regan, presented CT scan evidence showing a "widening" of sulci in Hinckley's brain in support of a diagnosis of schizophrenia. This was admitted notwithstanding the fact that there is no known "normal" figure of sulcal width.[26] The jury found the defendant not guilty by reason of insanity.

4.3 In Sentencing

The use of neuroscientific evidence in the formulation of claims for mitigation of penalty in the United States during capital sentencing cases is apparently widespread.[27] The aim in using neuroscience is to "bolster the defendants' mitigation claims with cutting-edge neuroimaging research that demonstrates a biological disposition to criminal violence".[28]

---

[22] *United States v Erskine* (1978).

[23] Rosen (2007).

[24] Pettit (2007).

[25] *United States v Hinckley* (1981).

[26] Perlin (2009), pp. 885–916.

[27] Brookbanks (2008), p. 623.

[28] Gazzaniga (2005).

5   Potential Problems with Neurolaw

Chief among concerns over using neuroscience information as evidence include the effect on juries, sentencing practices and the manner in which offenders – or even potential offenders – are viewed.

5.1   Human Rights Issues

From a human rights perspective, there are issues surrounding whether a brain scan could be obtained without something akin to a "search warrant", and the potential clashes with the New Zealand Bill of Rights Act.

In India during June 2008, evidence derived from a Brain Electrical Oscillations Signature (BEOS) was admitted during the murder trial of Aditi Sharma. During a police interrogation, the defendant, who was suspected of poisoning her former fiancé, had thirty two electrodes placed on her head while police read her their account of the murder. Despite the fact that the defendant made no verbal responses during the test, and without reference to specific evidence of the test's scientific validity, the judge concluded that the test proved she had "experiential knowledge" of the crime.[29]

Being compelled to be a witness against yourself arguably violates the New Zealand Bill of Rights Act, the primary legislation protecting civil rights in New Zealand. If applied to the above example of Aditi Sharma, this could even amount to an unreasonable search or seizure. Civil rights protection in New Zealand is weaker than in the United States where most neuroscience evidence is most frequently being used, so the position is unclear.

5.2   Efficacy of Evidence

The efficacy of such evidence is also of concern. Common law courts appear universally suspicious of such things as polygraph evidence, and it is suggested that some standard of accuracy would need to be established before neuroscience evidence can be regularly introduced as supporting evidence.

The Evidence Act 2006 provided that an opinion that is part of expert evidence offered in a proceeding is admissible if the fact finder "is likely to obtain substantial help from the opinion in understanding other evidence in the proceeding or in ascertaining any fact that is of consequence to the determination of the

---

[29] Giridharadas (2008).

proceeding".[30] This rule is applicable to neuroscience experts, but the question is the permissible scope of such evidence, and whether it tends to support or undermine institutions of justice.

Warren Brookbanks argues that while neuroscience may be able to supply evidence of a person's cognitive capacities for learning, memory and attention, it may be beyond scientific justification to infer anything about culpability.[31] In a similar line of thinking, Stephen Morse argues that although neuroscientific evidence can provide assistance in evaluating the relation of self-control to human responsibility, it does not answer the question of how much control ability is required for responsibility.[32] This is still a normative, moral and ultimately legal question.[33]

6 Effect on Juries

There is reportedly a large scope for such evidence to be misunderstood or misused, as happens with DNA probability evidence. Given the complex nature of neuroimaging, juries may not understand the true limits of the technology, tending to fall to the "lure of the mechanism", and take such evidence as an infallible truth.[34]

The nature of the adversarial process may lead to juries misinterpreting many complex findings, and as such it is widely recommended that any neuroimaging evidence be accompanied by a cautionary note. Individual brains can be affected by numerous external factors or circumstances causing that individual to perform mental or behavioural tasks in a manner different to another person. There is no reliable "normal" brain functioning, so while functional neuroimaging provides information about brain function in general, it cannot fully explain individual brain function.[35] There still remains a deductive gap between the hard science of the image and the subjective clinical interpretations admissible as evidence in courts. Jurors would have to be given a cautionary note to avoid this danger.

6.1 Abolition of Juries

Some concern exists that juries may not even be involved in the trial process,[36] even to the point that jury trials are to be abolished after lawyers begin insisting that

---

[30] New Zealand Evidence Act 2006, s 25(1).

[31] Brookbanks (2008), p. 623.

[32] Morse (2004).

[33] Morse (2004), p. 158.

[34] Morse (2008).

[35] Morse (2008).

[36] Erickson (2010).

jurors' brains be scanned for prejudice, antipathy or characteristics similar to the defendant.

7   Impact upon Offenders and Potential Offenders

7.1   Neuropersonal Model Shift

There are arguable concerns that neuroimaging technology and other such "brain scan" technology would cause a shift from judging the offender as a whole person to simply pointing at the brain as responsible for what the body has done due to an inherent mechanical and determined nature. This neuropersonal model promotes the idea that individuals only commit crimes because they possess brains that are unwell.[37]

If neuroscience and behavioural genetics can go some way to explaining and predicting behaviour, with particular reference to issues such as free will, determinism and their effect on the extent to which criminal culpability is likely to be undone by new scientific discoveries, then neuroscience may have an important role to play in assessing offenders.[38] Social agencies may learn how to best handle criminal behaviour accompanying drug addiction, given an apparent genetic predisposition towards the same. It is suggested by Garland and Frankel that by changing the way society views and understands addiction, drug use, and treatment, neuroscience has the potential to reshape our policies on criminalisation and incarceration as they pertain to drug-related offences.[39]

7.2   Uses in "Counterterrorism"

Recently, the United States Department of Homeland Security began testing a new "mind-reading" device to purportedly screen people at security checkpoints.[40] The device works by remotely measuring body temperature, heart rate and respiration and then measures this against physiological norms to generate conclusion about a subject's potential dangerousness.

---

[37] Pockett (2010), pp. 281–293.

[38] Garland and Frankel (2006), pp. 101–103.

[39] Garland and Frankel (2006), p. 104.

[40] Moreno (2009).

8 Conclusion

It is generally held that neuroscience, with particular reference to fMRI scans that attempt to find a causal nexus between brain pattern and behavioural function, are too nascent to require a redefinition of legal concepts, or have much drastic effect on legal systems.[41]

Although neuroscience could have a large effect on the legal system, it is unlikely that this will happen any time soon. Neuroscience is incapable of changing moral intuitions. Science is not concerned with values, but facts. Law, on the other hand, is fundamentally value based and operates on the basis that values both matter and are determinative.[42] In the words of Waldbauer and Gazzaniga, a decision about an individual's culpability is ultimately a judgement about a person made by another person. Unlike genetic paternity, questions such as diminished culpability cannot be handed to science for final arbitration. Responsibility is ultimately a judgement made within a social and legal framework.[43]

Where neuroscience will have the most impact is in developing understanding of the different ways that brains work and how this impacts on our ability to make decisions. The law's assumption that we are all equally competent will be challenged by neuroscience as it develops. We then have to face the difficult moral choice as a society that if there are quite different capacities between people because of how their brains function, then how should the law adjust accordingly?[44] The challenge for neuroscience is to prove that the way our brains function defines who we are. This is a chicken and egg dilemma. Do we make the choices and our brains follow, or does our brain make the choice and we follow? If the latter, then our concepts of responsibility will need radical revision. Pardo and Patterson argue that "The mind is not an entity or substance at all (whether non-physical or physical). To have a mind is to possess a certain array of rational powers exhibited in thought, feeling and action".[45] In essence "it is people who think, feel, intend and know (not parts of their brains)".[46]

References

Annas G (2007) Imaginaing a new era of neuroimaging, neuroethics and neurolaw. Am J Law Med 33:163–170
Brookbanks W (2008) Neuroscience, 'folk psychology' and the future of criminal responsibility. New Zealand Law Rev 2008:623–637

[41] Brookbanks (2008), p. 632.
[42] Brookbanks (2008), p. 636.
[43] Waldbauer and Gazzaniga (2001), pp. 357–364.
[44] Annas (2007).
[45] Pardo and Patterson (2010), p. 1249.
[46] Pardo and Patterson (2010), p. 1250.

Choong R (2007) Free nil – reflections on freedom, neurobiology and sin. Transdisciplinarity and the unity of knowledge conference, Philadelphia

Compton E (2010) Not guilty by reason of neuroimaging: the need for cautionary jury instructions for neuroscience evidence in criminal trials. Vanderbilt J Entertain Technol 12:333–355

Erickson S (2010) Blaming the brain. Minnesota J Law Sci Technol 11:22–77

Garland B, Frankel M (2006) Considering convergence: a policy dialogue about behavioural genetics, neuroscience and law. Law Contemp Probl 69:101–104

Gazzaniga M (2005) The ethical brain. Dana Press, New York (cited in Snead O (2007) Neuroimaging and the complexity of capital punishment. NY Univ Law Rev 82:1265–1339)

Giridharadas A (2008) India's novel use of brain scans in court is debated. New York Times, 14 September 2008

Grey B (2011) Neuroscience and emotional harm in tort law: rethinking the American approach to freestanding emotional distress claims. Current Legal Issues: Law and Neuroscience 13:203–229

Jones O et al. (2009) Brain imaging for legal thinkers: a guide for the perplexed. Stanford Technol Law Rev 5 (3):1–11

Libet B (1999) How does conscious experience arise? The neural time factor. Brain Res Bull 50:339–340

Merriman T, Cameron V (2007) Risk-taking: behind the warrior gene story. New Zealand Med J 120(1250):62–67 (citing Caspi A et al. (2002) Role of genotype in the cycle of violence in maltreated children. Science 297(5582):851–854; Kim-Cohen J et al. (2006) MAOA, maltreatment, and gene environment interaction predicting children's mental health: new evidence and a meta-analysis. Mol Psychiatry 11:903–913; Spatz-Widon C, Brzustowicz LN (2006) MAOA and the 'cycle of violence': childhood abuse and neglect. MAOA genotype and risk for violent and antisocial behaviour. Biol Psychiatry 60:684–681)

Moreno J (2009) Future of neuroimaged lie detection and the Law. Univ Akron Law Rev 42:717–738

Morse S (2004) New neuroscience, old problems. In: Garland F (ed) Neuroscience and the law – brain, mind and the scales of justice. Dana Press, New York

Morse S (2008) Neuroscience increasingly presented as evidence for trials in US Courts. Fox News, 3 March 2008 (quoted in Associated Press)

Pardo M, Patterson D (2010) Philosophical foundations of law and neuroscience. Univ Akron Law Rev 4:1211–1250

Perlin M (2009) His brain has been mismanaged with great skill: How will jurors respond to neuroimaging testimony in insanity defence cases. Akron Law Rev 42:885–916

Pettit M (2007) FMRI and BF meet FRE: brain imaging and the federal rules of evidence. Am J Law Med 33:319–340

Pockett S (2010) The concept of free will: philosophy, neuroscience and the law. Behav Sci Law 25:281–293

R v Dixon (2008) 2 NZLR 617 (CA)

R v Porter (1933) 55 CLR 182

Rosen J (2007) The brain on the stand. New York Times, 11 March 2007

Shafi N (2009) Neuroscience and the law: the evidentiary value of brain imaging. Graduate Student J Psychol 11:27–39

State v Anderson (2002) 79 S.W.3d 420 (Mo. 2002)

United States v Erskine (1978) 588 F.2d 721 (9th Cir. 1978)

United States v Hinckley (1981) 525 F. Supp. 1324 (D.D.C. 1981)

Waldbauer J, Gazzaniga M (2001) The divergence of neuroscience and law. Jurimetrics 41:357–364

Yang Y et al (2008) Brain abnormalities in antisocial individuals: implications for the law. Behav Sci Law 26:65–83

Switzerland: Brain Research and the Law

Rainer J. Schweizer and Severin Bischof

Abstract Many of the so far discussed methods of neurosciences have not yet reached a state in which they can be taken as reliable. Thus, the discussion on their application is still at a very early stage. However, it is the task of law to provide a legal compass showing the direction in which new technical developments have to be measured. In this context, one should not only see the risks of neuroscientific developments, but also bear in mind the new opportunities and chances.

1 Introduction

1.1 Recent Neuroscientific Insights

Various insights of the neurosciences within the field of brain research also require a discussion regarding their legal consequences. In what follows some of these research results will be mentioned; a more extensive presentation is already given elsewhere.[1]

1. According to the experiments by Libet as well as to the subsequent works by Haggard and Eimer an intentional action does already begin split seconds before it enters into the consciousness.[2]

---

[1] See Schleim 2008 or Roth 2003.

[2] Pauen/Roth, p. 72 et seq.

R.J. Schweizer (✉) • S. Bischof (✉)
Forschungsgemeinschaft für Rechtswissenschaft der Universität St. Gallen, Bodanstrasse 6,
St. Gallen 9000, Switzerland
e-mail: rainer.schweizer@unisg.ch; severin.bischof@unisg.ch

T.M. Spranger (ed.), *International Neurolaw*,   269
DOI 10.1007/978-3-642-21541-4_15, © Springer-Verlag Berlin Heidelberg 2012

2. In the field of criminology certain connections – which are yet to be taken with caution – between criminal behavior and dysfunctions or damages in the brain structure can be detected.[3]

3. Improved imaging procedures for the brain such as fMRI (functional magnetic resonance tomography or nuclear spin tomography),[4] PET (positron emission tomography),[5] EEG (electroencephalography[6]), etc. can be employed in order to detect lies (measurement of the so called P300-signals by EEG),[7] and they also provide first results in "reading" concrete *contents* of thoughts.[8]

4. First approaches of therapeutic stimulations through TMS (transcranial magnetic stimulation) or TDCS (transcranial direct current stimulation), for instance in order to cure depression, are very insecure so far.[9]

1.2   Evaluation

The insights of the neurosciences that have shortly been outlined above are by no means uncontroversial. Frank Rösler, for instance, identifies principal methodological problems in the construction of the experiments by Libet, Haggard, and Eimer.[10] Herbert Helmrich even claims contentual deliquencies of the experiments.[11] With regard to a juridical discourse this means in any way that biomedical conclusions of brain research shall still be judged with caution.[12]

Regarding the detection of thought contents it can justifiably be stated that the corresponding research is still in its infancy. To date, not thoughts are read, but rather brain-physiological activities, which do not sufficiently explain the functioning of the brain.[13] Although, for instance, it can be identified with a strike rate of approximately 85% whether a test person is thinking of a house or a face,[14] the insight into the world of thoughts of the human brain does not (yet) reach much further than that. Also, measurements of brainwaves do not yet show the reliability

---

[3] See Kiehl, Piefke/Markowitsch 2008, Neuroanatomische und neurofunktionelle Grundlagen gestörter kognitiv-emotionaler Verarbeitungsprozesse bei Straftätern, in: Grün et al., p. 96 et seq. and Roth, p. 115 et seq.

[4] See Schleim 2008, p. 39 et seq., p. 65.

[5] Ibid., p. 65.

[6] Ibid., l.c.

[7] Ibid., p. 34 et seq.

[8] Ibid., p. 79 et seq.

[9] Schleim 2008, p. 66 et seq.

[10] Rösler 2008, Was verraten die Libet-Experimente über den "freien Willen"? - Leider nicht sehr viel!, in: Lampe/Pauen/Roth, p. 145 et seq.

[11] Helmrich 2004, Wir können auch anders: Kritik der Libet-Experimente, in: Geyer, p. 94 et seq.

[12] See Bennet/Hacker 2010, passim.

[13] Mahlmann 2010, §18, margin number 22.

[14] Schleim 2008, p. 79 et seq.

that is necessary in order to examine the verisimilitude of statements. Accordingly, P300-signals, for instance, can be altered lastingly through "countermeasures", which makes it extremely difficult to prove lies.[15]

Thus, although the latest neuroscientific insights do not seem to be sufficiently secure at all, today certain brain researchers such as Wolf Singer[16] and Gerhard Roth[17] refer to exactly these insights when they plead in favor of a paradigm shift for society as whole, in the centre of which is the negation of the freedom of the human will. In the Swiss (legal) discourse such far reaching opinions are predominantly rejected in a critical manner. Marcel Senn, for example, judges that methodologically, brain research can currently not honor the advocated pretension to form a new image of humanity or society.[18] At most, brain research might contribute to the humanization of single cases in partial areas such as criminal law.[19] In a way that is symptomatic for the discussion Daniel Hell warns against an overinterpretation or absolutization of neuro-scientific results, namely when consequences in the realm of law are involved.[20] From the self-evaluation which Benjamin Libet draws from his own experiments Brigitte Tag concludes that the far-reaching deterministic consequences of brain research are misleading. According to her, the question why we act the way we do is in the end still unanswered.[21] The list of such objections could be continued.

Finally, it has also to be taken into consideration that every science – and as such the neurosciences, too – stand under the reserve of future insights. Already for this reason a scientific statement or thesis can hardly claim universality within a science which is strongly developing.[22]Nevertheless, jurisprudence must occupy itself with the exciting insights of the neurosciences, and it must discuss possible consequences for the legal system and the legal doctrine.

1.3   Possible Consequences For The Law

1.3.1   With Regard To The Present State Of Research

Which consequences do the recent results of brain research have for the law? Namely the following aspects appear to be of relevance:

---

[15] Ibid., p. 35.

[16] See Singer, p. 24 et seq.

[17] See Roth, p. 166 et seq.

[18] Senn 2008, Grenzen und Risiken der Hirnforschung: Folgerungen für die Rechtsordnung, in: Fachtagung Zürich, p. 11.

[19] Ibid., p. 11 et seq.

[20] Hell 2008, Das Gehirn ist kein Agent – Konsequenzen der Hirnforschung für das Recht (aus psychiatrischer Sicht), in: Fachtagung Zürich, p. 13.

[21] Tag 2008, Schnittstellen zwischen Recht, Gehirn und Technik, in: Fachtagung Zürich, p. 62.

[22] Senn 2008, Grenzen und Risiken der Hirnforschung: Folgerungen für die Rechtsordnung, in: Fachtagung Zürich, p. 7 et seq.

1. Improved procedures of brain imaging promise a more precise diagnosis of pathological results. This makes it possible to evaluate the power of judgment as well as culpability with greater precision, too.[23]
2. Besides more exact methods of diagnosis, it seems possible to substantially improve the therapies of diseases such as dementia, depression, fear, Alzheimer's, Parkinson's disease or multiple sclerosis in considerable time. Thus, beyond the evaluation of the power of judgment and culpability, one might possibly *give back* legal capacity to the persons concerned.

1.3.2   With Regard To A Possible Future State Of Research

Already today brain researchers proclaim a new era of right that supposedly looms attendant on neuroscientific insights. Thus, it makes sense to shortly realize the consequences in an anticipative manner.

1. In the field of criminal law much discussed at least the guilt concept would have to be reconsidered. The direction of the approached reform that is being discussed clearly points to preventive approaches, which eventually would make possible to already detect (so called) "dangerous subjects" by an early genetic diagnosis and to submit them to therapeutic and/or security-measures after their dangerousness has been prognosticated.[24] The highly problematic nature of such a qualification with respect to basic and human rights shall only be mentioned here so far.
2. Already today the surveillance of public places by cameras, metal detectors at airports, etc. is an everyday reality. With its so-called "brainscanning" brain research promises another kind of surveillance, as it will make possible to identify terrorists, for instance, by "scanning" their thoughts.[25]
3. The detection of lies by measuring brainwaves has already been addressed above. It actually is imaginable that such methods will be refined in the future and that they would be applicable in a more or less reliable way to juridical procedures of evaluation and decision.
4. Finally, one has to be aware of the fact that a legal-cultural decision to abandon the acknowledgment of the freedom of the will of the natural person will carry inestimable, fundamental consequences in all realms of law. To only mention a few: One has to think of the right to exercise ones political rights, of the right to

---

[23] See Tag 2007, Neurowissenschaft und Strafrecht, in: Holdegger et al., p. 350 as well as Gschwend 2006, Konsequenzen aus den Erkenntnissen der Hirnforschung für Straf- und Privatrecht, in: Senn/Puskás, p. 149.

[24] See e.g. Bommer 2007, Hirnforschung und Schuldstrafrecht, in: Stichweh/Bommer, p. 31 or Hillenkamp 2007, Das limbische System: Der Täter hinter dem Täter?, in: Hillenkamp, p. 98 et seq.

[25] Schleim 2008, p. 9.

marriage or to divorce, of the ability to testify or, last but not least, of the responsibility of the judges to the legal judgment.[26]

2 Approach to Neuroscientific Insights By The Swiss Constitutional Law

2.1 Fundamental Liberal Conception Of Humanity In The Constitution

Within the legal system of Switzerland (as well), one of the first questions of law is in how far the Swiss constitutional law can or shall adopt the neuroscientific results; furthermore, how namely the basic and human rights constitute a framework, which limits the neuroscientific research and its potential implications. As the central argument of brain researchers is in the end aiming toward a negation of the freedom of the will, it is reasonable to first analyze the constitutional law with regard to its statements about the "freedom" of humans. The purpose of this is to find a legal concept of "freedom" and to compare it with the neurological non-existence of "freedom of the will".

Already the preamble of the Swiss federal constitution (BV[27]) which introduces the BV with great symbolic power[28] conveys certain conclusions about the constitutional conception of freedom. The invocatio dei (appeal to God) at the beginning of the preamble is not only a Christian-religious confession; rather, by pointing to a transcendental power, it is also a reminder that the state is in the end an imperfect "work of man" and that it does not have the last command over its inhabitants.[29] In the last section the preamble explicitly mentions the freedom of the individual as a fundamental value, which, however, is not to be seen statistically, but as an everlasting task of the individual within the community.[30] It is important to note that this is not about an absolute freedom of action but about an *obligatory* freedom that is pitted against the public good.[31] Article 2, the article on purposes,[32] thus consistently mentions the protection of freedom and the rights of the people as the first duty of the state.[33]

---

[26] See also Hillenkamp 2006, Das limbische System: Der Täter hinter dem Täter?, in: Hillenkamp, p. 95.

[27] In Switzerland: Bundesverfassung, from now on cited as BV.

[28] See Ehrenzeller 2008, Zur Präambel, in: Ehrenzeller et al., margin number 7.

[29] Ibid., margin number 17.

[30] Ibid., margin number 26.

[31] Ibid., margin number 26.

[32] Art. 2 para. 1 BV: The Swiss Confederation shall protect the liberty and rights of the people and safeguard the independence and security of the country.

[33] Compare Ehrenzeller 2008, on art. 2, in: Ehrenzeller et al., margin number 15.

Article 6 BV[34] gets to the heart of what *freedom* in the sense of the BV means with regard to individual and societal responsibility. As a fundamental-values-article, it is preamble-like and in its normativity casts forth onto the entire constitution.[35] The article gains its concrete normative content only with regard to its particular context[36]; still, in principal two different norm-contents can be identified. On the one hand, Article 6 BV demands that everybody takes *responsibility* for herself or himself – a claim that is closely related to the concepts of "personhood, to human dignity, individuality and freedom".[37] The capability to take responsibility is thereby presupposed; it thus conforms with the image of humanity of the BV. On the other hand, Article 6 BV asks the individual person to take *co-responsibility* within the community, to participate in the tasks of the community.[38] According to Jörg Paul Müller, taking co-responsibility is one of the essential contents of freedom in the sense of the Swiss constitution. The individual freedom shall be secured beforehand the so called freedom rights by the status of the citizen and his contribution to shape the constitution.[39]

If one wants to elicit the conception of the constitution regarding human beings, then the fundamental rights, namely the *liberty* rights are of central importance as "the most fundamental norms and the most important organizing principles",[40] including the fact that they are strongly influenced by the international and European human rights.[41] The liberty rights are characterized by the very fact that they antagonize threats to freedom and personal development by state protection duties and by doing so account for the natural vulnerability and need for protection of the human being in light of economic, social and state power.[42] Thus, they presuppose a considerable degree of freedom in shaping one's own life, as otherwise the qualified objects of protection from the liberty rights would be omitted in each case.

The BV thus draws a picture of *freedom* in which freedom is closely connected to *duty*; the human being is not simply free for his own sake, but must time after time work on his freedom again, and as a social being he is thereby explicitly involved in *duties with respect to the community*, too. In contrast to the neurosciences the BV does not speak of determined or undetermined conceptions of (non-)freedom. Neither does it assume that human beings are in their actions *free*

---

[34] All individuals shall take responsibility for themselves and shall, according to their abilities, contribute to achieving the tasks of the state and society.

[35] This is the opinion of Häberle 2008, on art. 6 BV, in: Ehrenzeller et al., margin number 10.

[36] Ibid., margin number 11.

[37] See ibid., margin number 13.

[38] Ibid., margin number 14.

[39] Müller 2007, Geschichtliche Grundlagen, Zielsetzung und Funktionen der Grundrechte, in: Merten et al., margin number 22.

[40] Müller 2001, Allgemeine Bemerkungen zu den Grundrechten, in: Thürer et al., margin number 6.

[41] Schweizer 2008, St. Galler Kommentar zur Bundesverfassung, Vorbemerkungen on art. 7–36 BV, margin number 5 et seq., 15 et seq.

[42] Müller 2001, Allgemeine Bemerkungen zu den Grundrechten, in: Thürer et al., margin number 6.

from coercions.[43] However, it is at least being presupposed that human beings can consciously direct their actions to higher goals and values, because only then is the performance of duties and thus freedom in the sense of the BV possible.[44] With this, the constitutional law creates a (legal) construction of freedom, which is indispensable as the foundation of a legal system that is based on human dignity and self-determination[45], but which so far could not (yet) plausibly be called into question by the neurosciences (see above).

2.2   Intraconstitutional Reception of The Neuroscientific Results in Realms Which are Relevant to Fundamental Rights?

2.2.1   In The Realm Of Biomedical Application

Aspiration Of The Neurosciences

In the course of the current debates in the field of interfaces between brain research and law brain researchers like Wolf Singer and Gerhard Roth[46] readily take the liberty of presenting the normative consequences of their research onto the field of (criminal) law.[47] They fail to recognize, however, that even an incontrovertible scientific proof that refutes the existence of a free will would not be able to overthrow the liberal constitutional order of the judicial system, because due to its normativity the judicial system would continue to hold.[48] However, this does not yet mean that the judicial order wants to ignore scientific insights – nor that it does so. In what follows the constitutional law shall therefore be analyzed with respect to the question which guaranties of "listening" the constitutional law does give to research. However, at the same time it must be examined what kind of protection humans have as "guinea pigs" for brain research and, finally, as possible users of biomedical and other applications.

The BV grants qualified protection not least to neuroscientists, too. It does so by protecting the possibility to get "into contradiction with established conceptions,

---

[43] On that note also see Gschwend 2006, Konsequenzen aus den Erkenntnissen der Hirnforschung für Straf- und Privatrecht, in: Senn/Puskás, p. 148.

[44] See the Kantian conception of freedom as it has been presented by Höffe (Höffe 2004, Der entlarvte Ruck. Was sagt Kant den Hirnforschern?, in: Geyer, p. 181).

[45] About the "legal value of freedom" see Mahlmann 2010, §27, margin number 1 et seq.

[46] For instance Roth/Merkel 2008, Bestrafung oder Therapie? Möglichkeiten und Grenzen staatlicher Sanktion unter Berücksichtigung der Hirnforschung, in: Fachtagung Zürich, p. 21 et seq.

[47] See also the critique of Senn (Senn 2008, Grenzen und Risiken der Hirnforschung, in: Fachtagung Zürich, p. 4).

[48] See Möllers 2008, Willensfreiheit durch Verfassungsrecht, in: Lampe/Pauen/Roth, p. 270 et seq.

views or ideologies",[49] by protecting the freedom of research in article 20 BV[50] as a basic right of communication. It namely protects learning and passing on scientific findings as well as scientific research as a "method-directed search for universalizable knowledge".[51/52] Every natural and juridical person, Swiss and foreigners just as research facilities can refer to the freedom of science.[53] If the research activity also serves purchase purposes – for instance within the scope of industrial research – the economic freedom[54] offers protection, too.[55]

Finally, people who arrive or already did arrive at a different image of the world and of humanity due to the recent results of the neurosciences may in our opinion also be protected by the freedom of belief and the freedom of conscience[56] in their world-view as an "encompassing life-plan without a reference to transcendency".[57]

Thus, particularly for neuroscientists, too, the BV offers a distinct scope of "listening", in the bounds of which it can be researched, and the resulting insights can be passed on and discussed.

Protection Of The Individual Addressees Of Law

In The Field Of Research

To be sure, the scope of the freedom of research outlined above does not hold unlimitedly.[58] First, the freedom of science can be restricted by opposing predominantly public interests.[59] Van Spyk distinguishes between the relevant public interests according to the societal, the scientific and the personality-related perspective. The societal perspective embraces legally protected goods of the general public, in particular the classic police areas, as well as the basic rights of third parties. The second perspective refers to the scientific integrity of research, i.e.

---

[49] Meyer/Hafner 2008, on art. 20, in Ehrenzeller et al., margin number 1 et seq.

[50] Art. 20 BV (Freedom of research and teaching is guaranteed.) and also art. 15 subpara. 3 UNO-Treaty I, art. 19 UNO-Treaty II and art. 10 ECHR on the international scale.

[51] Botschaft HFG, p. 8092.

[52] Meyer/Hafner 2002, on art. 20, in: Ehrenzeller et al., margin number 3 et seq., art. 3 FIFG (Federal Act of October, 7th 1983 on the promotion of research and innovation (SR 420. 1)) and Schweizer 2007, Wissenschaftsfreiheit und Kunstfreiheit, in: Merten et al., margin number 12.

[53] Meyer/Hafner 2002, on art. 20, in Ehrenzeller et al., margin number 7 and Schweizer 2007, Wissenschaftsfreiheit und Kunstfreiheit, in: Merten et al., margin number 18 et seq.

[54] Art. 27 BV: Economic freedom is guaranteed (para. 1).

[55] See Vallender 2008, on art. 72, in: Ehrenzeller et al., margin number 18 with reference to Swiss Federal Court Decision BGE 125 I 277.

[56] Art. 15 BV and art. 9 EMRK and art. 18 UNO-Treaty II.

[57] See Cavelti/Kley 2008, on art. 15, in: Ehrenzeller et al., margin number 3 et seq.

[58] See for the further limitations of the freedom of research: Schweizer 2007, Wissenschaftsfreiheit und Kunstfreiheit, in: Merten et al., margin number 25.

[59] Schweizer/Hafner 2008, on art. 20, in: Ehrenzeller et al., margin number 19 et seq.

qualitatively insufficient studies or researches that are in a blatant discrepacy between the expected benefit and the intended violation of the integrity of the test person are not protected by the freedom of research. The freedom of research prohibits research in the case that it contradicts duties of protection of the state in favor of the right of self-determination or of the personal integrity of the test person regarding his or her fundamental rights.[60] Especially under the personality-related perspective, the concrete restriction of the freedom of research due to a collision with other basic rights is being determined by balancing legally protected interests against conflicting fundamental rights or constitutional principles.[61] Thereby particularly in the field of reproductional medicine and gene technology – in principle, however, in the field of neurosciences, too – an eminently important role is played by human dignity[62] as well as the right to live and the right to personal freedom[63] – here namely the protection of the psychological-mental integrity.[64] However, under certain conditions relatively far-reaching restrictions of, for instance, the right of mental integrity in favor of research are possible, *in case* a consent of the test person in the form of an "informed consent" is given.[65] However, only little importance should be given to the „informed consent" with reference to an assessment, in case the intervention into the integrity which the research intends is needless or purposeless[66]; such a kind of research would reduce the human being to a pure object, a means to an end, and would thus not be compatible with human dignity.[67] Even illegitimate are studies which aim towards intentional killing of the test person or which intend to violate the content of the dignity of the human species – for instance in the case of a xenotransplantation aiming at a change of the germline.[68] When conducting research with minors especially the state duty to protect and to promotion of art. 11 BV[69] is to be observed – besides increased requirements regarding the "informed consent"[70]–, which is complemented on the level of international law by

---

[60] Van Spyk, p. 315.

[61] Meyer/Hafner 2008, on art. 20, margin number 13, as well as van Spyk, p. 313.

[62] Art. 7 BV: Human dignity must be respected and protected.

[63] Art. 10 BV: Everyone has the right to life. The death penalty is prohibited (para. 1). Everyone has the right to personal liberty and in particular to physical and mental integrity and to freedom of movement (para. 2).

[64] Meyer/Hafner 2002, on art. 20, in: Ehrenzeller et al., margin number 11 and Tag 2007, Neurowissenschaft und Strafrecht, in: Holdegger et al., p. 352 et seq.

[65] See on this van Spyk, p. 328 et seq.

[66] Ibid., p. 325.

[67] Schweizer/van Spyk, p. 572 et seq, Mastronardi 2008, on art. 7, in: Ehrenzeller et al., margin number 42 et seq., 49 and Schweizer/Hafner 2008, on art. 20, in: Ehrenzeller et al., margin number 24 et seq.

[68] Van Spyk, p. 327. Schweizer/Hafner 2008, on art. 20, in: Ehrenzeller et al., margin number 27.

[69] Children and young people have the right to the special protection of their integrity and to the encouragement of their development (para. 1). They may personally exercise their rights to the extent that their power of judgement allows (para. 2).

[70] See on this Schweizer/van Spyk, p. 563 et seq.

the programmatic right of an maximum of health of art. 24 CRC[71] and the child's entitlement of hearing and voice in its own affairs stated in art. 12 CRC.[72]

The sketchy exposition of the protective mechanisms of fundamental rights just given shall suffice at this point; an outline of the Swiss law regarding research with humans in force will be given in chapter 3.

In The Field Of Additional Medical Applications

With a view to future in what follows it shall be focused on possible medical applications that have already transcended the stadium of research. Namely, future discussions will presumably focus on the diagnosis of pathological results in the brain based on improved procedures of brain imaging, and furthermore on enhancement, i.e. the increase of brain activity beyond the standard measure.[73]

With the latest procedures of brain imaging – so-called neuroimaging – it is already possible today to diagnose disorders in the awareness of the self like in the case of schizophrenia[74] or other mental limitations, but also damages in brain-regions which are responsible for the repression of violent behavior.[75] A further development of these technologies – as is to be expected – carries consequences for the current order of law, but does not shake it to its very foundations. The subjective content of the equality of rights[76/77] already that manifests itself in the established wording of the equal treatment of what is equal according to its equality and the unequal treatment of what is *unequal according to its unequality*, which again has manifested itself in constant jurisdiction of the Swiss Federal Court,[78] is affected by improved diagnosis techniques insofar as its validity can be accounted for in a better way. Doubtlessly, it would run contrary to our sense of justice if, for instance, a – as far as one can see – psychologically healthy delinquent would suffer the same penologic treatment as a delinquent who is psychologically deficient and whose deficiency played a major role when committing the crime. The principle of equality alone calls for a differenciation here. Precisely at this point of intersection between medicine and law, when assessing the legal culpability in criminal law[79]

---

[71] Convention of November 20th, 1989 on the Rights of the Child.

[72] See in detail: Sprecher 2007, p. 95 et seq.

[73] See Tag 2007, Neurowissenschaft und Strafrecht, in: Holdegger et al., p. 350 et seq.

[74] Hell 2007, Moderne Erkenntnisse verändern das Ich-Bewusstsein, in: Holdegger et al., p. 370 et seq.

[75] Tag 2007, Neurowissenschaft und Strafrecht, in: Holdegger et al., p. 350.

[76] Art. 8 BV: Everyone shall be equal before the law (para. 1). No one may be discriminated against, in particular on grounds of origin, race, gender, age, language, social position, way of life, religious, ideological, or political convictions, or because of a physical, mental or psychological disability (para. 2).

[77] See Schweizer 2008, on art. 8, in: Ehrenzeller et al., margin number 20.

[78] Ibid. margin number 22.

[79] See Kiesewetter 2010, p. 2 et seq.

and the sagacity in civil law,[80] the neurosciences promise for more precise forensic expertise and thus for a stimulation of the law. At the same time, however, it has to be prevented that a neurological "handicap" that has been diagnosed in a special way does not lead to a medical or social discrimination against the person in question (see art. 8 para. 4 BV).[81]

Due to the fast developing techniques of diagnosis, there will likely also be a breakthrough of special therapies both for yet unknown and already known diseases. Already today the treatment with psychotropic drugs is a common way of healing, while the number of drugs is strongly correlated with the growing knowledge about the structure and the chemicals of the brain.[82] The neurorecovery after a brain damage is a special filed in this context. It is the declared aim here to substitute the functions of damaged brain areas through other brain areas. This is to be achieved by a specific training program, which is structured in an interdisciplinary way.[83] Further rapid improvements have recently been made in the field of neuroprosthetics, in which electric signals within the brain are being used, for instance, to steer wheelchairs.[84]

In the fields of therapies and neuroprosthetics as well as deep brain stimulation, the neurosciences thus promise to resocialize people suffering, for instance, from Parkinson [85] or the locked-in syndrome, by giving back to them their communicational abilities, so that they can be subjects to legal rights and duties again. The state has to generally foster such developments. Namely article 2 para. 3 BV[86] explicitly demands from the state to secure a maximum degree of same chances to every citizen in Switzerland.[87] Furthermore, the principle of same chances for anyone is recognized as a social-political basis according to the nondiscrimination percept of article 8 BV. The aim is to create the same "starting conditions" for everyone and thus to enable the effective use of freedom rights, election rights, and procedural rights.[88] The program is being concretized by the social aims of article 41 BV.[89/90]

---

[80] Gschwend 2006, Konsequenzen aus den Erkenntnissen der Hirnforschung für Straf- und Privatrecht, in: Senn/Puskás, p. 150 et seq.

[81] See Eggenberger-Bigler 2008, on art. 8, in: Ehrenzeller et al., margin number 103 et seq.

[82] Helmchen 2007, Therapiemöglichkeiten in der Psychiatrie: Modellfall Psychopharmakotherapie, in: Holdegger et al., p. 381.

[83] Kesselring 2007, Neurologische Aspekte des Lernens – Lektionen aus der Neurorehabilitation, in: Holdegger et al., p. 417 et seq.

[84] See Pousaz 2011, Neuroprosthetics: the mind is the pilot (EPFL – news), <http://actu.epfl.ch/news/neuroprosthetics-the-mind-is-the-pilot/>, accessed on 13/03/2011.

[85] Also see Hildt 2007, Ethische Überlegungen zur Tiefenhirnstimulation, in: Holdegger et al., p. 457.

[86] It [The Swiss Confederation] shall ensure the greatest possible equality of opportunity among its citizens.

[87] See Ehrenzeller 2008, on art. 2, in: Ehrenzeller et al., margin number 22.

[88] Schweizer 2008, on art. 8, in: Ehrenzeller et al., margin number 32.

[89] See also art. 12 UNO-Treaty I.

[90] See Schweizer 2008, on art. 8, in: Ehrenzeller et al., margin number 33.

The provision states inter alia the protection against social risks such as severe or chronicle diseases, accidents (para. 1 lit. a)[91] or the promotion of appropriate care for health (para. 1 lit. b).[92] The relevant competences for this can be found in articles 117 et seq. BV[93], from which article 118b BV covers research with human beings[94]. In addition to this, the far-reaching federal competence in the field of civil law has to be mentioned, which the civil protection of the personality[95] and the provisions on the protection of children and adults[96] are based on[97]. Special protection is given to for the personality of children and juveniles according to article 41 para. 1 lit. g BV.[98/99] In this context, special attention is to be given to the protection of challenged children or juveniles,[100] which is granted in particular by the regulations on health insurance and invalid insurance.

Last but not least the issue of the so-called enhancement shall be addressed. Hanfried Helmchen defines enhancement as "medical treatment, in particular the use of psychotroph drugs by healthy people, with the aim to compensate slight physcial or mental disadvantages and to improve human talents and abilities, such as the cognitive abilities or creativity".[101] An already current example is the use of methylphenidate by school children suffering from ADHS (in German: Aufmerksamkeitsdefizit-Hyperaktivitätsstörung), or the use by students in order to enhance their cognitive abilities.[102] However, such constellations seem to be rather an ethical than a legal problem.[103] Nevertheless, on a legal level questions of equality before the law and the access to same chances will likely come into play, if it turns out that for example only rich persons are able to gain such enhancement-technologies.[104]

---

[91] Bigler-Eggenberger 2008, on art. 41, in: Ehrenzeller et al., margin number 29.

[92] Ibid., margin number 37 et seq.

[93] Ibid., margin number 39.

[94] Art. 118b BV: The Confederation shall legislate on research on human beings where this is required in order to protect their dignity and privacy. In doing so, it shall preserve the freedom to conduct research and shall take account of the importance of research to health and society (para. 1).

[95] Art. 11 et seq. Swiss Civil Code.

[96] Child protection: art. 307 et seq. Swiss Civil Code, guardianship: art. 360 et seq. Swiss Civil Code.

[97] Art. 122 BV: The Confederation shall be responsible for legislation in the field of civil law and the law of civil procedure (para. 1).

[98] The Confederation and the Cantons shall (. . .) endeavour to ensure that: (. . .) children and young people are encouraged to develop into independent and socially responsible people and are supported in their social, cultural and political integration (art. 41 para. 1 lit. g BV). See also art. 11 and art. 10 para. 2 BV.

[99] Bigler-Eggenberger 2008, on art. 41, in: Ehrenzeller et al., margin number 76.

[100] Ibid., margin number 80.

[101] Helmchen 2007, Therapiemöglichkeiten in der Psychiatrie: Modellfall Psychopharmakotherapie, in: Holdegger et al., p. 392.

[102] Ibid., p. 392.

[103] See ibid., p. 392.

[104] See also ibid., p. 392.

2.2.2   In The Field Of Police- And Security Law

Especially within the field of criminal law the accusation of being able to also act differently is of utmost importance for the connection of responsibility and punishment.[105] If this precondition is no longer existent – as some brain researchers already state – the existing system of criminal responsibility would have be at least to be rethought. As a consequence, one could think of an increasing application of therapeutic measure or extended terms of imprisonment.[106] If it became possible to connect abnormalities in the brain to criminal action, even preventive measures on the basis of danger prognosis could become an issue.[107] On the long-term even the search for terrorists or other special perpetrators via controls of the public area by mind-reading devices could become possible.[108] Following such scenarios, Felix Bommer speaks of a general tendency toward a more preventive criminal legal system.[109] From a constitutional and a public international law perspective, such demands have to face some serious points of criticism, because they see the individual as foreign to himself and unable to act.[110] Not only would there be a tension with view to the principle of legality[111] – in particular to the principle of legal security, according to which the basis of state actions have to be sufficiently determined, so that the individual is able to relate his own actions to them[112] – and the principle of appropriateness[113], but also it comes to tensions with constitutional positions such as the right to personal freedom according to article 10 para. 2 BV[114] or the protection of a private sphere according to article 13 BV.[115] A preventive observation on the basis of neurological findings could in the end lead to a stigmatization of the concerned group of persons, which would conflict with the principle of equality before the law.[116]

---

[105] Bommer 2007, Hirnforschung und Schuldstrafrecht, in: Stichweh/Bommer, p. 25.

[106] Ibid., p. 26; as well as Gschwend 2006, Konsequenzen aus den Erkenntnissen der Hirnforschung für Straf- und Privatrecht, in: Senn/Puskás, p. 148 et seq.

[107] Tag 2007, Neurowissenschaften und Strafrecht, in: Holdegger et al., p. 351.

[108] Schleim, p. 9.

[109] Bommer 2007, Hirnforschung und Schuldstrafrecht, in: Stichweh/Bommer, p. 30.

[110] Mahlmann 2010, §25, margin number 6 et seq.

[111] Art. 5 BV. All activities of the state shall be based on and limited by law (para. 1). State activities must be conducted in the public interest and be proportionate to the ends sought (para. 2).

[112] Hangartner 2008, on art. 5, in: Ehrenzeller et al., margin number 11.

[113] See art. 5 para. 2 BV and art. 36 para 3 BV (Any restrictions on fundamental rights must be proportionate.)

[114] See art. 5 ECHR and art. 9 UNO-Treaty II.

[115] Everyone has the right to privacy in their private and family life and in their home, and in relation to their mail and telecommunications (para. 1). Everyone has the right to be protected against the misuse of their personal data (para. 2). And also art. 8 EHCR and art. 17 UNO-Treaty II.

[116] In particular art. 8 para. 2 BV and also see Schweizer 2007, on art. 8, in: Ehrenzeller et al., margin number 70.

2.2.3   In The Field Of Procedural Law

As for the field of the fundamental procedural rights,[117] there arise quite a few problems to be solved in connection to neurosciences. First of all, one may raise the question, which scope of content remains left for the prohibition of formal negation of rights, especially the right to be heard, if a prognosis on the basis of neuroscientific surgery would be seen as sufficient in order to ground preventive measures or even an extended term of imprisonment[118] on this. In this context, it seems to be certain that a negation of free will would possibly also bring forward a loss of the rights of the concerned persons in court trials.

As already mentioned above, also the decision of the judge would have to be faced with some suspicion under neuroscientific evaluation. The juristic method of connecting an actual situation in life to and evaluating it under the legal provisions could no longer be seen as the free will of the judge, as one had to assume that automatic brain procedures would in some way or the other preclude the result of the assessment. With regard to the determination of guilt or the capacity of an accused person, this would consequently result in an increasing influence of neurological experts, which would take the genuine responsibility of the judge.[119] In its result, this would not be in compliance with the fundamental principle of the independence of the judge according to article 30 BV[120] and article 6 of the European Convention on Human Rights. Last but not least, preventive measures taken by the police, which would be based on neurological findings at a certain person, would not be in accordance with article 32 BV[121] and article 6 para. 2 ECHR,[122] in which the general presumption of nonguilt is laid down.

2.2.4   In The Field Of Data Protection

A pivotal role is reserved for data protection, which holds true not only for the field of research, but also for the collection and usage of personal data in the field of police and security law. The discussion regularly focuses on so-called sensitive personal data, which are worth being especially protected[123] according to art. 3 lit. c

---

[117] Art. 29–32 BV, art. 6 ECHR.

[118] In principle the preventive custody of a mentally ill person is indisputable nowadays. However, the ECtHR has ruled on this, that even in such cases access to a court and fair proceedings according to art. 5 para. 4 ECHR and art. 6 para. 1 ECHR must be granted (Winterwerp v. The Netherlands (no. 6301/73), §67 and §75).

[119] Compare for instance Hell 2008, Das Gehirn ist kein Agent, in: Fachtagung Zürich, p. 19.

[120] Anyone whose case falls to be judicially decided has the right to have their case heard by a legally constituted, competent, independent and impartial court. Ad hoc courts are prohibited (para. 1).

[121] As well as art. 6 ECHR and art. 14 UNO-Treaty II.

[122] Grabenwarter 2009, §24, margin number 120 et seq.

[123] See Schweizer 2008, on art. 13, in: Ehrenzeller et al., margin number 42.

of the Swiss Act on Data Protection.[124] The right to privacy[125] in its core protects private persons from being registered in public through state authorities with words, pictures, or sounds.[126] Although the assessment of contents of thoughts is not explicitly named in this context, it would probably be content of art. 13 para. 1 BV[127] according to its ratio.

In its content, data protection according to art. 13 para. 2 BV[128] stretches to the basic right to determine the use of one's own personal data. The individual shall be free in determining, if, and if yes to which degree and for which purpose, state authorities are allowed to collect and process individual personal data.[129] The scope of protection by the basic rights in this context stretches to any kind of processing of the personal data.[130] The most important consequence of this is the general right to knowledge about the content of a data collection,[131] which shows a strongly basic right–like character.[132]

Besides the general rules of data protection, there are more specific provisions on data protection, for instance in the field of biomedical research, which are mainly influenced by the European Convention on Human Rights and Biomedicine[133], which was ratified by Switzerland in 2008, and its Additional Protocol on Biomedical Research[134], which has not yet been ratified by Switzerland. For the field of police and security law, the specific requirements of data protection are mainly based on specific state contracts and Federal Acts or Acts of the cantons.[135] For the field of biomedical research, for instance the disclosure of a professional secret, which

---

[124] Federal Act of 19 June 1992 on Data Protection (FADP, SR 235.1).

[125] Art. 13 para. 1 BV as well as art. 8 para. 1 ECHR and art. 7 of the Charter Of Fundamental Rights Of The European Union.

[126] Breitenmoser 2008, on art. 13, in: Ehrenzeller et al., margin number 13 et seq., Schweizer 2008, on art. 13, in: Ehrenzeller et al., margin number 41 et seq.

[127] Everyone has the right to privacy in their private and family life and in their home, and in relation to their mail and telecommunications.

[128] Everyone has the right to be protected against the misuse of their personal data.

[129] Schweizer 2008, on art. 13, in: Ehrenzeller et al., margin number 39.

[130] Ibid., margin number 41.

[131] Art. 8 FADP, compare also art. 8 of the draft for a Federal Act on Research on Human Beings (BBl 2009 8163).

[132] See Schweizer 2008, on art. 13, in: Ehrenzeller et al., margin number 45 et seq.

[133] Convention for the Protection of Human Rights and Dignity of the Human Being with regard to the Application of Biology and Medicine of April, 4th 1997 (SR 0.810.2/ETS 164).

[134] Additional Protocol to the Convention on Human Rights and Biomedicine concerning Biomedical Research (ETS 195).

[135] Schweizer 2008, on art. 13, in: Ehrenzeller et al., margin number 49, 53. See the Council Framework Decision 2008/977/JHA of 27 November 2008 on the Protection of Personal Data Processed in the Framework of Police and Judicial Cooperation in Criminal Matters (OJ L350/60) and also the Federal Act of March, 19th 2010 on the Implementation of the Council Framework Decision 2008/977/JHA on the Protection of Personal Data Processed in the Framework of Police and Judicial Cooperation in Criminal Matters (AS 2010 3387) or e.g. art. 4, 12 et seq., 15 et seq. of the Federal Act of June, 13th 2008 on the Police Information Systems of the Federal Government (SR 361).

has been gained via biomedical research, is prohibited by punishment.[136] In the field of police law, special restrictions of data protection can be justified with measures of preventive state protection.[137] However, bearing in mind the development of control systems, such restrictions have to be seen from a very critical viewpoint.[138]

3   The Existing Legal Framework On Research

Just in short terms the currently existing legal framework of Swiss research shall be lined out.[139] The basic right protection of the freedom of research and science will have to be accurately shaped due to the risks of biomedical research with regard to the patients' rights or the rights of other concerned persons. Currently, beside the international standards[140] – for instance in the European Convention on Human Rights and Biomedicine – the field of biomedical research is in particular regulated by the provisions of health law and patients' rights of the individual Swiss cantons.[141] However, in the near future a harmonization shall be achieved on the Federal level,[142] when enacting an Act on research with human beings, based on the Federal competence of article 118b BV.[143] Similar to article 2 of the Convention on Biomedicine, the planned Act clearly points out that the protection of dignity, personality and health of the human being is higher-ranking than the sole interest of the society in research and progress.[144] In the named Act itself, this fundamental decision for instance becomes clear, when it is regulated that the health of the involved persons has to be protected in cases where unexpected complications arise in the course of the research project and these complications can only be ceased by ending the project. In such cases, the project has to be aborted and must not be brought to an end.[145]

---

[136] Art. 121 Swiss Criminal Code and art. 35 of the Federal Act on Data Protection.

[137] See art. 14 para. 1 sentence 2 and art. 18 of the Federal Act of March, 21st 1997 on Measures of Ensuring Inner Security (SR 120) and also the decision of the Federal Commission on Data Protection, ZBI 108 (2007), p. 392–401, in special: considerations 5b and 5c.

[138] Schweizer 2008, on art. 13, in: Ehrenzeller et al., margin number 53.

[139] As far as it hasn't been subject to chapter 2.2.1 yet.

[140] E.g. the "Declaration of Helsinki", the "International Guidelines for Biomedical Research involving Human Subjects" (CIOMS-Guidelines), the "Guideline for Good Clinical Practice" (GCP) and so on (see in detail Schweizer/van Spyk 2007, p. 548 et seq.).

[141] See Tag 2007, Neurowissenschaft und Strafrecht, in: Holdegger et al., p. 352.

[142] Botschaft HFG, p. 8046 et seq.

[143] The Confederation shall legislate on research on human beings where this is required in order to protect their dignity and privacy. In doing so, it shall preserve the freedom to conduct research and shall take account of the importance of research to health and society (para. 1).

[144] See art. 4 of the draft of the Federal Act on Research on Human Beings. On this Botschaft HFG, p. 8046.

[145] See art. 15 of the draft of the Federal Act on Research on Human Beings. On this Botschaft HFG, p. 8097.

As an outflow of a compromise between the protection of the human welfare and the necessity of research involving human beings in order to cure diseases, research with human beings is still devoted to a row of legally binding research principles.[146/147] In the context of the latest developments in the field of neurosciences, one has to especially mention the legal nature of medical interventions connected to medical treatment and the principle of informed consent. According to the Swiss understanding, any medical intervention is at first hand a violation of the personal integrity and thus also violates the personal rights of the concerned person. Any medical intervention thus is in its origin illegitimate according to Swiss national law.[148] The most important reason of justification in this context is of course the prior informed consent of the concerned person.[149]

In particular, this issue could consequently be subject to some sceptical review by brain researchers. How should a free informed consent be construed, if the result is merely based on neuronal force? However, the legal order recognizes that the principle of informed consent is rather based on an ideal than on the reality. Medical interventions and also research regularly are connected to difficult situations demanding decisions. Difficult or new situations often demand a variety of medical treatment measure, which sometimes even cannot be overlooked by qualified experts. The standards of informed consent are thus considered to be very high and set as a precondition an enormous level of understanding by the concerned persons, which is not always given in reality. Especially in cases of serious diseases, the desire for treatment and a better physical or mental condition will more likely affect on the decision of the concerned person than a rational evaluation of arguments pro and contra.[150] Consequently, the Swiss legal order also provides special regulations on the informed consent of incapable persons or persons with diminished capacity.[151]

Thus, in the field of medical law and research law, the argumentation that the individual is not free in his will, is not new at all. Nevertheless, the image of the human being behind this demands to stick to the principle of informed consent.[152] Without mentioning the consequences one more time, it shall be mentioned at this point that the legal basis for brain research is grounded on the legal construct of free will.

---

[146] In detail: Schweizer/van Spyk 2007, p. 556 et seq.

[147] See e.g. Sprecher 2007, p. 15 et seq., Botschaft of September, 12th 2001 concerning the European Convention on Human Rights and Biomedicine (SR 01.056), p. 313 et seq., art. 16 of the European Convention on Human Rights and Biomedicine and also art. 4 et seq. of the draft of the Federal Act on Research on Human Beings.

[148] Sprecher 2007, p. 200.

[149] Ibid., p. 200 et seq.

[150] See on this the critique on the principle of informed consent by Sprecher (Sprecher 2007, p. 232 et seq.).

[151] See art. 17 of the European Convention on Human Rights and Biomedicine and art. 21 et seq. of the draft of the Federal Act on Research on Human Beings.

[152] See van Spyk, p. 108, 111.

4 Conclusion

If one tries to take the neurological developments for granted and also the connected demands of some brain researchers, the consequences for the law would be enormous. The constitutional analysis has shown that freedom is not understood in an absolute way, which would mean that there were no kinds of constraints or commitments, but that the constitutional freedom is a fundamental principle throughout the whole Swiss legal order. A complete recognition of all neurological findings would mean serious reformations of the legal order. However, already in this regard there would be arising serious legal concerns. How shall a constitution be renewed, which demands the free election of the people's representatives as well as the free consent of those entitled to vote.

Furthermore, one would have to raise the question about possible alternatives. For sure this has been discussed intensively for the field of criminal law and in parts also for the field of private civil law. In which way a legal order, which would not be built on the concept of free will, could be designed, remains an open question. Such demands are thus object to severe criticism by the jurisprudence and are commonly rejected due to problems of legitimating the constitutional order and also for practical reasons.

However, one should not generally take a hostile view toward methods of neurosciences. Deeper insights into the inner sphere of the human being are also to be seen as a chance, and not just a risk. This holds true in particular in areas where the legal order ties consequences to the pathological state of a person. In this context, especially the individual evaluation of criminal responsibility and the civil capacity are to be mentioned.

Concerning possible future developments in the field of neurosciences and law in particular the basic rights and the human rights serve as a precious compass for the question in which direction these developments should run or where possible boundaries may be situated. The legal system does not only provide the boundaries for critical and risky research areas, but it is also to be seen as a bridge and a forum for an interdisciplinary discourse between neurosciences and other branches, such as psychology, psychiatry or other fields of social or health politics.

References

Bennet, MR/Hacker, PMS (2010) Die philosophischen Grundlagen der Neurowissenschaften, Darmstadt

Botschaft vom 21. Oktober 2009 zum Bundesgesetz über die Forschung am Menschen (BBl 2009 8045), cited: Botschaft HFG

Ehrenzeller, B/Mastronardi, P/Schweizer, RJ /Vallender, KA (eds.) 2002, 1th edition (used where marked), 2008, 2th edition Die schweizerische Bundesverfassung – Kommentar, Zürich

Geyer, C (2004), Hirnforschung und Willensfreiheit: zur Deutung der neuesten Experimente, Frankfurt a.M.

Grabenwarter, C (2009) Europäische Menschenrechtskonvention, 4th ed., München/Basel/Wien

Grün, K/Friedman, M/Roth, G (2008) Entmoralisierung des Rechts. Massstäbe der Hirnforschung im Strafrecht, Göttingen

Hillenkamp, T (2006) Neue Hirnforschung – neues Strafrecht? (Tagungsband der 15. Max-Alsberg-Tagung am 28.10.2005 in Berlin), Baden-Baden

Holdegger, A et al. (2007) Hirnforschung und Menschenbild, Fribourg

Kiehl, KA et al (2001).Limbic Abnormalities in Affective Processing by Criminal Psychopaths as Revealed by Functional Magnetic Resonance Imaging. Biological Psychiatry,50: pp. 677–684

Kiesewetter, M (2010) Schuld, Delinquenz und forensische Psychiatrie. ZStrR, 128(3): pp. 318–332

Lampe, EJ/Pauen, M/Roth, G (2008) Willensfreiheit und rechtliche Ordnung, Frankfurt a.M.

Mahlmann, M (2010) Rechtsphilosophie und Rechtstheorie, Baden-Baden

Merten, D/Papier, HJ/Müller, JP/Thürer, D (eds.) (2007) Handbuch der Grundrechte, Bd. VII/2 Grundrechte in der Schweiz und in Liechtenstein, Heidelberg/Zürich/St. Gallen

Pauen, M/Roth, G (2006) Grundzüge einer naturalistischen Theorie der Willensfreiheit, Frankfurt a.M.

Rechtswissenschaftliche Fakultät der Universität Zürich (ed.) (2008) Hirnforschung – Chancen und Risiken für das Recht (Fachtagung Zürich), Zürich, cited: Fachtagung Zürich

Roth, G (2003) Aus Sicht des Gehirns, Frankfurt a.M.

Schleim, S (2008) Gedankenlesen – Pionierarbeit der Hirnforschung, Hannover

Senn, M/Puskás, D (eds.) (2006) Gehirnforschung und rechtliche Verantwortung, Fachtagung der Schweizerischen Vereinigung für Rechts- und Sozialphilosophie, 19. Und 20. Mai 2006, Universität Bern (Archiv für Rechts- und Sozialphilosophie, Beiheft 111)

Schweizer, RJ/van Spyk, B (2007). Arzt und Forschung. In: Kuhn, Moritz W. / Poledna, Tomas (eds.), Arztrecht in der Praxis. Zürich/Basel/Genf: pp. 535–595

Singer, W (2003) Ein neues Menschenbild?, Frankfurt a.M.

Sprecher, F (2007) Medizinische Forschung mit Kindern und Jugendlichen (Diss. St. Gallen 2007), Berlin/Heidelberg

Stichweh, R/Bommer, F (eds.) (2007) Die zwei Kulturen? Gegenwärtige Beziehungen von Natur- und Humanwissenschaften (Luzerner Universitätsreden, Nr. 18), Luzern

Thürer, D et al. (2001) Verfassungsrecht der Schweiz, Zürich

van Spyk, B (2010) Das Recht auf Selbstbestimmung in der Humanforschung (Diss. St. Gallen 2010), Zürich/St. Gallen

Neuroethics and Neurolaw in Turkey

Berna Arda and Ahmet Acıduman

Abstract This section is dedicated to examining the subject "Neuroethics and Neurolaw in Turkey". The development of medicine and related branches in Turkey generally demonstrates a parallelism with the examples from the similar countries in the world. In brief, the contemporary criterions are applicable to both education and daily practices of these fields. In this context, the headlines under the disciplines of neurology and neurosurgery shall be evaluated from the medical ethics and medical law points of view under the heading of scientific neurothics and neurolaws.

Today, the worthiness problems related with the end of life constitute one of the most important subjects of discussion in medical ethics. In the neurology area, where this problem frequently arises, the commands: do not apply euthanasia and do not resuscitate are two important phenomena to be studied closely. A very crucial subject from the organ transplantation point of view is making the decision on the "brain death". The specialists on neurosurgery and neurology in Turkey are legally tasked among the decision-making doctors in this subject. Therefore, this is one of the headings that will be discussed in the text from both ethical and deontological or medical law points of view. Thus, referring to an eternal problem of medicine "terminating a life" and also to a new concept the organ trade, belonging to the 21st century and which is the result of modern economical and political factors.

The "clinical researches" shall be discussed as a rather discrete dimension of the daily doctor – patient relations under the heading of neurological sciences as a separate subsection within the text. Here, we shall discuss how the concept of informed consent may be applied to the patient and subject groups which the neurological sciences deal with in the normal daily medical applications and in research phases and the potential problems related with it. Another concept to be scrutinized here is how experimental treatments may be turned out to be a subject of hope trade in some communities.

B. Arda (✉) • A. Acıduman
Department of Medical Ethics and History of Medicine, Ankara University, Faculty of Medicine, Sıhhiye 06100, Ankara, Turkey
e-mail: Berna.Arda@medicine.ankara.edu.tr

T.M. Spranger (ed.), *International Neurolaw*,
DOI 10.1007/978-3-642-21541-4_16, © Springer-Verlag Berlin Heidelberg 2012

Finally, this section is a response given from a geography on the junction of Asia and Europe to the query of how different nations may develop different approaches to similar subjects from a "neuroethics and neurolaw" point of view.

1 A Few Related Examples from the Medical History

Neurology is an area where chronic, progressive, and incurable sicknesses are frequently seen. The diseases called "Degenerative Diseases" such as multiple sclerosis (MS), amyotrophic lateral sclerosis (ALS), and Alzheimer disease (AD) represent a disease group which also conceives ethical problems. On the other hand, usually all a doctor can do is limited with maintaining the quality of life of the patient at a certain level in the cases related with neurosurgical discipline such as crainocerebral or spinal traumas and serious acute neurological function losses observed following the surgeries of tumors in the neural system and neurological function losses connected with the tumors in the neural system and related with vascular incidents (ischemic or hemorrhagic, requiring a surgical intervention or not). From this point of view, neurology and neurosurgery is an area where ethical problems related with termination of life are frequently confronted with and these inevitably will have reflections on the doctor and patient relations. In this context, some examples from the historical dimension of medicine may be impressive to the reader.

The famous surgeon of Andalusia Al-Zahrawi (936–1013) in section 30 of his famous art piece *Al-Tasrif* which is related with surgery has emphasized that deficiencies or mistakes in anatomy training may lead to injuring the patient or the patient even may die due to these reasons. Zahrawi, in the prolog of the second section of his book related with incision, perforation and venesection he emphasizes that surgical operation is mainly separated into two parts where the first part deals with the health of the patient while the second part bearing many threats and in order to avoid these threats one has to study and learn anatomy very well. The reason behind Zahrawi's insistence in surgeons to learn anatomy very well is an almost universally accepted principle in training of doctors and surgeons and it is also meant for the well-being of the patients. Once again according to Zahrawi, in order to obtain a positive result from the surgical operation of a patient is to examine the patient closely prior to the attempt and not to conduct the operation if there is no positive information about it. These approaches are in consistent with the "primum non nocere" ethical principle, which is extremely important from the patients' point of view and well known all the way from the times of Hippocrates (Acıduman 2010).

Zahrawi advises that a surgeon must demonstrate care and attention for himself and sympathy and permanence to his patients. In a case where there is no hope for a patient's disease to get cured and there is no indication for a surgical treatment, he warns the surgeons not to conduct surgical operation to such patient with greediness and ambition to make money. Zahrawi's order of "illness that are very threatening or difficult to cure leave alone!" reminds us of the "do not resuscitate!" order of our modern times (Acıduman 2010).

The above are some ethical examples of a surgeon who lived more than a thousand years ago. The expression "my sons" he used while calling the other surgeons shows us that Zahrawi accepts himself as the authority and admits that he was in charge of warning the young surgeons about the threats of surgery and to training them. This situation is also in consistent with the paternalistic medicine and ethic concepts of that era (Acıduman 2010).

As a situation that is interesting for the contemporary neurosurgery area, in the part with the heading "On setting the vertebrae of the back and the neck" we see an example of Zahrawi's "do not treat" understanding:

> When a fracture occurs in the bones of the neck, which is rare, as mostly they suffer contusion, as do the spinal vertebrae –when it happens to anyone and you want to know whether it will heal or not, then look and if you see both his hands relaxed and numb and dead he has no power to move or stretch or close them, and when you pinch them or prick them with a needle he does not notice it or feel any pain in them, you may know, as a general rule, that it will not mend, for he is doomed. But if he moves them both and feels in them the pinching and pricking, you may know that the spinal medulla is still intact and that under treatment the patient will recover. If anything of this nature happens to the vertebrae of the back and you wish to know if he will recover or not, then pay attention to his feet. If you see them relaxed and in the situation we described in the case of hands, and when he lies on his back he passes flatus and faeces involuntarily, and lying prone he passes water involuntarily, and laying on his back he cannot pass water if he wishes, then you may know his case is hopeless, so do not concern yourself with his treatment. But if nothing of this kind occur then the case is easier. . . . (Albucasis 1973)

We see Esmail Jorjani, an Iranian, who was one of the most important physicians of 12th Century (1042–1137) has a similar approach in his famous study Zakhire-i Khwarazmshahi (The Treasure of the Khwarazm Shah) in the section where he handles spinal fractures: "dorsal vertebra is rarely broken, but the edges may be crushed. Spinal cord and membrane and the nerve squeezes between two vertebra and quickly dies. In case this phenomenon takes place around the neck level, do not attempt to treat it . . . (Jorjani)"

It is interesting to see that some subjects within the area of the contemporary neurological sciences were defined as headings carrying certain sensitiveness in medical applications even in 10th and 12th centuries.

2   Neuroethics and Neurolaw in Daily Medical Applications

Informing the patient and patient's relatives about the structure and prognosis of neurological diseases may often create problems in daily medical applications. Transferring them accurate and realistic information about the treatability of the disease, life expectancy of the patient, and quality of life of the patient is the obligation of the doctor and it is necessary. In Turkey socioculturally not only the patient but also the relatives of the patient are also members of this group; therefore, the doctors must take them into consideration throughout the process too. It is stated that in many cases the patient is being informed about the disease and process following a consultation with the patient's relatives about the level of information to be given to the patient (Kızıltan and Kızıltan 2001). It is possible to state that the principle of doctor being honest to his patient is being forced to convert to "treating the patients in a paternalistic manner" and "sharing the truth in a covered manner."

This section is aimed to study the commands "don't apply euthanasia" and "don't resuscitate" and the picture that chronic and mostly incurable diseases create, from the neuroethic point of view.

2.1 Euthanasia from Neuroethics and Neurolaw Point of View

As known, euthanasia is a term created from Latin words "good death". Even if it has been transformed into various forms through time, it is a subject of debate being discussed from the ethical point of view for long eras starting from the philosophers of the ancient period. Death of course is not a phenomenon that we have chance to practice. Which states may be the ones that we wish death consciously and insistently? Is it suffering greatly from pain, or inexpressible deficiency – feeling of impotency, or the incontestable frustration caused by requiring the attention of others to live, or the despair caused by finding ourselves in a state of remoteness to our former life and estranged to it? Therefore deciding on what "good death" or "bad death" can be is a conclusion that we may reach through our individual, social, and cultural perceptions.

Euthanasia is classified into three distant categories: the doctor's action, the patient's willingness, and the nature of the action. The euthanasia done according to the patient's willingness is divided into three groups: "voluntary euthanasia", "unwillingly euthanasia", and "involuntary euthanasia". In case of voluntary euthanasia, a written or verbal request of the patient is the base. In the cases of unwillingly euthanasia, the decision comes up through relatives, persons who the patient formerly stated his/her opinion about euthanasia or a court decision. Involuntary euthanasia is the one conducted against the will to live of the patient or without learning the patient's will to live or not.

A well-described right must create equal opportunities to everyone who has the potential to exploit that right. Therefore in this context, being a request that may only be brought up by a limited number of patients in certain conditions, we may conclude that in the hypothetical level euthanasia may not be defined as a right. Briefly, the diversity of the cases in medicine where obtaining the consent of the patient is not always possible, creates a question mark on the euthanasia's nature of being a right. Some examples from the local legislation are related with this subject; these are mainly hinting in the direction that euthanasia cannot be depicted as a patient right in Turkey. However, this situation does not totally eliminate the chance of discussing the subject from the ethics point of view. Nevertheless, we may state that at least mentioning that there is no such a right is not considered as a crime.

Euthanasia is defined as "active" or "passive" according to the role of the doctor in this action. Passive euthanasia means shutting down the system that maintains the life artificially and leaving the disease to its natural course; the active euthanasia is terminating the life of the patient by the doctor himself. Doctor-assisted suicide is the help given by the doctor to the patient who wishes to kill himself or herself.

Euthanasia may be considered as a phenomenon far away or even against the purposes of medicine, which has a field of occupation that placed "life" and "life support" to the main purpose of the profession. This heading which constitutes a very hot subject of debate from the medicine point of view brings forth the arguments such as "we cannot be considered as an executioner even if we approve euthanasia under certain conditions", "even when the role of a doctor in an execution is merely limited to certifying the death, active euthanasia certainly cannot be a doctor's practice", "medicine science that strive to return the people who committed suicide back to life, should be against euthanasia in order to protect its internal consistency".

Here, the question to which extent the effort for "trying to keep alive with insistence, by any means" can ignore the patients' sovereignty is an important heading of debate. Euthanasia is defined as a crime in the local legislation of Turkey. In this context, what shall we name the action in the cases of shutting down the life support and cutting the artificial feeding of a patient with brain death, or cutting of the medical treatment and leaving the patient out of hospital due to the request of the relatives... or similar practices? On the one hand looking from the decision-maker side, the weight of giving an irremeable decision, and on the other hand the worries that emerge from the possibility that these decisions may lead to stigmatization on the professional identity of the doctor, carries the subject to a very challenging situation from the medicine point of view.

Questions such as "permitting" the death of a patient who has no chance to live humanly has any complying point with the most respectful principle of this profession "respect to life"? Or at least are there any cases that match with it? Is it possible that doctors' undertaking this duty may rumple the doctor – patient relations which has to be built on a confidence base? constitute the main headlines of neuroethics debates in Turkey.

The legal system in Turkey forbids euthanasia and defines openly that it considers any action that may be related with euthanasia as "intentional killing".

The article 13 of the Patients Rights Regulation put into force in 1998 (Hasta Hakları Yönetmeliği 1998) is dedicated to illicitness of euthanasia: "regarding the medical requirements it is not permissible to abandon the right to live. No body's life may be terminated even if that person or others request so". According to article 24, "drawing back the consent after the commencement of a treatment, is subject to nonexistence of a condition of medical inconvenience." It is understood from this statement that the legislation vest this right in only a limited manner, and does not consider a patient's refusal to treatment as a patient right under conditions that threaten the life of that patient.

2.2   The "DNR Order" in Daily Life of Medicine

"Do not resuscitate (DNR)" is a concept defined back in the 1970s. Doctor's consent is a must in DNR. It comes up in the cases of patients in the terminal phase of diseases, prolonged cardiac arrests or due to patients' own consent. This order means that any treatment attempt to be done on such a patient is considered as a futile treatment.

In a limited study conducted in Ankara, the 17% of the patients died in neurology clinics, verbally requested DNR. The number of DNR requests occurred in the highest rate in the cases of cerebrovascular diseases among the neurological diseases (Büyükkoçak et al. 1998). İyilikçi et al. 2004 starting from the point that there are numerous information in the literature related with life support and DNR decisions; yet, there is no survey report in Turkey about how the doctors decide on end of life of a patient, and if there are any social factors that influence their decisions or not, have prepared a questionnaire questioning the euthanasia and DNR order experiences of the anesthetist, how they place the DNR order, and their knowledge level about the related articles in the Turkish Crime Law and sent to the 439 members of the Turkish Society of Anesthesiology and Reanimation; 369 (84%) of the members whom the questionnaire was sent replied; 65.5% of these anestheists mentioned that they have placed verbal or written DNR order. It is quite meaningful that 94% of these DNR orders were verbal and only 6% were written. Another interesting finding is that the doctors prefer to conduct consultation with their friends about the DNR order rather than taking it to an ethical committee prior to final DNR decision (82.7 %). About determining who must decide on giving the DNR order, contributors in excess of 58% of the total number stated that hospital administration, ethical committee, the patient, relatives of the patient and doctor must collectively give the order, 31.4% of the contributors stated that the patient, relatives of the patient and the doctor must collectively give the order and only 10% of the contributors stated that according to their idea, the doctor should give this order alone. Even if there are ethical committees established in many universities and ethical consultations being carried these committees are not being consulted neatly during the DNR order placing phases in Turkey. Most of the contributors (58.5 %) have answered that ethical consultation is helpful in giving these decisions and they receive assistance from the ethical committee during the decision-making phase.

There is a small number of studies about the financial strengths of the families and the regulations of the Social Security Organizations that may provide a clear picture about the results of practicing or not practicing the attempts that may prolong the life of the patient in spite of the family's objection due to high cost, and therefore it is obvious that it is not possible to make a generalization about this subject (Uysal 1998a). The doctors are under the influences of many factors including the patients and the relatives of the patients; besides their medical knowledge and their capabilities, they are forced to decide on to apply resuscitation or not. It is obvious that the doctors who perform in the intensive care, anesthesiology and reanimation, neurology departments require a common tendency system in this subject (Uysal et al. 1995).

2.3   Brain Death Decisions and Organ Transplantations

Brain death decisions are especially important from the organ transplantation point of view. Organ transplantation law in Turkey was put into force in the year 1979 with the name "The law related with organ and tissue explantation, preserving, inoculation and transplantation". With this law, the organ and tissue trade was prohibited, and all the advertisement activities other than the ones that have a scientific, statistical and informative nature are also prohibited. According to the law, explanting organs or tissue from the young individuals who are under 18, which is accepted as the maturity age in Turkey, and noncompus mentis are deemed inappropriate and banned. It is obligatory that a donor must fully consciously and without being under any pressure declare in written and signed form before two witnesses that he intends to be a donor and this protocol must be approved by a medical doctor. It is mandatory that the donor shall be informed about the complications or dangers of the action, in case of married donors the spouse shall be informed about the situation, the identities of the donor and receiver shall be kept secret in the transplantations other than the ones that are made between the relatives and the required medical examination shall be done prior to transplantation.

The third section of the law is dedicated to explanting organ or tissue from the dead bodies. In this part, it is stated that, as a rule, the death state of a person shall be certified unanimously by four expert doctors from cardiology, neurology, neurosurgery and anestheology fields. It is also forbidden that the doctors who will explant the organ and the treating doctor of the patient shall not be among the group that will decide on the brain death of the patient. The law that enables explanting tissue or organs such as cornea which causes no visual difference of the dead, also clearly states that no tissue or organ can be explanted from the dead bodies of the individuals who has stated while he/she was alive that he/she refuses such a practice on his/her dead body.

Presidency of Religious Affairs makes an evaluation from the Islamic point of view about the subjects where public opinion is deemed to be important. This Presidency, which had declared in 1960s that blood transfusion was appropriate from religious point of view, had made an explanation in 1980 that organ donation was "the greatest favour a human being can do to another human being." Presidency defines the certain conditions where organ and tissue donorship creates no trouble from the religious point of view. These are medical necessities, the donor being a dead person, and in the case of alive donor the subject tissue or organ not being an essential one, no payment to be made against the organ or tissue, prior consent of the donor and also the consent of the receiver for the said transplantation.

The number of organ donorship following a brain death cases is rather low in Turkey. In a limited survey conducted related with the subject, it was observed that the number of donorship was getting higher proportional to the education level of the individuals; however, the tendency to donate was decreasing for other people if the donor is a relative. Medical doctors constitute the profession group that stated they will be donors with the highest rate such as 72%; however, the rate of donation dropped to the lowest level when the donor shall be the relatives of the doctors. Taking this point into consideration, we may conclude that systems that depend on people donating their organs is more realistic than the systems that depend on organ requesting from the donors (Uysal 1998b).

It is obvious that the decisions on brain death are sensitive and very crucial from the results of this decision. Protecting the patient and doctor relations which is built on mutual trust is not easy in cases like these. Recently in Turkey in a case brain death of a young patient had occurred due to a traffic accident. Upon the neurosurgeon's statement that took place in the media *"never give up hope from Allah"* Turkish Neurosurgical Society had made an official statement declaring that this attitude was "out of science" and blamed the action to be a kind of "hope trade." They also stated that an action like this, aimed to mistaken the public opinion is unacceptable:

"...Mistaken perceptions that will effect the public opinion adversely following this situation, may lead the relatives of the patient whom the brain death was confirmed by related committees to sceptisizm about the system. Such attitudes may jeopardize the treatment of thousands of patients who are expecting organ transplantation from the patients having brain death and spoil their hopes. In case of brain death the patient is inviable (Türk Nöroşirürji Derneği 2010a).

The brain death diagnosis defines certainly the death level today. Without the confirmation of this diagnosis, the life support systems must not be shut off. However, the patients' request not to be connected to the ventilator or to be separated from the ventilator seems to be open to a lot of discussion. The doctors are continuously facing the insisting requests and expectations coming not only from the patients but also from the relatives of the patients about the state of the treatment. Taking the unique sociocultural structures of the countries into consideration, we may conclude that giving brain death decision is an extremely delicate subject on doctors level and also social level which requires great care.

Today, we may easily state that there is a great number of organ and tissue traffic on the world, we may even state that "body parts trade" has made a peak. This traffic where generally "rich" buyers and "poor" sellers take place is also called as a special type of tourism activity. According to the numbers issued by WHO, 10% of the organ transplantations are being conducted in illegal ways in the world. Kidney is somewhat "star product" in this trade. Two reasons, namely that there are two kidneys in every person and humans may continue to live with only one kidney, carry the kidney to top of the list in the organ trade. In this "market," there is a buying cost of the kidney that differs from one country to other; for example in South Africa it is 470 Euros, in USA 20200 Euros, in Moldova 1800 Euros, in Turkey it is 6800 Euros. In Pakistan, only 1/3 of the kidney transplantations were done for the Pakistani citizens in 2006 (Rugemer 2010).

In short, the deepening of poverty in especially these countries creates new potential donors for the organs. On the other hand, developments in the medicine and treatment methods that reduce the transplant rejection possibility is another important factor in this subject. Finally, ethical attempts and legal regulations are important factors in preventing organ traffic but the main determinant in this context is the economical conditions.

3 Neuroethics and Neurolaw in the Research World

In general using the human subjects in the experiment must be considered only in the cases where there is no other chance to obtain the expected scientific development from such experiment. And when the decision is given in that direction, the human subject number must be kept as low as possible, their biological, psychological, sociological and legal beings must be preserved and most importantly the basic ethical rules must be carefully followed in the cases where human subjects are used. Not causing harm, privacy, respect to sovereignty – informed consent and honesty seems to be the prominent ethical principles in the cases of scientific research ethics. In many countries, there are ethical committees established and performing according to the legal arrangements in order to preserve the well-beings and rights of the human subjects that take place in such research activities. Obtaining the approval of an ethical committee is a legal obligation in the cases where experiments will be made on humans. Research Ethics Committees constitute the basic assurance of safety, health and protection of human rights of the subjects, and assures that the worries about the health of the subjects will be above the scientific and social concerns and prevents the exploitation of the subjects for other purposes other than the one intended. This assurance is provided by evaluating the research protocols by taking the valid legislation and regulations into consideration.

In Turkey, the experimental researches to be conducted on human beings are defined by laws, bylaws and regulations. One of the earliest regulations in Turkey that monitor the scientific research practices on human beings is 1960 dated Medical Deontology Regulation (MDR). The article 10 of MDR states that "the doctor or dentist who deal with research activities shall not apply or advise the method he develops without practicing the same for enough number of times and prove that it is helpful, or at least does not have adverse effects. However, he may advise the new method together with the warning that it is not practiced in adequate number of times and its in the test phase and also mentioning about all of the precautions to be taken during the practice. It is forbidden to express unfair statements that will create misperception about an invention." Article 11 states that "no experimental surgical attempts may be conducted on human beings, also no experimental chemical, physical or biological treatments may be done on humans. In case it is determined during the clinical and laboratory examinations that classical methods do not have a positive effect on the patient, then conducting new treatment methods may be permissible, provided that the new treatment is experimented on animals for adequate times and the favorable results are proved, prior to conducting on humans. To the extent that, it will be clear that the result obtained by conducting this new treatment, shall not be worse than expectations from the results of classical methods. New and unexperienced treatments may be conducted on the patient when there is no prossibility cause any harm to the patient and there is the possibility to cure the patient from the disease (Tıbbi Deontoloji Nizamnamesi 1960)."

The article 17 of the 1982 dated constitution of Turkish Republic states that consent of the related human being must be obtained prior to any medical or scientific experiments on human beings: "Every citizen has the right to live, protect and develop their material and moral being. Unless there is a medical necessity or ordered by law, the completeness of no citizen's body may be spoiled, no medical or scientific experiment may be conducted on his body without the consent of the related citizen; nobody may be tortured; nobody may be held subject to a punishment that is against human dignity (Türkiye Cumhuriyeti Anayasası 2010)."

In the year 1993 by the development and putting into force of the "Regulation Related With Drug Researches" by the Ministry of Health, it became mandatory to obtain the approval from ethical committees prior to experiments to be conducted on humans by using drugs, and it also became mandatory to carry such experiments under the supervision of the ethical committees. Local ethics committees and a Central Ethics Committee was established within the Ministry of Health in order evaluate the experiments to be conducted on human beings, from ethics point of view (İlaç Araştırmaları Hakkında Yönetmelik 1993).

"The Convention for Protection of the Human Rights and Dignity of the Human Beings With Regard to the Application of Biology and Medicine, and Human Rights and Biomedicine Convention, Oviedo" which Turkey signed on the date of 04.04.1997 was approved by the Turkish Grand National Council on the date of 03.12.2003 and "The Law Related to Approval of The Convention for Protection of the Human Rights and Dignity of the Human Beings With Regard to the Application of Biology and Medicine", which was issued in the 09.12.2003 dated and 25311 numbered Official Gazette, was put into force. As a consequence of this, the Oviedo Convention has become an internal part of Turkish Legislation (Tıbbi Deontoloji Nizamnamesi 1960). The articles 15, 16, 17, and 18 of the section 5 of this Law is about the conditions of conducting "Scientific Researches", protection of human subjects and protection of individuals who do not have the capability to state his/her consent on the research to be conducted on him/her and researches to be conducted on in vitro embryo (Avrupa Antlaşmaları 2010).

Article 90 of the section with the heading of "Offences Against Physical Integrity" of Turkish Criminal Law (TCL) is about "Experiments on Human Beings" and states that anyone who conducts scientific experiment on human beings shall be imprisoned for one to three years and also lists the conditions where scientific experiments shall not be considered as a crime (Türk Ceza Kanunu 2010).

In the section six of the Patient Rights Regulation with the heading "Medical Researches", obtaining consent of, informing and protecting the volunteer in the scientific research cases and the way to obtain the said consent and the procedure related with the noncompus mentis and infants are defined (Hasta Hakları Yönetmeliği 1998). In this context, it is clear that most of the patients that are in the group that neurological sciences deal with are somehow easily open to be hurt and dealing with them requires a higher level of attention.

The above legal arrangements deal with the well beings of humans that are being subject to scientific researches and cover the general rules. In Turkey, from the medical experimental researches point of view, there is no special legal arrangement related with "Activities for Stemcell," which is also related with neural sciences, and this situation makes it very hard even to draw a frame from the ethical discussions point of view (Arda and Aciduman 2009). Treatment Services General Directorate of Ministry of Health has issued a circular about the subject "Embryonic Stemcell Researches" in the year 2005. There are intense researches about the usage of stem cell in the medical applications is being widely carried in the world and in our country, the results of these researches are potentially great hopes for the treatment of various diseases for the future, but however it is emphasized that the source of the stem cell is continuing to be a subject of debate. Even if researches related with treatments through somatic stem cell transplantation is generally approved in the world, the usage of stem cells obtained from embryo causes a lot of debate from especially the ethical and legal points of view. The Ministry of Health continues its efforts in order to finalize the legal arrangements in a manner that will satisfy the public conscious and according to the contemporary scientific norms. It is noted that the above researches are being handled from ethical, legal, and cultural aspects and within the scope of compliance works to EU legislation and until this work is completed the stem cell researches should not be carried in Turkey (Sağlık Bakanlığı 2005).

The second circular issued by the Ministry of Health related with stem cell and its annex was aiming to regulate the studies being carried for this subject. This second circular issued by Treatment Services General Directorate of Ministry of Health in the year 2006 related with the Stemcell Studies, it was stated that stemcell transplantation is being used for years as a means of treatment in the cases of especially hematological, oncological patients and recently usage of stem cell in the treatment of other diseases has commenced too. There are very intensive studies being carried about treatment through stem cell transplantation both in Turkey and in the world, but it is also emphasized that the number of patients that stem cell transplantation is done is very small, and there is not enough information about the possible complications this practice will cause on the patients. Ministry of Health notes that "Stemcell Transplantations Scientific Advisory Board" is being established and *Clinical Purposed Nonembryonic Stemcell Study Guide* is being prepared in order to establish the required infrastructure within the organization and to enable work conditions according to requirements of contemporary science, and also emphasizes that all of the studies should be carried in accordance with this guide (Sağlık Bakanlığı 2006).

In the prolog of the 2006 dated *Clinical Purposed Nonembryonic Stemcell Study Guide*, it is stated that the guide is prepared especially for those patients who cannot be cured with the currently available medical and surgical methods in Turkey and the guide highlights the rules to be obeyed during the human origin clinical purposed nonembryonic stem cell applications. It is also explained that these applications shall be evaluated by the "Stemcell Transplantations Scientific

Advisory Board" established within the organization and permissions shall be given to the scientific centers, which has the approval of the Ministry to work in this field, the board shall evaluate the following aspects related with the center; the quality and quantity of the specialist staff to be hired in these centers and the infrastructure requirements of a center, the previous researches of these centers conducted on animals, scientific studies they produced, scientific publications and the accumulated knowledge of the scientists who perform in these centers. The study application file shall be submitted promptly to the Stemcell Transplantations Scientific Advisory Board following the approval obtained from the "Local Ethics Committee" that is established within the organization which the clinical purposed study is being planned to be carried and only upon evaluation of the application file by Stemcell Transplantations Scientific Advisory Board the Ministry may grant permission to work. During the application phase, among the detailed information about the study, the draft of the "informed volunteer consent form" shall also be included in the file and also the measures to be taken in the cases of unintended or unexpected effects on the patient and guarantees in order to cure/recover the complications (insurance). It is defined that studies may only commence upon completion of above necessities (Sağlık Bakanlığı 2006).

In order to commence a clinical stem cell study: a) Similar study must be conducted on the subjects other than human beings or on animals in an adequate number of times; b) The studies done on nonhuman subjects or animals demonstrate that in order to reach to the desired result, it is necessary to conduct the same study human beings; c) The study shall not have a predictable permanent harm on the health of human subjects. It is mandatory that progress reports in 6 months periods and the final report at the end of the study shall be submitted to the Ministry of Health. Again the organization that performs the study is obliged to inform the Ministry of Health together with all the details and complications within 7 days in case an unexpected severe side effects appear or death occurs. Following the completion of the clinical stem cell study, it is prohibited to announce the results of the study to public in a manner to direct/mislead the public opinion. The guide also states the rules for securing the privacy rights of the patients and obeying the medical ethics in accordance with patient rights and respect to human dignity are among the indispensable rules of the guide.

Meanwhile, the Ministry of Health, which searches to cease the embryonic stemcell researches in order to complete the legal arrangements, has permitted the adult stem cell studies and decided that the rules about how these studies shall be carried are issued with a circular and its annex. The expectation of usage in cell renewal and reuse of the cell or tissue and regaining the functionality has brought together the thought that stem cell treatment which is successfully applied especially in many hematological diseases' treatments may also be helpful in the diseases where neural cells are damaged. Recently, the stem cell applications aiming to bring cure for the neurological diseases have become a source of hope for many patients in Turkey. With this respect, the expectations of the patients and their relatives have far passed the scientific facts or current capabilities of medicine and even reached to a level that may be considered as forcing the boundaries of hope trade.

Therefore, the Turkish Neurosurgical Society had to issue a "Stemcell Newspaper Announcement" in order to enlighten the public. In order to put the stem cell transplantation into public service, all of the tests to ensure the success of the method should be completed according to the valid norms first, then these must be published and the predicted benefits and harms of the treatment should be proved. In the final part of the announcement, it was stated that there were some news on the newspapers about stem cell (!) transplantations to the patients who has cerebral and spinal cord damages or diseases. It is emphasized that these types of information about the success of treatment in such cases should be evaluated by objective specialists and should also be accepted and approved verbally and in written form by the experts before they are brought in front of the public. Announcing the stem cell applications to Turkish people in such a manner is nothing but hope trade (Türk Nöroşirürji Derneği 2010b).

On the other hand, in an article issued in the web site of Turkish Neurological Society it was stated that the stem cell applications for some genetical disorders such as Duchenne Muscular Dystrophy (DMD), which are observed in young males and has no real treatment yet, are still in the research phase, test conducted on animals having DMD model have not demonstrated satisfactory; and there is only one case where the study was made on human, and there is suspect that the diagnosis and evaluation methods for this case may be insufficient. While it was being mentioned that studies made on animals for cure of DMD, usage of inappropriate stem cell type and applications may be the reason for unfavorable results and they emphasized that even if these studies gave favorable results this does not guarantee that studies on human beings will have favorable results too. In the article where it was clearly stated with ethical worries, it must be known that currently there is no DMD patient cured from the disease with stem cell applications. And it is emphasized that insisting on stem cell treatment methods to cure the patients with DMD will lead to great expenditures for the families of the patients with vain hopes. Finally, it is stated that at this stage it will be better to concentrate on the experiments on animals (Serdaroğlu and Topaloğlu 2010).

4 Conclusion

Neurological sciences are one of the areas of medicine which is widely open to progress yet at the same time requiring much more attention. The reason that makes this area so special is the weight of the ethical problems related with the end of life and appearance of this situation in different types in both daily activities of medicine and during the research phase (8). The news taking place in the written and visual media about the experimental treatments, especially stem cell treatments and servicing this news as proved and treatments with high potential of success, even introducing these treatments as "Lazarus Miracles", creates a great threat from the ethical point of view.

There is no doubt that each country has to make its own legal arrangements in order to meet the requirements of their own legislations. Moreover, it is expected that this legal process shall also cover the ethical principles and should be discussed deeply prior to putting the legislation into force. However, even if these requirements are met, solely depending on the laws in order to obtain a qualified medicine environment is not enough. Some of the major determinants of this profession is the ethical dimension in the continuous medical education and researcher education prior to graduation, professional activities being open to auditing or not, functionalities of the ethical committees and the appearance of the malpractices in the neurological sciences field. Along with the above, the field being open to hope trade or not, the limits of the explanations that can be made under the modest nature of science, the importance given by the society to the nonscientific explanations are also important headings. This situation is related to the question to which extent the scientific environment can express itself to the society and also the extent of health of the relation between the society and the scientific environments. Therefore, we have to emphasize that several different factors play a role for better medical applications, and it is not possible to establish a qualified environment by only making legal arrangements.

References

Acıduman A (2010) Al-Zahrawi and his ethical principles in his surgical practice. Turkiye Klinikleri, J Med Ethics 18(2):109–112

Albucasis (1973) On Surgery and Instruments [A Definitive Edition the Arabic Text with English Translation and Commentary by M. S. Spink and G.L. Lewis]. London: The Wellcome Institute of the History of Medicine, pp. 734–737

Arda B, Aciduman A (2009) An evaluation regarding the current situation of stem cell studies in Turkey. Stem Cell Rev and Rep 5:130–134

Avrupa Antlaşmaları, Biyoloji ve Tıbbın Uygulanması Bakımından İnsan Hakları ve İnsan Haysiyetinin Korunması Sözleşmesi Tasarısı: İnsan Hakları ve Biyotıp Sözleşmesi [Oviedo - Convention for The Protection of Human Rights and Dignity of the Human Being with regard to The Application of Biology and Medicine: Convention on Human Rights and Biomedicine]. [Internet] [cited 2010 Dec 13]. Available from: http://www.avrupakonseyi.org.tr/antlasma/aas_164.htm [in Turkish]

Büyükkoçak Ü, Uysal H, Ertürk Ö, Bilgin S, Ketene, A, İnan S (1998) Resüsite edilmeyecek kararının etik yönden içeriği [Nature of do not resuscitate orders from ethical point of view]. In: Şahinoğlu- Pelin S, Arda B, Özçelikay G, Özgür A, Çay Şenler F, editors. 3. Tıbbi Etik Sempozyumu Bildirileri [3 rd Medical Ethics Symposium Proceedings]. Ankara: YÖK Matbaası, pp. 59–64 [in Turkish]

Hasta Hakları Yönetmeliği [Patient Rights Regulations]. Resmi Gazete [Official Gazette], Tarihi [Date]: 01.08.1998, Sayısı [Number]: 23420. [Internet] [cited 2009 Apr 14]. Available from: http://www.saglik.gov.tr/TR/BelgeGoster.aspx?F6E10F8892433CFFAAF6AA849816B2EF46148DEDD773827B [in Turkish].

İlaç Araştırmaları Hakkında Yönetmelik [The Regulation Concerning Research On Drugs]. Resmi Gazete [Official Gazette], Tarihi [Date]: 29.01.1993, Sayısı [Number]: 21480. [Internet] [cited 2009 Mar 30]. Available from: http://www.bsm.gov.tr/mevzuat/docs/13052005_y_16.pdf [in Turkish]

Iyilikçi L, Erbayraktar S, Gökmen N, Ellidokuz H, Kara HC, Günerli A (2004) Practices of anaesthesiologists with regard to withholding and withdrawal of life support from the critically ill in Turkey. Acta Anaesthesiol Scand 48:457–462

Jorjani E, Kitab-i Zakhire-i Khwarazmshahi [The Treasure of the Khwarazm Shah]. Istanbul: Suleymaniye Manuscript Library, Fatih Collection, Nr: 3551, p. 400b

Katoğlu T (2005) Türk hukukunun bir parçası olarak Avrupa Konseyi İnsan Hakları ve Biyotıp Sözleşmesi [European Council Human Rights and Biomedicine Convention as a part of Turkish Legislation]. Ankara Üniversitesi Hukuk Fakültesi Dergisi 55(1):157–193 [in Turkish]

Kızıltan G, Kızıltan M (2001) Klinik Nöroloji Uygulamaları ve Etik Sorunlar [Clinical Neurology Applications and Ethical Problems]. In: Demirhan-Erdemir A, Oğuz NY, Elçioğlu Ö, Doğan H, editors. Klinik Etik [Clinical Ethics]. İstanbul: Nobel Tıp Kitabevi, pp. 334–352 [in Turkish]

Rugemer C (2010) Body parts. Special Report: Transplantation. Research EU, No: 62, pp. 8–9

Serdaroğlu P, Topaloğlu H. Duchenne kas distrofisi (DMD) ve kök hücre [Duchenne Muscular Dystrophy (DMD) and stemcell]. Türk Nöroloji Derneği [Turkish Neurological Society]-Duchenne Kas Distrofisi (Dmd) Ve Kök Hücre [Duchenne Muscular Dystrophy (DMD) and stemcell]. [Internet] [cited 2010 Dec 7]. Available from: http://www.noroloji.org.tr/page.aspx?menu=522 [in Turkish]

T.C. Sağlık Bakanlığı, Tedavi Hizmetleri Genel Müdürlüğü [T.R. The Ministry of Health, the General Directorate of Treatment Services]. Embriyonik Kök Hücre Araştırmaları (2005/141) [Embryonic Stem Cell Research (2005/141)], Tarihi [Date]:19.09.2005, Sayısı [Number]: 17972. [Internet] [cited 2009 Mar 31]. Available at: http://www.saglik.gov.tr/TR/BelgeGoster.aspx?F6E10F8892433CFF7A2395174CFB32E164B5516ED5B497B2. [in Turkish]

T.C Sağlık Bakanlığı, Tedavi Hizmetleri Genel Müdürlüğü [T.R. The Ministry of Health, the General Directorate of Treatment Services]. Kök Hücre Çalışmaları (2006/51) [Stem Cell Studies (2006/51)], Tarihi [Date]: 01.05.2006, Sayısı [Number]: 8647. [Internet] [cited 2009 Mar 31]. Available at: http://www.saglik.gov.tr/TR/BelgeGoster.aspx?F6E10F8892433CFF7A2395174CFB32E1D0652C231336A42D. [in Turkish]

Tıbbi Deontoloji Nizamnamesi [Medical Deontology Regulations]. Resmi Gazete [Official Gazette] Tarihi [Date]: 19.02.1960, Sayısı [Number]: 10436. [Internet] [cited 2009 Apr 14]. Available from: http://www.mevzuat.adalet.gov.tr/html/5044.html [in Turkish]

Türkiye Cumhuriyeti Anayasası. Kanun No: 2709 [Constitution of Turkish Republic] [Internet] [cited 2010 Oct 5]. Available from: http://www.tbmm.gov.tr/anayasa.htm [in Turkish]

Türk Ceza Kanunu. Kanun No: 5237 [Turkish Criminal Law]. [Internet] [cited 2010 Oct 5]. Available from: http://www.tbmm.gov.tr/kanunlar/k5237.html [in Turkish]

Türk Nöroşirürji Derneği (2010a) [Turkish Neurosurgical Society]-Haberler [News]. [Internet] [cited 2010 Dec 7] Available from: http://www.turkishneurosurgicalsociety.org/detay.php?id=149 [in Turkish]

Türk Nöroşirürji Derneği (2010a) [Turkish Neurosurgical Society]-Hastalar için Bilgiler [Information for the Patients]. [Internet] [cited 2010 Dec 7] Available from: http://www.turkishneurosurgicalsociety.org/bilgiler.php?content=5 [in Turkish]

Uysal H, Ertürk Ö, Üçkardeşler L, İnan S, Bilgin S (1998a) Terminal dönemdeki ALS'li hastaya yaklaşım [Approach to the patient with ALS on the terminal phase]. In: Şahinoğlu-Pelin S, Arda B, Özçelikay G, Özgür A, Çay Şenler F, editors. 3. Tıbbi Etik Sempozyumu Bildirileri [3 rd Medical Ethics Symposium Proceedings]. Ankara: YÖK Matbaası, pp. 65–69

Uysal H, Ertürk Ö, Üçkardeşler L, Evrenkaya T (1998b) Otopsi ve organ nakli; ortak etik yaklaşımlar yapılabilir mi? [Autopsy and organ transplantation; may there be common ethical approaches?] In: Şahinoğlu-Pelin S, Arda B, Özçelikay G, Özgür A, Çay Şenler F, editors. 3. Tıbbi Etik Sempozyumu Bildirileri [3 rd Medical Ethics Symposium Proceedings]. Ankara: YÖK Matbaası; pp. 145–147 [in Turkish]

Uysal H, İnan LE, Kuli P, Yurdakul M (1995) Nörolojide etik sorunlar [Ethical problems in Neurology]. T Klin Tıbbi Etik 2–3:72–5 [in Turkish]

Neuroscientific Evidence in the English Courts

Lisa Claydon and Paul Catley

Abstract This chapter examines the use of neuroscientific evidence in the courts of England and Wales. It considers the breadth of use which has been made of this evidence. In particular it examines the use of this evidence in cases where the capacity of the legal actor has been questioned. This may apply in evaluations of criminal responsibility and in a civil context in assessing capacity to perform legally meaningful actions such as the making of wills. Consideration is given to what this evidence adds to determinations of whether individuals are in a persistent vegetative state in particular in relation to the withdrawal of treatment. This chapter looks at the use of expert evidence in court and briefly considers proposed changes. Finally the chapter considers how neuroscientific evidence may be used in the future and also whether it has wider application in the criminal and civil justice systems.

1 Introduction

The use of neuroscientific evidence in court has increased over the last decade as neuroscientific brain imaging and other techniques have improved. The purpose of this chapter is to explore the impact of the introduction of this evidence in the courts of England and Wales.[1] There is a potential wide range of uses for neuroscientific evidence in court, within both the criminal and the civil justice systems. It provides a means of evaluating degrees of damage caused to plaintiffs in civil cases, degrees of injury to victims in criminal cases and may shed some light on claims

---

[1] It is common to refer to English law and not to English and Welsh Law. Accordingly, references to English law should be read as applying to the law of England and Wales.

L. Claydon (✉) • P. Catley (✉)
Department of Law, Bristol Law School, University of the West of England, Frenchay Campus, Coldharbour Lane, Bristol BS16 1QY, UK
e-mail: Lisa.Claydon@uwe.ac.uk; Paul.Catley@uwe.ac.uk

T.M. Spranger (ed.), *International Neurolaw*,
DOI 10.1007/978-3-642-21541-4_17, © Springer-Verlag Berlin Heidelberg 2012

that actors lacked capacity when taking actions which would normally engage legal consequences.

As this book is to be read by an international audience perhaps some comment about how the law develops and is applied in England and Wales is required. This will be kept as brief as possible in order to avoid obscuring the main investigation which is to look at the effect of neuroscientific advances on English law. The law is found both in statute and in case law and though statute law takes precedence where it determines an issue, case law will normally be utilised to interpret the meaning of particular sections of statutes or of particular words in particular sections. The Human Rights Act 1998, by virtue of Section 3, imposes on all courts the duty to interpret case and statute law and to apply it, so far as is possible, in accordance with the European Convention for the Protection of Human Rights and Fundamental Freedoms (ECHR).[2] Certain law is entirely case law based. For example, much of the law relating to injuries caused by negligence or the law relating to homicide offences is not statute based. This law is based on the common law as declared by judges hearing cases argued before them. The system of precedent means that a higher court's decision is binding if the issues in a case before a lower court are the same as a previously decided case in a higher court.

The use of neuroscientific evidence in court will normally be to provide an explanation of human behaviour or to assess damage suffered by a claimant. As such it provides one explanation to be placed before the judge or jury dependent on the court or subject matter of the issue being adjudicated. Normally such evidence will be presented by an expert giving oral evidence in court or in the form of an expert's written report.

1.1   Rules of Evidence Including Expert Evidence

Under English law there are different rules regarding the admissibility of evidence in criminal and civil courts. The general rule is that evidence is admissible if it is relevant.[3] However, there are further rules which exclude potentially relevant evidence; there are rules which allow judges to exclude evidence at their discretion, for example in a criminal case a judge can exclude prosecution evidence which 'would have such an adverse effect on the fairness of the proceedings that the court ought not to admit it'.[4]

The English judicial system is an adversarial system with each side usually being represented by their own legal advocates. The calling of evidence is normally the

---

[2] By virtue of Section 2 of the 1998 Act courts are obliged, where appropriate, to take into account the jurisprudence of the European Court of Human Rights.

[3] Goddard LJ stated: 'generally speaking, all evidence that is relevant to an issue is admissible, while all that is irrelevant is excluded.' *Hollington v. F Hewthorn & Co. Ltd.* [1943] KB 587, 594.

[4] Police and Criminal Evidence Act 1984 s.78.

exclusive preserve of the two contesting parties. The judge acts as arbiter determining what evidence is to be allowed and what is to be excluded. Where a jury of lay people is present, as in more serious criminal trials, it is the judge who advises the jury on the law, but it is the jury who determine the facts in issue. Normally, witnesses can only give evidence of what they said, did, saw or heard and cannot give evidence as to the inferences or conclusions that they drew. However, where a witness has expertise, which is not possessed by the court, the witness may be deemed an expert witness and, as such, can give evidence of her opinion. In terms of the admissibility of neuroscientific evidence in a court, whether civil or criminal, then almost certainly such evidence will be introduced by an expert and that expert will be called upon to give her opinion as to the interpretation that should be given to the evidence. Experts' overriding duty is to the court, however, it is the parties who select and pay for their experts. In civil cases there is a power for the court to appoint its own expert,[5] but the exercise of this power is rare. In criminal cases such an approach has been considered, but was rejected on the grounds that 'there would, however, be no guarantee that he or she was any nearer to the truth of the matter than the expert witnesses for the parties'.[6] It is therefore likely that in cases involving contested neuroscientific evidence both parties will seek to call expert witnesses to give their opinion as to the proper interpretation that should be given to the evidence before the court. There used to be a rule that an expert witness could not give his or her opinion on 'the ultimate issue' – in other words, the issue which the court had to decide upon. This rule has now been abolished for civil cases.[7] In criminal cases it is less clear cut. In *Stockwell*[8] the Lord Chief Justice described the rule as more a matter of form than substance, whereas in *Doheny*[9] Lord Justice Phillips spoke of the need for an expert not to 'overstep the line which separates his province from that of the jury'. In this case, which related to DNA evidence, the court also advised that expert witnesses should not express a view on the probability of guilt given the DNA match.

1.2   The Recording of Decisions

In England and Wales, the cases that appear in the official law reports tend to be cases that are appealed from the lower courts. Whilst some first instance judgment are recorded and transcripts are taken of court proceedings, the immediately accessible reports of cases are largely from the High Court, Court of Appeal and

---

[5] Civil Procedure Rules r.35.7 courts may direct evidence be given by 'a single joint expert'.

[6] Report of the Royal Commission on Criminal Justice, Cm 2263 (1993) ch.9, para 74.

[7] Civil Evidence Act 1972 s.3.

[8] (1993) 97 Cr. App. R. 20, 374.

[9] [1997] 1 Cr. App. R. 369 CA.

the Supreme Court, formerly known as the House of Lords. This means that the precise number of cases in which neuroscientific evidence is used in court cannot be known. However, by using legal databases such as Lexis and Westlaw it is possible to identify and read decisions in the higher courts where neuroscientific evidence has been utilised.

The grounds of appeal against the decision of the court of first instance are limited. Appeals do not lie as of right, it is generally necessary to gain permission to appeal. The rules are detailed and do not need to concern us here. In brief, appeals from civil law courts are permitted to remedy serious procedural or other irregularity. For example an appeal might lie where neuroscientific evidence has been improperly excluded at first instance. Similarly in criminal cases appeals lie to remedy procedural or other irregularity. One example would be a judge misdirecting a jury with respect to the application of law to the facts of the case. One of the means of appeal in criminal cases has been to request the consideration of new evidence, sometimes brain scan or neurological evidence that was not available at the time of the trial, or was not disclosed to the defence at the time of the trial. The Criminal Appeal Act 1968 allows a court to hear such evidence where it is 'necessary or expedient in the interests of justice'[10] so to do.

2 Criminal Law and Neuroscientific Evidence

If a search for MRI, CT or EEG scans is carried out in the legal databases, then many criminal law cases will be found. The vast majority of the references will relate to evaluations of the extent of medical injury to victims of crime or to causes of death. These serve a useful purpose for example in providing evidence of the severity and brutality of an attack.[11]

When reviewing the new neurological evidence which is the subject of the appeal, the continued existence of pictures of sections of the brain and/or scans from the first trial enables any appeal to re-examine the conclusions drawn by medical experts at the time of the trial. The new interpretation of the existing evidence in the light of modern scientific advances in the understanding of brain states may of itself provide new evidence to support the appeal. Additionally, the advances achieved by greater use of scan techniques to study diseased and healthy brains also adds to the accuracy of determinations of causes of death or injury and may cast doubt on previously held theories. Our knowledge of brains and how they react to injury has changed and is changing rapidly. Evidence for this is to be found

---

[10] Section 23(1).

[11] For example *R v H* [2007] EWCA Crim 2330 a case where scan evidence of an injury to the eye socket of the victim of a serious assault was used on appeal against sentence to assist the consideration of whether the appellant posed a 'danger to society'.

in the reconsideration of cases where death or serious injury is said to have been caused by the shaking of a baby or young child.

In 2005 the Court of Appeal revisited four cases where the appellants' convictions which ranged from murder to serious assault were based, in part, upon scan evidence of injuries suffered by children.[12] What is clear from the reasoning in respect of each case is that there was huge disagreement amongst those giving expert evidence in the appeals as to the causes of the injuries. The court accepted that a previously held theory in relation to brain damage in shaken baby cases was flawed.

Interestingly for the use of neuroscientific evidence in court, the Court of Appeal felt that the admission of scientific evidence in trials – even where it was based on a developing theoretical approach was essential: 'developments in scientific thinking should not be kept from the Court, simply because they remain at the stage of hypothesis'. The court went on to add that the evidence supporting the hypothesis should be made clear. 'Obviously, it is of the first importance that the true status of the expert's evidence is frankly indicated to the court'.[13]

2.1   General Issues Relating to Criminal Capacity

The issue of capacity bears on the law in a number of ways. There are two presumptions made of the mental condition of a defendant who is arraigned for trial in a criminal court. One is that the defendant's criminal act(s) were voluntary and the other is that she was sane at the time of the crime. Where a defendant wishes to challenge one of these assumptions she must provide evidence which satisfies the court that there is a basis for the assertion that her acts were involuntary or that she was not sane at the time of the criminal act.[14] Absence of sufficient capacity is likely to be problematic to establish in court, especially where the loss of capacity is temporary. Responsibility for criminal acts in English law rests on establishing that the defendant committed the criminal act in the relevant circumstances and that she had the required mental element at the time of the crime.

Capacity issues may also arise in relation to the denial of prescribed mental elements required by criminal offences: intention or recklessness being examples of two mental states that may render a defendant liable for potentially criminal

---

[12] R v Harris; R v Rock; R v Cherry; R v Faulder [2005] EWCA Crim 1980.

[13] R v Harris; R v Rock; R v Cherry; R v Faulder [2005] EWCA Crim 1980, para 270.

[14] Bratty v Attorney General for Northern Ireland [1963] AC 386, 413.

'Whilst the ultimate burden rests on the Crown of proving every element essential in the crime, nevertheless in order to prove that the act was a voluntary act, the Crown is entitled to rely on the presumption that every man has sufficient mental capacity to be responsible for his crimes: and that if the defence wish to displace that presumption they must give some evidence from which the contrary may be reasonably inferred.' Per Lord Denning.

actions. The argument here would be that in view of the neuroscientific evidence the defendant was unable to act with the requisite mental state. In 1985 the results of an EEG scan were introduced in court in the case of *R v Birmingham Justices ex parte Lamb*.[15] The expert interpreting the EEG evidence suggested that it provided proof that Lamb could not have had the mens rea of theft, which is the intention to permanently deprive. Whilst this was not central to the determination of the case, this assertion was not questioned in the written court report.

Additionally, there are degrees of capacity recognised by the criminal law in terms of the trial process. A defendant must be fit to plead. They should have the mental capacity required to be a participant in the trial process. This would include having the ability to comprehend questions and instruct their legal team. Partial loss of capacity may also establish a mental condition sufficient to support a partial defence to murder in the form of diminished responsibility.

2.2 Early Judicial Approaches to the Use of Neuroscientific Evidence

An early use of scientific testing to establish whether a defendant's acts were involuntary occurs in the case of *Hill v Baxter*.[16] The appeal was heard in 1957. The case concerns a car accident where the defendant failed to stop at a junction. The explanation proffered by the defence and accepted by the magistrates was that the driver was unconscious at the time of the accident. This raised the issue that his actions were not voluntary, challenging the presumption of voluntariness. The plea was rejected as inapplicable to the offence with which he was charged. The magistrates had found that the defendant was not 'capable of forming any intention as to the manner of his driving'.[17] This was said by the appeal court to be irrelevant because the statute[18] placed an absolute prohibition on dangerous driving – therefore the mental state of the defendant was irrelevant.

However, the case is of interest because an EEG test was carried out on the defendant and the consultant neurologist's report was submitted for consideration by the court. It showed no abnormality which could explain the defendant's loss of consciousness. The appeal court notes its disapproval of the use of this written medical evidence commenting 'medical reports so often used in civil actions have

---

[15] Transcript (QBD, 4th July 1985).

[16] [1958] 1 All ER 193.

[17] [1958] 1 All ER 193, 194.

[18] At that time the Road Traffic Act 1930.

no place in criminal courts',[19] though they would have accepted oral testimony which could then be subject to cross examination.[20]

However, medical reports now play a far greater role than they did in 1957. Additionally, the role of medical explanations of behaviour has become more generally accepted by the courts. This raises interesting questions with regard to the usefulness of such reports when they are produced a considerable time after the criminal act occurred.

2.3 Fitness to Plead

An interesting case where the Court of Appeal was willing to examine neurological evidence suggesting that a conviction was unsafe: and to use its powers under Section 23 of the Criminal Justice Act 1968 to review neuroscientific evidence is the case of Mohammed Sharif.[21] This case concerned Mohammed Sharif's fitness to plead at the time of his trial.[22] The ability of a defendant to understand the trial process is assessed at the time of the trial.[23] The jury hearing the evidence with regard to fitness to plead found Sharif fit to plead.[24] A trial of the facts was then held.

Sharif was charged with conspiracy to defraud the Criminal Injuries Compensation Board. A claim had been entered in respect of alleged injuries he had suffered. These injuries were said to be caused by an assault which resulted in a head injury. In December 1998 before Sharif's trial, but after he had been found fit to plead, an MRI scan was taken of his brain. One of the experts, Professor Deakin, who had given evidence at the fitness hearing altered his view slightly in the light of the MRI scan evidence. He still said he thought that Sharif was 'malingering' though he did say that he could not express this view with certainty[25] and he continued to express the view that Sharif was fit to plead. The 1998 MRI scan was interpreted as showing an 'enlargement of the extra cerebral spaces' and another medical expert, Dr Forbes, presenting evidence at trial suggested this 'indicated a mild generalised

---

[19] [1958] 1 All ER 193, 194 per Goddard LCJ.

[20] In the English adversarial system both the defence and prosecution have the right to question each other's expert witnesses. The judge also has the right to question witnesses.

[21] R v Mohammed Sharif [2010] EWCA Crim 1709.

[22] Criminal Procedure (Insanity) Act 1964 s 4. A person may be found unfit to plead where they are charged with a criminal offence and the court finds they are 'under a disability'.

[23] The process by which a defendant is found unfit to plead is set out in the Criminal Procedure (Insanity) Act 1964. As amended by subsequent legislation.

[24] The Criminal Law Act 1967 Section 6(c) provides 'if he stands mute of malice or will not answer directly to the indictment, the court may order a plea of not guilty to be entered on his behalf ...'. This applies where a defendant refuses or is unable to speak in court.

[25] R v Mohammed Sharif [2010] EWCA Crim 1709, para 14.

atrophy of the brain'. EEG evidence was also referred to at trial. This was said, by another expert, Dr Launer, to indicate the possibility that Sharif was 'suffering from a conversion syndrome or longstanding functional psychosis'.[26] At his trial Sharif did not give evidence. Evidence was presented by medical experts on his behalf and on behalf of the prosecution. Sharif was convicted by the jury hearing the case in 1999.

The appeal focussed upon Mohammed Sharif's fitness to plead at the time of his trial. The arguments made by the medical experts were based in part on the MRI scan evidence. Two scans were considered: the 1998 scan and a scan from January 2000, which was taken after Sharif's trial and conviction. The later scan suggested that deterioration had occurred in the appellant's mental condition since 1998. The January 2000 scan was said to demonstrate a 'significant degree of cerebral atrophy' and to suggest that Mohammed Sharif was suffering from 'progressive organic brain damage'. Two medical experts made comments to this effect, the second suggesting that Sharif was suffering from a 'chronic degenerative disorder … and some sort of autosomal disorder'.[27]

The length of time taken to prepare the appeal is noteworthy. The case report describes this process as a 'medical snowball which becomes very large indeed as it is rolled down the hill'.[28] The appeal court also refers to the need to use one expert to mediate the other evidence and refers to this expert as the 'ringmaster of the subsequent investigations'.[29] The conclusion of those writing the medical reports was that Sharif's mental condition had a genetic cause.[30] The experts giving evidence were drawn from backgrounds in neurology, neuroradiology, neuropsychiatric genetics, ophthalmology, psychiatry and learning difficulties. The degree of care with which the evidence is presented is impressive. The appeal court accepted the submission, taking the view that the original conviction was unsafe because Sharif may have been unfit to plead at the time of the trial. This conclusion was based on the 2000 MRI scan evidence and new interpretations of evidence existing at the time of the original trial of fitness to plead.

The case is intriguing for a range of reasons. First, it reflects a changing attitude to the use of scan evidence and to the value of that evidence. This is evidenced in the case reports by the change in the approach taken by the medical experts. Second, it emphasises the need for a broad range of expert opinion in interpreting scan evidence data. In this case, the court refers to the need for someone who will take the lead as expert in pulling all this information together. It is also remarkable as a case where the scan evidence suggests that the cause of the degenerative

---

[26] *R v Mohammed Sharif* [2010] EWCA Crim 1709 para 9 all quotations.

[27] *R v Mohammed Sharif* [2010] EWCA Crim 1709, para 17.

[28] *R v Mohammed Sharif* [2010] EWCA Crim 1709 para 18.

[29] *R v Mohammed Sharif* [2010] EWCA Crim 1709 para 19.

[30] *R v Mohammed Sharif* [2010] EWCA Crim 1709 para 19 ' the most likely explanation for the appellant's medical problems is a previously unrecognised autosomal recessive disorder occurring as the result of multiple consanguineous marriages in his family.' Similar comments were made about Mohammed's blindness which were said to be related to an inherited genetic disorder.

condition suffered by the defendant is genetic and ongoing. This raises interesting issues with regard to whether the evidence could actually have determined Mohammed Sharif's fitness to plead at the time of the trial. The case demonstrates the power of scan evidence where convincingly presented to make the appeal court question the safety of convictions.

Clearly in Sharif's case there was sufficient visual scan evidence supported by medical reports to cast doubt on the safety of his conviction. Use of scans has also been made by appellants claiming that they were suffering from mental conditions at the time of their first trial, which were not considered and which if they had been considered would have permitted the courts to form a different view of their criminal responsibility.

2.4 Diminished Responsibility

There has been a growth over the last 10 years of the use of scan evidence to support appeals against sentence and against conviction in cases where the plea of diminished responsibility has been raised. Diminished responsibility is a very specific defence. It provides a partial defence to murder, but does not apply to other offences. Where it succeeds it reduces the murder charge to a conviction for manslaughter. This means that the defendant will not receive the mandatory life sentence. All murder convictions carry a mandatory life sentence, whereas those convicted of manslaughter may receive a shorter sentence.

The law has recently been changed and the 2009 statutory definition of diminished responsibility is set out in the footnote below.[31] It bears some relation to the

---

[31] Coroners and Justice Act 2009

52 Persons suffering from diminished responsibility (England and Wales)

(1) In Section 2 of the Homicide Act 1957 (c. 11) (persons suffering from diminished responsibility), for subsection (1) substitute -

'(1) A person ('D') who kills or is a party to the killing of another is not to be convicted of murder if D was suffering from an abnormality of mental functioning which—

(a) arose from a recognised medical condition,

(b) substantially impaired D's ability to do one or more of the things mentioned in subsection (1A), and

(c) provides an explanation for D's acts and omissions in doing or being a party to the killing.

(1A) Those things are—

(a) to understand the nature of D's conduct;

(b) to form a rational judgment;

(c) to exercise self-control.

(1B) For the purposes of subsection (1)(c), an abnormality of mental functioning provides an explanation for D's conduct if it causes, or is a significant contributory factor in causing, D to carry out that conduct.'

earlier law.[32] The earlier law required that the defence establish evidence of abnormality of mind. That abnormality having to be linked with 'a condition of arrested or retarded development of mind or any inherent causes or induced by disease or injury'. Medical evidence was usually required to establish this linkage. The partial defence to murder included cases where the defendant suffered a loss of control when killing. This loss of self control had to be shown to be linked by medical evidence with the abnormality of mind. If defendants could show this abnormality of mind had substantially impaired their responsibility to the satisfaction of a jury, they would receive a conviction for manslaughter.

Appeal cases demonstrate the use of brain scan evidence to establish brain damage was extant at the time of the crime. In the case of *R v Hanson*,[33] Hanson's appeal was based upon the dispute between the medical experts at his trial as to whether the abnormality of mind from which all the experts agreed he suffered was sufficient to substantially impair his responsibility for the killing. Two experts were cited in court as concluding, after considering MRI scans of Hanson's brain taken in 1994 and 1996, that his responsibility was impaired. Three experts were cited in court as concluding his responsibility was not impaired. What was not disclosed by the prosecution at the time of Hanson's first trial was that there was an opinion of a further medical expert, Dr Lewis. He said 'that he could not accept that the 1996 scan was normal. That was possible, but the findings could be indicative of temporal lobe epilepsy'.[34] The Court of Appeal gave careful consideration to this evidence which was not considered at the first trial and to new neurological evidence based upon this and further brain scans. However, they did not accept that the conviction for murder was unsafe. They instead concluded that the defendant's criminal behaviour suggested that his responsibility was not substantially impaired.[35]

The courts tend to be sceptical when reviewing brain scan evidence in cases of diminished responsibility. A problem area for the law in relation to claims of this type is the interaction of alcohol and prescription or illegal drugs with underlying medical conditions. In trying to analyse whether the capacity for control is diminished by the alcohol, drugs or mental condition suffered by the defendant the courts

---

(2) In Section 6 of the Criminal Procedure (Insanity) Act 1964 (c. 84) (evidence by prosecution of insanity or diminished responsibility), in paragraph (b) for 'mind' substitute 'mental functioning'.

[32] Homicide Act 1957 s 2

(1) Where a person kills or is a party to the killing of another, he shall not be convicted of murder if he was suffering from such abnormality of the mind (whether arising from a condition of arrested or retarded development of mind or any inherent causes or induced by disease or injury) as substantially impaired his mental responsibility for his acts and omissions in doing or being a party to the killing ....

[33] [2005] EWCA Crim 1142.

[34] [2005] EWCA Crim 1142, para 8.

[35] For a fuller examination of this case and its use of neuroscientific evidence see Claydon L. Law Neuroscience and Criminal Culpability in Current Legal Issues 2010 vol 13 (2011, Oxford University Press) 141–169, 152.

have encouraged juries to try to disentangle from the factual and medical evidence the most likely cause of the defendant's loss of capacity to control her actions.

The House of Lords considered this issue in the case of *R v Dietschmann* where their Lordships formulated the question to be considered by the jury as follows: 'has the defendant satisfied you that, despite the drink, his mental abnormality substantially impaired his mental responsibility for his fatal acts, or has he failed to satisfy you of that'? [36] How this ruling was precisely to be applied in cases of diminished responsibility was considered by the Court of Appeal in the case of *R v Wood (Clive)*.[37] In this case the issue was whether the accused was suffering from alcohol dependency syndrome. He killed his victim after a day and an evening spent drinking. In *R v Tandy*[38] it was stated that was that the defence had to provide evidence that the accused was so addicted to alcohol that he could not voluntarily stop drinking alcohol. The question in *Wood* was did the reasoning in *Dietschmann* permit the Court of Appeal to interpret s2 Homicide Act 1957 so that evidence of alcohol dependency syndrome might be sufficient for a jury to find that the defendant's responsibility for the killing was diminished? In other words was evidence of alcohol dependency syndrome sufficient to establish abnormality of mind?

The appeal was allowed and the reason the Court of Appeal gave was that the strict dividing line in *Tandy* was not really appropriate. Most alcoholics could stop drinking at some time. In a sense the jury is being asked to determine how voluntary the drinking is given the evidence and it is open to a jury to conclude having heard the scientific evidence that there is an abnormality of mind. This is an area where brain scan evidence may in future be used to establish whether the consumption of alcohol or drugs has resulted in identifiable damage pointing to the possible conclusion that the defendant was suffering at the time of the killing from abnormality of mental function. In the case of *R v McCann (Patrick)*[39] reference is made to the comments of a medical expert who expressed the opinion that MRI or CT scan evidence would have assisted in assessing the reliability of a key witness at McCann's trial. According to this witness the scans would have assisted in determining the extent of her incapacity which may have been caused by 'Korsakoff's syndrome', 'alcoholic dementia' and/or 'other cognitive problems'.[40]

---

[36] [2003] UKHL 10 para 41.

[37] [2008] EWCA Crim 1305.

[38] (1998) 87 Cr App R 45 (CA).

[39] 2000 WL 1918514.

[40] This raises a further interesting questions namely whether in the future neuroscientific evidence will be admitted which relates to the reliability of witness' evidence.

2.5 Claims That Actions Are Not Voluntary

Perhaps even more problematic are claims of lack of capacity in the form of an assertion that the defendant's acts were not voluntary. Where the presumption of voluntariness is challenged, the defendant must establish that there is evidence to support the assertion. The defence of automatism is case law based[41] and may be asserted as a defence to any crime. In *R v Roach*[42] the claim made by the defendant, who had committed a serious assault on a work colleague, was that he had no recollection of the events. The claim he wished to assert was that he was 'not conscious of what he was doing' at the time of the crime. A great deal of medical evidence was available at trial and Roach was convicted of the crime and sentenced to 4 years imprisonment. One of the grounds of his appeal was that the judge had refused to leave the issue of non-insane automatism to the jury.

In English law the distinction between non-insane automatism and insane automatism is fairly arbitrary. If the cause of the automatism is viewed as internal, for example a pre-existing mental disorder, then the M'Naghten[43] rules will be applied and if satisfied a finding of insanity will result. If however, the cause of the automatism is legally deemed to be external, then the jury may find that the defendant has the defence of non-insane automatism with the result that the defendant will be acquitted of all charges.

To say the least, the evidence to support Roach's claim of non-insane automatism was mixed. There was EEG and CT scan evidence which according to one expert indicated 'activity in normal limits'. Such evidence clearly was useful to a plea of non-insane automatism as it might suggest the cause of the unconscious behaviour was external. However, the expert went on to say that these scans were 'fairly insensitive investigations' and further asserted that he could not rule out the possibility of 'temporal lobe epilepsy'. The further assertion by the expert of the possible existence on a more sensitive brain scan of signs of epilepsy would have impeded a non-insane automatism claim. Epilepsy has been identified, by the courts, as an internal cause of automatism and therefore insane automatism.[44] Presumably, the expert knew this and that is why he presented the evidence as he did.

---

[41] Bratty v Attorney General for Northern Ireland [1963] AC 386

'No act is punishable if it is done involuntarily: and an involuntary act in this context – some people nowadays prefer to speak of it as 'automatism' – means an act which is done by the muscles without any control by the mind such as a spasm, a reflex action or convulsion; or an act done by one who is not conscious of what he is doing such as an act done while suffering from concussion or while sleepwalking.' Per Lord Denning.

[42] [2001] EWCA CRIM 2698.

[43] M' Naghten s Case All ER Rep. (1843) 10 C & F 200, 8 Eng Rep 718, [1843–60] All ER Rep 229 at 233. The M'Naghten Rules '... to establish a defence on the ground of insanity it must clearly be proved that, at the time of committing the act the party accused was labouring under such a defect of reason, from disease of the mind, as not to know the nature and quality of the act he was doing, or, if he did know it, that he did not know he was doing what was wrong.'

[44] *R v Sullivan* [1983] 2 All ER 675 (HL).

The medical evidence on appeal was to say the least diverse, a whole variety of causes being suggested by different medical experts, most of which pointed to internal causes of the behaviour. There was, however, some evidence of potential external causes of the automatism in the form of alcohol and prescription drugs. The Court of Appeal accepted that the conduct of the trial by the trial judge had contained irregularities with regard to the admission and exclusion of certain evidence. The Court of Appeal also noted that the jury had been misinformed as to the burden of proof on the defence in cases of non-insane automatism. The conviction was declared unsafe and Roach's conviction was quashed. The role of the scan evidence is interesting. It does not appear to support a case for an internal cause of the defendant's alleged automatism. The evidence was taken from EEG and CT scans which were said to be "insensitive". This raises the intriguing question - would more sensitive brain scan evidence which indicated the absence of evidence of a mental condition - assist the defence in making a plea of non-insane automatism?

3   Civil Law and Neuroscientific Evidence

3.1   Testamentary Capacity

In English law a testator must be of sound mind to make a valid will. The test of testamentary soundness of mind has developed over many years and encompasses the testator's memory and understanding. In cases where a testator's capacity is questioned, medical evidence is commonly adduced. In recent years, evidence from brain scans has increasingly been used.

In *Carr and another v. Thomas*[45] the issue was whether Christopher Ward had capacity when he made a will in 2006 whilst terminally ill with bowel cancer and secondary brain cancer. Mr. Ward died 9 days after making the 2006 will. There had been no contemporaneous medical assessment of the testator's capacity at the time of the 2006 will – a will which changed the primary beneficiaries from those named in the testator's 2002 will. In assessing the testator's mental capacity, evidence of his mental state in the period before he wrote the 2006 will was given by people who visited him during the later stages of his illness. Perhaps unsurprisingly, the person who was the primary beneficiary under the 2006 will observed no evidence of confusion, but those people who benefitted most from the 2002 will said that Christopher was unable to sustain a conversation and that his attention wandered. Evidence was also adduced from those treating him. This was inconclusive: some of the evidence indicated that he was 'muddled' and that he had 'memory impairment', other evidence described him as 'alert'. The day before he made his will he was described as 'unable to assimilate advice or instructions',

---

[45] [2008] EWHC 2859 (Ch).

however on the day on which he made his will he was described by his medical carers as being alert and 'seeming brighter'.[46]

The main medical evidence came from Dr. Campbell, a Consultant Psychiatrist in Forensic Psychiatry, and from Professor Trimble, a Professor of Behavioural Neurology and Consultant Physician of Psychological Medicine. Both experts relied on the evidence of a CT scan conducted 3 days after the testator's death. The experts agreed that there 'was gross swelling and a large area of abnormality in the left hand side of the brain' and that this could have led to 'aphasic disturbance'.[47] However, they disagreed as to the likely impact of the aphasia. Both experts also related their view of the brain abnormality to the evidence of the other witnesses, finding evidence of the testator's conduct which fitted their interpretation of the likely impact of the aphasic disturbance.

The case illustrates the willingness of the courts to utilise brain scan evidence to assist in determining capacity. However, whilst both experts relied on the same brain scan images, they disagreed as to the appropriate conclusions to be drawn from the evidence. This illustrates a common problem that will be faced by the courts. Brain scan images may appear useful in determining the issue before the court: in this case whether the testator had capacity. In one sense, the evidence may appear more 'real' than the impressionistic evidence of witnesses who observed the actions of the individual whose capacity is being questioned. Brain scans are tangible, visual evidence and may lead to agreement as to the nature of what is observed – in this case 'gross swelling and a large area of abnormality in the left hand side of the brain'. However, this evidence was not conclusive as to whether the testator had or had not got capacity. Even if the brain scan had been contemporaneous with the signing of the will, it would still not have been conclusive. The scan still needed to be interpreted. Did it mean that the testator did not have capacity as one expert concluded? Or did it mean that he could have had capacity at the time of making the will as the other concluded?

Faced with conflicting interpretations from the experts and conflicting evidence from the other witnesses the judge had in effect to decide which evidence he found most plausible. Interestingly, the witness on whose evidence the judge chose to rely most heavily was not one of the medical expert witnesses, but was the solicitor who had taken instructions from the testator and who had drawn up the will. As for the two experts: the trial judge expressed his preference for the evidence of Dr. Campbell as against that of Professor Trimble – favouring Dr. Campbell's reliance in drawing conclusions based on the evidence of the solicitor and dismissing Professor Trimble's contention that 'it requires a medical professional (or even a consultant neurosurgeon) to assess aphasia'. This perhaps illustrates an unwillingness of judges to hand over the determination of medico-legal issues entirely to the expert medical witness in court.

---

[46] [2008] EWHC 2859 (Ch), quotations taken from paras 23–35.

[47] [2008] EWHC 2859 (Ch), quotations taken from paras 62–63.

Similarly in other cases the existence of brain scan evidence, whilst providing part of the picture, tends not to be determinative. In *Scammell and another v. Farmer*,[48] a CT scan which indicated 'the early stages of Alzheimer's disease of late onset' was viewed as less significant than the testator's performance in a series of memory tests[49] in evidencing capacity to testate. However, neither was conclusive. The CT scan evidenced that the testator was suffering from Alzheimer's disease, but did not evidence the extent of the impact. Similarly, the memory tests provided evidence as to the state of the testator's memory, but this was not the same as the testator's capacity to make a valid will. This test of capacity included the ability to comprehend the extent of her estate, where arguably memory was relevant, but also included a moral responsibility to consider those such as her immediate family who might expect to inherit. As in *Carr* the final determination as to whether the testator had capacity was based on all the available evidence including the medical evidence, but also relying on the evidence of relatives, friends, carers and the solicitor who drew up the will.

In both the case of *Carr* and the case of *Scammell,* there was no medical examination of the testator at the time of making the will to assess the testator's capacity. In *Sharp and another v. Adam and others*,[50] the Court of Appeal noted the so-called 'golden rule' that the making of a will by an old and infirm testator ought to be witnessed and approved by a medical practitioner who satisfies himself as to the capacity and understanding of the testator and makes a record of his examination and findings'.[51] This was done in the case of Mr. Adam. Indeed, the court went on to note that the solicitor responsible for drawing up the will did 'everything conceivably possible, short of submitting Mr. Adam to a wholly impracticable full scale series of neuro-physiological tests and examinations, to satisfy herself that Mr Adam had testamentary capacity'. This is interesting for two reasons. First, despite the contemporaneous medical evidence to the contrary, both the trial court and the Court of Appeal concluded that Mr. Adam did not have capacity. Second, the judge's reference to the possibility of neuro-psychological tests indicates a judicial awareness that such tests exist and might be employed.

In fact in *Sharp and another v. Adam and others*, there was neuroscientific evidence. The testator had undergone MRI tests whilst in Germany. The expert witnesses who gave evidence in the case had not met the testator and had not seen the scans, but had read the detailed notes provided by those who had treated Mr. Adam including their notes relating to the scans.[52] There was argument in

---

[48] [2008] EWHC 1100 Ch.

[49] Mini Mental State Examination.

[50] [2006] EWCA Civ 449.

[51] [2006] EWCA Civ 449, para 27.

[52] The notes included the following: '*MRI head* demonstrates extensive decreased signals in the periventricular areas especially near anterior and posterior horns but also in the basal ganglia and the brainstem. Noted also advanced supratentorial atrophy. All this is in keeping with the diagnosis of longstanding disseminated encephalomyelitis.' [2006] EWCA Civ 449, para 44.

the case as to whether expert evidence should be allowed where the experts had not seen the scan, but only the report on the scans. However, the judge concluded that the expert evidence interpreting the scans should be admitted.

The experts agreed that the German report indicated 'damage to the brain' and that this was 'likely to impair his cognitive functions, in particular his memory and his executive functions'.[53]

The MRI scans were from 1996, 5 years before the date of the will. It was agreed that the testator, Mr. Adam, was suffering from secondary progressive multiple sclerosis. The expert witnesses, in accordance with approved practice, produced a joint report on those matters on which they agreed – for example that this condition was likely to have worsened in the intervening 5 years. However there were areas of disagreement – in particular on the crucial issue as to whether the testator had capacity at the time that the 2001 will was made.

As stated earlier, notwithstanding the contemporaneous medical evidence, the trial judge held that the testator did not have capacity. In arriving at this conclusion, the judge cited a number of reasons: the first of which was 'the expert evidence as to the severity of the damage to Mr. Adam's mental faculties'. This indicates a very high judicial regard being given to the interpretation of the scan evidence. The other reasons given by the trial judge related to Mr. Adam's unexplained change in attitude to his daughters. His daughters were the main beneficiaries under his two earlier wills, but were excluded in his 2001 will.

The Court of Appeal's decision is most interesting. They disagreed with the trial judge's reasoning and did not accept that the evidence of Professor Maria Ron, the expert called by those challenging the 2001 will, could be interpreted in the way it had been by the trial judge. The Court of Appeal went to great efforts to understand fully the expert testimony, including going to the length of obtaining a transcript of the trial in order to examine in detail the expert testimony. Having done this, the Court of Appeal concluded that whilst the trial judge could not properly have arrived at his conclusion on the basis of Professor Ron's testimony, it was possible to draw the same conclusions from a part of the evidence of Dr. Hawkes, the expert who had been called by those seeking to uphold the 2001 will.

The Court of Appeal noted that both experts focussed primarily on the testator's cognitive abilities However, as noted by both the trial judge and the Court of Appeal, the test of testamentary capacity also includes a requirement that 'no disorder of the mind shall poison his affections, pervert his sense of right, or prevent the exercise of his natural faculties – that no insane delusion shall influence his will in disposing of his property and bring about a disposal of it which, if the mind had been sound, would not have been made'.[54] The Court of Appeal viewed this requirement as being 'concerned as much with mood as with cognition'.[55] The

---

[53] [2006] EWCA Civ 449, para 53.

[54] *Banks v. Goodfellow* (1870) 5 Q.B. 549, 565.

[55] *Sharp and another v. Adam and others* [2006] EWCA Civ 449, para 93.

court felt this was an area in which neuroscience could assist. 'Our general understanding is that modern neurology and neuro-psychology is capable of addressing affections of mood in scientific terms'.[56]

The Court of Appeal did not seek to set aside the trial judge's conclusion that Mr. Adam had the requisite cognitive capacity. However, whilst that Court of Appeal concluded that almost all Professor Ron's evidence had focussed on cognition, they did find a passage in Dr. Hawkes' evidence that was relevant to the issue of the testator's affections and sense of right. This passage the Court concluded 'provided the evidential lifeline for Grace and Emma Adams' case'.[57] This is doubly interesting. Not only is Dr. Hawkes' evidence being used against the interests of the side who called him to give expert evidence, but it is being used in an area where he claimed no special expertise.[58]

In considering the appeal the Court of Appeal were limited in their options. The Court could only allow the appeal if satisfied that the first instance decision was wrong.[59] Although they rejected the reasoning of the trial judge, they considered that there was evidence on which he could have arrived at the conclusions that he did. In doing so, they were heavily reliant on the neuroscientific evidence and were ready to apply their own understanding of scientific developments in order to draw their conclusions.

3.2   Other Issues of Capacity

In civil cases capacity will not only be an issue with regard to capacity to testate. It can also arise in relation to capacity to contract and capacity to make gifts. However, examples of such cases being reported are less common. This may indicate that there are fewer such cases or that such cases are generally not reaching the level of court where decisions will be reported. An example of a case where neuroscientific evidence was admitted to help answer a question as to whether a person had the requisite capacity to make a gift is *Sutton v. Sutton*.[60] In this case brain scan evidence was produced and this showed 'some shrinkage of the brain (atrophy) but no other abnormality'.[61] Whilst this evidence clearly contributed to the conclusion that the benefactor did not have capacity, it was only a small part of the medical and other evidence on which the conclusion was based.

---

[56] *Sharp and another v. Adam and others* [2006] EWCA Civ 449, para 93.

[57] *Sharp and another v. Adam and others* [2006] EWCA Civ 449, para 93.

[58] 'Although Dr Hawkes had referred to himself in his written report in this respect as a lay observer, the deputy judge was, we think, entitled to take this part of his cross-examination as expert opinion.' *Sharp and another v. Adam and others* [2006] EWCA Civ 449, para 93.

[59] *Sharp and another v. Adam and others* [2006] EWCA Civ 449, para 95 citing CPR 42. 11 (3).

[60] [2009] EWCA 2576 (Ch).

[61] [2009] EWCA 2576 (Ch), para 24 (v).

3.3 Personal Injury

Probably the most widespread use of neuroscientific evidence and in particular brain scan evidence in English courts arises in cases of personal injury and medical negligence. As early as 1969 there is reference in the case of Jones v. Griffith[62] to the claimant having produced seven EEG reports in support of her claim to have developed grand mal epilepsy as a result of a motor accident. The case was one in which the medical evidence was agreed and an agreed medical statement was submitted to the Court of Appeal. However, notwithstanding the absence of medical disagreement the Court of Appeal commented on the desirability of such experts being called to give evidence in person so that they could be questioned by the court. This highlights a tension within the system. To speed up the trial process and save costs, it is generally considered desirable that evidence should, wherever possible, be agreed before the hearing and submitted in written form. On the other hand, the opportunity to question witnesses may assist the judiciary in gaining as full an understanding of the evidence as possible.

As stated, whilst it cannot be accurately quantified, because of the limitations of court reporting it is clear that the use of scan evidence in personal injury cases has become quite widespread.

The case of *Meah v. Mcreamer*[63] illustrates an interesting area on which there is currently very little caselaw, but which may become more common as further advances in neuroscience take place. Meah was a passenger in a car which was involved in a motor accident. As a result, Meah suffered head injuries. Dr. Gooddy, a consultant neuro-physician gave evidence on the basis of brain scans that Meah had sustained 'a clear cut area of left anterior frontal damage extending from the inner surface of the skull to or almost to the anterior end of the left lateral ventricle'.[64] Further medical evidence was provided by two consultant neuro-surgeons, two consultant psychiatrists, and by Mr. Offen a hospital consultant who reported that a month after the accident 'was still a little aggressive but this is common after such a severe head injury'.[65]

All the doctors agreed that the accident had affected Meah's personality. However, there was disagreement as to the extent of the personality change. Prior to the accident Meah had done badly at school, had numerous convictions for largely low level offences[66] and had on numerous occasions been involved in and injured in

---

[62] [1969] 2 All ER 1015.

[63] [1985] 1 All ER 367.

[64] [1985] 1 All ER 372.

[65] [1985] 1 All ER 372.

[66] [1985] 1 All ER 374 He had convictions for theft, house breaking, burglary, several driving offences and one conviction for assaulting a police officer.

fights.[67] However Meah's convictions were, with the exception of an assault which arose when resisting arrest, for non-violent offences and his last reported fight[68] had been in 1971, during what the judge, Woolf J., described as his 'skinhead phase', 7 years before the car accident.

Prior to the crash he had no convictions for any sexual offences and the evidence of witnesses including his former wife was that he had not been sexually violent. However, Meah became generally more aggressive after the accident and there was evidence that he was more callous in his attitude to his wife and following the end of that relationship with his subsequent sexual partners.

Dr. Noble, one of the consultant psychiatrists involved in the case, commented that the type of injuries would be 'characterised by blunting of emotional and moral sensitivity. There is a tendency to be bland and even callous. There is often apathy and irresponsibility. Control is weakened and there may be aggressiveness. The person becomes irresponsible and without adequate regard for the welfare of others or their own welfare, except on an immediate basis. Any anti-social tendency is likely to be enhanced'.[69]

Three and a half years after the car crash, Meah carried out three violent sexual attacks. He was convicted of these offences and sentenced to life imprisonment. Meah's civil case arising from the car accident was remarkable in that he was seeking damages not simply for the injuries sustained and the loss of earnings, but also for the fact that he had committed offences which he would otherwise not have committed. All the experts agreed that his personality had been changed by the accident. However, there was disagreement amongst the experts as to whether it could be said that Meah would not have committed the sexual attacks but for the injury he had sustained.

Both Dr. Noble and Dr. Gooddy were of the opinion that Meah would not have committed the sexual offences had he not suffered the brain injury. The nature of the brain injury sustained was likened by Dr. Noble to a leucotomy: an operation which the consultant psychiatrist explained had been known to cause disinhibition and could 'bring people with sexually abnormal tendencies into conflict with society'.[70] The trial judge found these experts' evidence most compelling and awarded Meah damages. Quantifying the damages was acknowledged to be

---

[67] [1985] 1 All ER 373 According to the evidence at the age of 17 he had received hospital treatment for cuts following an attack with a bottle. The following year he had been hit over the head with a poker. That same year he received further injuries to the head as a result of either being kicked or hit by someone using a knuckle duster. A year later he head butted someone who was the other side of a plate glass window and received cuts to his head as a result. That year he also sustained a back injury.

[68] If you exclude the assault conviction for head butting a police officer in order to resist arrest which had taken place in 1973.

[69] [1985] 1 All ER 373, 379.

[70] If you exclude the assault conviction for head butting a police officer in order to resist arrest which had taken place in 1973, 380.

difficult and interestingly was calculated by the judge on the basis of the approach adopted in cases of wrongful imprisonment.[71,72]

The personal injury cases considered above all involved neuroscientific and brain scan evidence to support the claim. As scans have become a more regular part of the medical professions armoury there has developed an expectation that scans will be undertaken when they are necessary. This has, unsurprisingly, led to the possibility of medical negligence claims being made when a scan does not take place or is delayed.[73]

3.4   Withdrawal of Treatment from Those in Persistent Vegetative States

It is well established in English law[74] that it is lawful to cease treatment for a person in a persistent vegetative state who has no prospect of recovery. Neurological evidence is likely to be key in determining whether a person is in a persistent vegetative state and whether they have any prospect of recovery. In the landmark case of *Airedale NHS Trust v Bland*,[75] CT and EEG scans showed no evidence of cortical activity and that there was 'more space than substance in the relevant part of Anthony Bland's brain'.[76] Experts called to give evidence in the case included a number of distinguished medical and neuroscience experts.[77] Again, the expert witness evidence was treated with great care by the courts.

---

[71] Meah was awarded £60,000. The award was quite low in comparison to the awards typically given in miscarriages of justice because given Meah's previous employment history meant that his loss of earnings were viewed as negligible. Two of Meah's victims subsequently sued him and were awarded damages (*W v. Meah; D v. Meah and another* [1986] 1 All ER 935). An attempt by Meah to recover these awards against the driver and his insurers failed, the damage being held to be too remote (*Meah v. Mcreamer and others No.2* [1986] 1 All ER 943).

[72] Subsequently the courts have had to consider a similar scenario to Meah's case. In *Gray v. Thames Trains and another* [2009] UKHL 33, [2009] 4 All ER 81 it was accepted that Gray had suffered a personality change as a result of a train accident caused by the defendants' negligence. It was accepted that but for the personality change Gray would not have committed the killing for which he was subsequently convicted. However, in this case the House of Lords rejected the claim for compensation; taking the view that to allow such a claim would be to allow the claimant to profit from his criminality.

[73] In *Rehman v University College London and another* [2004] EWHC 1361 (QB) experts agreed that a CT scan should have been undertaken to identify whether the claimant's bowel was perforated. This examination was delayed by two days and as a result the claimant received damages for her pain and suffering during the days in question.

[74] See *Airedale NHS Trust v Bland* [1993] AC 789 and *NHS Trust (AVM)* [2001] Fam 348.

[75] [1993] AC 789.

[76] [1993] AC 789, 795.

[77] Dr. Keith Andrews, Director of Medical Research Services at the Royal Hospital and Home, Putney – a hospital with a large brain injury rehabilitation centre. Professor Bryan Jennett, former Foundation Professor of Neurosurgery at Glasgow University's Institute of Neurological Science.

More recently in *An NHS Trust v. J*[78] Sir Mark Potter, President of the Family Division, showed a similar commendable willingness to acquaint himself with the latest scientific developments. The evidence considered in court included a 2003 report of the Royal College of Surgeons and two articles on patients recovering from persistent vegetative states[79] published in the year of the trial. Again, the court was assisted by high profile expert witnesses.[80] The extent of care taken in preparing the case is evidenced by the fact that the Official Solicitor, who acts in such cases to safeguard the interests of the patient, not only entered into correspondence with the British experts, but also engaged in correspondence with the authors of the two articles published in the year of the trial, which had explored the issue of patients recovering from persistent vegetative states.

The possibility of an fMRI scan was considered, but on the facts it was considered unnecessary. However, the possibility of administering Zolpidem, a sleeping medication, identified in Clauss and Nel's work as having dramatic effect in some cases was considered at length. The administration of this drug was opposed by the patient's family, who did not want J to awake temporarily and discover the condition in which she was surviving, and was opposed by Professor Andrews who viewed the published research as flawed, but was supported by the Official Solicitor. The court held that the drug should be tried, but that if it had no positive or beneficial effect the parties should return to court without delay, the clear implication being that the order to withhold treatment would then be given.

4 The Possible Future Uses of Neuroscientific Technologies Inside and Outside the Courtroom

Pictures of our brains or, of blood oxygen activity levels in our brains, enable claims to be made by those who produce such evidence in court regarding the value and purpose of human actions.[81] This is particularly true where the evidence provides arguments for the existence of a mental condition affecting behaviour at the time

---

Professor Jennett was the person who with Professor Plum jointly coined the term persistent vegetative state. Dr. Cartlidge a consultant neurologist and university lecturer and Professor Behan from the Institute of Neurological Sciences at Glasgow Southern General Hospital.

[78] [2006] EWHC 3152 (Fam).

[79] R Clauss and W Nel, 'Drug Induced Arousal from the Permanent Vegetative State' (2006) 21 *Neuro Rehabilitation* 23–8 and AM Owen and others, 'Detecting Awareness in the Vegetative State' (2006) 313 *Science* 1402.

[80] Professor Keith Andrews, who had given evidence in the Anthony Bland case and Professor John Pickard, Professor of Neurosurgery, one of the authors of the Royal College of Surgeons 2003 report.

[81] For a review of some of the neuroscientific and philosophical arguments in relation to criminal responsibility see Claydon L, Mind the Gap, in Freeman M and Goodenough, O. R. (eds) Law Mind and Brain, (Ashgate, 2009) 55–80.

of the criminal act which may reduce culpability or be used to appeal for a lower sentence. This may also be true where neuroscientific evidence provides proof of damage to the brain where the claim is a civil action and again this may be a claim that the brain damage has altered behaviour. More contentiously, it may enable claims to be made in respect of personality traits, propensities to risky behaviour, dispositions and mental conditions which may have existed at the time of the crime. As the cost of brain scanning and other neuroscientific techniques become cheaper and more accurate so the attractiveness of the techniques to states in using them to identify those who are a risk to themselves or the general population increases. So to conclude this chapter it is worth noting that there are areas largely outside the remit of court discussion where this neuroscientific evidence may be used.

In England and Wales under a new offender management regime polygraph tests which are designed to test the truthfulness of statements about behaviour of sexual offenders after release on licence from prison are being introduced. The Offender Management Act 2007 enables the imposition on such offenders of a mandatory requirement to undertake polygraph tests to enable the risk they pose to the community to be better managed. Section 29 of the Act controls the manner in which the tests are conducted and Section 30 prohibits using the results of the tests taken by an offender as part of this regime being used in evidence 'against the released person for an offence'. Of course, this does not preclude the instigation of an investigation into the offender's activities using information gleaned from the offender during the test interviews. Polygraph evidence has been stated in a recent case from the Court of Appeal to be inadmissible.[82]

The Act makes the testing regime mandatory for certain sexual offenders.[83] The government are reviewing the use of the testing and the effectiveness of the tests as a management tool. What is interesting is the justification given for making such testing mandatory by Baroness Scotland, Minister of State for the Home Office in the House of Lords. She said 'We will commission a research study to run alongside the mandatory testing pilot, with a view to determining whether the polygraph test is efficacious in assisting the collection of useful evidence about offenders' behaviour and whether it genuinely facilitates effective offender management without disproportionally affecting the rights of those tested.[84] We think it is a sensible way forward'.[85]

---

[82] *R v Malcolm MacMahon* [2010] ECWA CRIM 1953.

[83] The English courts in relation to a particular applicant have accepted the mandatory testing regime does not contravene Article 8 of the ECHR see Corbett v Secretary of State for Justice and Another [2009] EWHC 2671 (Admin) – where the imposition of the regime was said to be a proportionate interference with the applicants rights in view of Article 8(2) in view of the risk posed by the applicant's release.

[84] The Human Rights Act 1998 Section 19 requires that the Government makes a statement of compatibility before the second reading of a new piece of legislation.

[85] Hansard, HL Vol.692, 5.45pm (June 12, 2007).

Such a view is interesting given the claims that are being made by neuro-scientists in America about the efficacy of fMRI testing of defendants. Websites such as http://noliefmri.com make the following offer to potential clients: 'The technology used by No Lie MRI represents the first and only direct measure of truth verification and lie detection in human history'![86] Such claims are hotly disputed.[87] Given the government view that where the collection of data is efficacious in managing behaviour then some interference with human rights is acceptable provided that it is not viewed as disproportionate then its introduction at sometime in the future cannot be ruled out.

In England, at present, there is no sign of the government looking to introduce such tests, particularly if they offer no particular advantage in terms of accuracy over polygraph tests. But none the less the use of the polygraph in such a manner suggests that where there is a test which appears to distinguish truth from lies then the government may be tempted to use such tests in the management of offenders, particularly where the risk posed by the release of those offenders is seen by the public as high risk. At present there is civil case law in England stating that a claimant in a civil case should not be required to undergo an MRI test,[88] there is however no such statement in the criminal law. In fact there is reference in *R v Gwaza*[89] to the comments from a trial judge that the absence of brain scan evidence may have had a negative impact upon the judge's decision in relation to sentencing Gwaza.

In reading the court decisions, it is hard not to be impressed by the degree of care taken in handling what is often complex neuroscientific evidence. There is an undoubted willingness to engage with difficult novel medical and scientific arguments. Clearly, such evidence needs to be carefully evaluated and the court needs to be sure that it can bear the weight of proof asserted by the expert witness.

One factor which will greatly affect the impact of neuroscientific evidence in criminal proceedings in the future are proposal to introduce changes to the Admissibility of Expert Evidence. The Law Commission has been reviewing the use of expert evidence in criminal proceedings. A consultation paper was issued in 2009

---

[86] Date of download 6th January 2011.

[87] See for example Greely H. T., and Isles J., Neuroscience Based Lie Detection: The Urgent Need for Regulation American *Journal of Law and Medicine, 33* (2007): 377–431.

[88] *Laycock v Lagoe* 40 BMLR 82, 89. In this case a defendant wanted the claimant to undergo a second brain scan. The court set out a two part test. Was it in the interests of justice to require the claimant to undergo a further scan? On the facts the answer to this was yes. The second question was whether the party who opposed the test had put forward a substantial reason for the test not to be undertaken? The court found that the danger to the claimant was not 'imaginary or illusory and therefore has to be regarded as substantial.' The court therefore ruled that the second scan should not be ordered.

[89] [2009] EWCA CRIM 1101 'We had to rule out any possible mental illness because you refused to be examined, and you refused to have a brain scan, or any other test which might show whether you were suffering from some real illness, so the only basis one can sentence you is on the basis that you suffered this episode because you were taking cannabis.'

and the Law Commission's recommendations are due to be published in spring 2011. A press release on the Law Commission's web site states that: 'The final report will recommend that a clear admissibility test for expert evidence should be created to replace the present application of the 'relevance and reliability' test. To accompany the test, we will also set out proposed guidelines to assist Crown Court trial judges (and magistrates' courts) to determine whether proffered expert evidence is sufficiently reliable to be considered'.

In the consultation paper reference was made to the Court of Appeal's deliberations in *Gilfoyle (No 2)*[90] as to the admissibility of expert evidence in new areas of scientific or medical knowledge. The court commented that such evidence 'is not admissible until accepted by the scientific community as being able to provide accurate and reliable opinion'. The consultation paper goes on to point out that such an approach is irreconcilable with another Court of Appeal decision that stated that it: 'would be entirely wrong to deny the law of evidence the advantages to be gained from new techniques and new advances in science'.[91] It also conflicts with the decision in *R v Harris; R v Rock; R v Cherry; R v Faulder.*[92] It should be noted that Law Commission recommendations are often substantially changed before enactment so there is no means of telling what the final statutory regime for the submission of evidence is likely to be.

For anyone who studies the law in this area, it is impossible not to feel that care should be and normally is taken when interpreting brain scan evidence in court. If the rules of evidence are to be changed and there are some excellent proposals in the consultation paper as to how the reliability of evidence should be assessed.[93] It is to be hoped that this if enacted, would enhance the present use of neuro-scientific evidence in court. Lawyers and scientists need to work together to build a mutual understanding and respect for each others' work. By these means a greater understanding may be achieved of both how the law should be applied and how it should be developed.

---

[90] [2001] 2 Cr App R.

[91] R v Robb (1991) 93 Cr App R 161,166 all quotations taken from Law Commission Consultation Paper 190 paras 3.5 and 3.6.

[92] See footnote 12 and the accompanying text.

[93] See Law Commission Consultation Paper No 190 Part 6.
For a fuller examination of the Law Commission's suggestions see footnote 35 page 167.

Neurolaw and UNESCO Bioethics Declarations

Darryl Macer

Abstract This paper presents an analysis of the bioethics Declarations agreed by all member countries of UNESCO with reference to neurosciences. The texts, the 1997 Universal Declaration on the Human Genome and Human Rights, and the 2005 Universal Declaration on Bioethics and Human Rights, provide a number of useful points for countries considering policy for use of knowledge of neurosciences, and in the education of society of the issues that arise from our increased understanding of neurosciences.

1 Introduction

UNESCO's ethics program started in 1993 with the establishment of the International Bioethics Committee (IBC), the first bioethics committee with a global scope and expert membership. As a forum for international reflection on bioethics, the IBC and the resultant bioethics Declarations provide an example of international reflection on these subjects, and the texts were unanimously endorsed by all member countries of UNESCO, making them useful texts for development of national laws, and more specific international guidelines.

This paper will analyze, in particular, the 1997 and 2005 Declaration, while noting that there are also useful references in the International Declaration on Human Genetic Data adopted by the General Conference of UNESCO on 16 October 2003 as well. I also refer readers to the text of the 1995 IBC report,

D. Macer (✉)
Regional Advisor on Social and Human Sciences for Asia and the Pacific, RUSHSAP, UNESCO Bangkok, Prakanong, Bangkok 10110, Thailand
e-mail: d.macer@unesco.org

T.M. Spranger (ed.), *International Neurolaw*,
DOI 10.1007/978-3-642-21541-4_18, © Springer-Verlag Berlin Heidelberg 2012

which described some of the ethics posed by neurosciences that also aided the work in drafting the 1997 Declaration.[1]

The Universal Declarations constitute nonbinding instruments in international law. However, the unanimous adoption by the member states gives the Declarations moral authority and creates a spirit of commitment to ethical principles in the use of applications of science and technology. All states of the international community are committed to respect and implement the basic principles of bioethics, set forth within these texts. The Declarations identify shared values and common principles. The formulation of these values and principles, however, is very general, allowing for flexibility in precise interpretation.

2 Analysis of the 1997 Declaration

Let us examine the insights of the 1997 Universal Declaration on the Human Genome and Human Rights for developments of law to govern neurosciences. (In each case, relevant text of the Declaration is presented in italics with accompanying commentary). Given that a significant part of neuroscience research is directly linked with genetics research, there are obvious applications of the Declaration to genetics research in any field. However, the principles of the Declaration can also be applied to scientific knowledge in general, and those broader issues were discussed in the International Bioethics Committee during the 5 years of the drafting of the Declaration.[2]

> The Preamble of UNESCO's Constitution refers to "the democratic principles of the dignity, equality and mutual respect of men", rejects any "doctrine of the inequality of men and races", stipulates "that the wide diffusion of culture, and the education of humanity for justice and liberty and peace are indispensable to the dignity of men and constitute a sacred duty which all the nations must fulfill in a spirit of mutual assistance and concern", proclaims that "peace must be founded upon the intellectual and moral solidarity of mankind",

The equality of people is not dependent upon all persons having equal abilities or equal capacities. Even before the development of neurosciences, it is clear that each person is unique. It is also apparent that even if we provide an identical and educational environment for humans, each will have different outcomes. The point is solidarity among varied individuals and communities. Neuroscience is showing us further explanation of the differences between people, and their psychological and intellectual abilities.

---

[1] **Ethics and Neurosciences** (reported by Mr. Jean-Didier Vincent). Document available on: http://unesdoc.unesco.org/images/0010/001051/105160e.pdf.

[2] Readers can refer to relevant reports of the UNESCO International Bioethics Committee (IBC), including a report on Ethics and Neurosciences. The document can be downloaded from http://unesdoc.unesco.org/images/0010/001051/105160e.pdf. The author was also a member of UNESCO IBC.

In the Preamble, there is also reference to "the Convention on the Prohibition of the Development, Production and Stockpiling of Bacteriological (Biological) and Toxin Weapons and on their Destruction of 16 December 1971".

Neuroscience research will also provide data that may be misused to develop targeted drugs and genetic treatments that could be used to harm certain individuals. Even if such interventions are directed against communities of individuals who share common genetic traits, rather than common nationality, it would still be against the principles of human rights if such acts were intended for harm. The ethical challenge, however, is to define what is "harm", since there are a variety of constructions of what is a good life. Although one community may not consider an act to be harm, the values of another may consider it a harm. We will need to know more clearly what each community values and does not value to be clear on what is good or harmful.

The Preamble reference to *the UNESCO Declaration on Race and Racial Prejudice of 27 November 1978* reminds us not to discriminate against racially common behavioral patterns. Although most genetic traits are common between races, there are some specific traits with increased tendency that could be associated with so-called race. Related to this are also the words:

> Bearing in mind also the United Nations Convention on Biological Diversity of 5 June 1992 and emphasizing in that connection that the recognition of the genetic diversity of humanity must not give rise to any interpretation of a social or political nature which could call into question "the inherent dignity and (...) the equal and inalienable rights of all members of the human family", in accordance with the Preamble to the Universal Declaration of Human Rights,

The Declaration is generally positive to research and thus promotes scientific research, as stated:

> Recognizing that research on the human genome and the resulting applications open up vast prospects for progress in improving the health of individuals and of humankind as a whole, but emphasizing that such research should fully respect human dignity, freedom and human rights, as well as the prohibition of all forms of discrimination based on genetic characteristics,

The specific articles of the text are quite useful guides for member states to consider development of laws not only for genetic studies but also for neuroscience. We could also reflect on them in the context of any future Declaration on Neurosciences, when research in neurosciences is developed to the extent that its applications threaten the humanity that the United Nations was established to protect.

First, the heritage (past, present and future) of humanity is centered on the human mind and its products, and relationships between minds that socially construct what we call humanity. While article 1 states:

> 1. The human genome underlies the fundamental unity of all members of the human family, as well as the recognition of their inherent dignity and diversity. In a symbolic sense, it is the heritage of humanity.

Although the genome is common between all human beings, we could be challenged to consider if the brain or mind is common between all. In the case of anencephalic humans, born with only a brain stem, but without most of the brain, the point may be made that they are not human minds. The more we focus on the "mind" as opposed to simply some part of a "brain", some will be tempted to consider some human individuals as nonpersons. A genome-centered definition of humankind is more inclusive that a one which is based on the mind. However for distinctions of some ethical issues, such as consent, a functional mind is necessary. In order to protect persons, we can refer to article 2, and replace "genetic" with "mental", as in:

2. a) Everyone has a right to respect for their dignity and for their rights regardless of their genetic characteristics.

This is a solid guarantee for the rights of all humans, regardless of their mind, brain, or personality.

2b) That dignity makes it imperative not to reduce individuals to their genetic characteristics and to respect their uniqueness and diversity.

We can expect greater understanding of neuroscience to add many insights into the prediction of the way people make decisions. Even if we can determine the moral choices people are likely to make, we should treat them as individuals with free choice. A similar tone is seen in the next article:

3. The human genome, which by its nature evolves, is subject to mutations. It contains potentialities that are expressed differently according to each individual's natural and social environment including the individual's state of health, living conditions, nutrition and education.

No matter how well we can predict a person's behavior from their mind, the mind is still subject to change and evolution. Already we can see in some legal jurisdictions the use of determinism (genetic environmental, fetal, social, etc.) as a defense for crimes that were committed. Depending on the location, judges may accept such ideas as a defense for what people do. Neurosciences will provide a greater scientific understanding of the mind and decision-making. One day it will allow prediction of actions, and challenge the concepts of free will and criminal responsibility.

Here lies a great challenge to the accepted norms of current criminal account-ability, as well as to methods for incarceration of persons for modification of their behavior. If we can manipulate the environment of persons to lower the chances of criminal acts, should not we do so? In fact, we already do in most societies protecting children as they grow up to lessen influences that will encourage crime and abuse. Ethically, this will be challenging with regard to the consent of persons, however, it may also provide good alternatives for return of convicted criminals (of any cause) to society, with lower chance of repeated crime.

4. The human genome in its natural state shall not give rise to financial gains.

The idea of this article is to avoid patents on genes. The same sentiment could be applied to mental characteristics, and we could imagine intellectual property rights (IPR) arriving from the use of neuroscience knowledge for patent.

The issues of consent are even more relevant to the question of neurosciences than to genetics, because consent requires an ability to consider and decide among several medical options. Given the complexity of neuroscience, it will be arbitrary to determine when sufficient assessment has been made of the risks and benefits of treatment. Even for diagnosis, when there is a chance of stereotyping, and prediction of choices, it will be difficult to decide what sufficient assessment is required. As stated:

B. RIGHTS OF THE PERSONS CONCERNED

5. a) Research, treatment or diagnosis affecting an individual's genome shall be undertaken only after rigorous and prior assessment of the potential risks and benefits pertaining thereto and in accordance with any other requirement of national law.

b) In all cases, the prior, free and informed consent of the person concerned shall be obtained. If the latter is not in a position to consent, consent or authorization shall be obtained in the manner prescribed by law, guided by the person's best interest.

The next article (5c) poses a number of challenges that will challenge the doctrine of consent. If we understand the mind of an individual, we may be able to assist in their decision making. However, if they decide they do not what to know the results, should we continue to respect their decision not to know? What if the results may provide an alternative to imprisonment?

c) The right of each individual to decide whether or not to be informed of the results of genetic examination and the resulting consequences should be respected.

The practice of neuroscience research should be established like genomics and other scientific research upon the laws of the country. As the other papers in this volume discuss, we need to clarify what these laws should be, and in the context of international research these should follow the generally accepted norms. A number of the research projects will also be subject to the guidelines of funding agents. It is important to have some common standards between countries.

5d) In the case of research, protocols shall, in addition, be submitted for prior review in accordance with relevant national and international research standards or guidelines.

We can expect neuroscience research to totally reconstruct the meaning of consent, and consent of those thought to be incapable of consent. We have seen this in the increasing awareness of the abilities of persons in so-called persistent vegetative state (PVS), who like some persons in "locked in syndrome" seem to be quite capable of consent if only we can communicate with them. Thus, as we consider:

5e) If according to the law a person does not have the capacity to consent, research affecting his or her genome may only be carried out for his or her direct health benefit, subject to the authorization and the protective conditions prescribed by law. Research which does not have an expected direct health benefit may only be undertaken by way of exception, with the utmost restraint, exposing the person only to a minimal risk and minimal burden and if the research is intended to contribute to the health benefit of other persons in the same age category or with the same genetic condition, subject to the conditions prescribed by law, and provided such research is compatible with the protection of the individual's human rights.

We will find ways to benefit persons through research, and they may be able to consent to the research or treatment. They also may be able to refuse such interventions, in new ways of communication, based on greater understanding of neurosciences. One could imagine challenges in countries, especially in those who have laws against suicide and assisted suicide, when consenting individuals refuse interventions. There will be numerous ethical issues raised for active euthanasia.

The following article is against genetic reductionism, and can be applied to any form of reductionism, including reducing people to merely their mental characteristics or psychological predispositions:

6. No one shall be subjected to discrimination based on genetic characteristics that is intended to infringe or has the effect of infringing human rights, fundamental freedoms and human dignity.

Confidentiality of medical information of any type, including genetic or neurological, is protected in article 7:

7. Genetic data associated with an identifiable person and stored or processed for the purposes of research or any other purpose must be held confidential in the conditions set by law.

Article 7 would also apply to psychological tests, such as intelligence tests and personality tests that are used in education and employment. The next article if applied to neurosciences would apply not only to medical injury, but also to social and educational actions that damage a person, such as incorrect schooling or social treatment:

8. Every individual shall have the right, according to international and national law, to just reparation for any damage sustained as a direct and determining result of an intervention affecting his or her genome.

Neurosciences will challenge some of the concepts of consent and confidentiality in national law, as further discussed in the conclusion to this paper. Some guiding principles for regulation are in article 9:

9. In order to protect human rights and fundamental freedoms, limitations to the principles of consent and confidentiality may only be prescribed by law, for compelling reasons within the bounds of public international law and the international law of human rights.

The third section of the Declaration promotes adherence to human rights standards for research on individuals as well as groups of persons:

C. RESEARCH ON THE HUMAN GENOME
10. No research or research applications concerning the human genome, in particular in the fields of biology, genetics and medicine, should prevail over respect for the human rights, fundamental freedoms and human dignity of individuals or, where applicable, of groups of people.
11. Practices which are contrary to human dignity, such as reproductive cloning of human beings, shall not be permitted. States and competent international organizations are invited to co-operate in identifying such practices and in taking, at national or international level, the measures necessary to ensure that the principles set out in this Declaration are respected.

Article 12 upholds the free sharing of research results for all, as well as emphasizing the freedom of scientific research as part of the freedom of human thought. There may be times in the future when neurosciences itself may challenge our concepts of what is freedom of thought. Whatever we learn however, it should be applied according to the principle of beneficence.

12.a) Benefits from advances in biology, genetics and medicine, concerning the human genome, shall be made available to all, with due regard for the dignity and human rights of each individual.

b) Freedom of research, which is necessary for the progress of knowledge, is part of freedom of thought. The applications of research, including applications in biology, genetics and medicine, concerning the human genome, shall seek to offer relief from suffering and improve the health of individuals and humankind as a whole.

The fourth section of the Declaration refers to ethical conduct of biomedical and scientific research and scientific responsibility, and this is particularly useful for scientific associations and scientists and professionals. Researchers have obligations to conduct research at all stages, and use it, ethically. There may need to be careful consideration of some types of research that challenge concepts of human identity. It will also be useful for enhancing professional standards of neurosciences and social work, so that human rights is applied after careful consideration of the implications of all of the research results:

D. CONDITIONS FOR THE EXERCISE OF SCIENTIFIC ACTIVITY

13. The responsibilities inherent in the activities of researchers, including meticulousness, caution, intellectual honesty and integrity in carrying out their research as well as in the presentation and utilization of their findings, should be the subject of particular attention in the framework of research on the human genome, because of its ethical and social implications. Public and private science policy-makers also have particular responsibilities in this respect.

14. States should take appropriate measures to foster the intellectual and material conditions favourable to freedom in the conduct of research on the human genome and to consider the ethical, legal, social and economic implications of such research, on the basis of the principles set out in this Declaration.

15. States should take appropriate steps to provide the framework for the free exercise of research on the human genome with due regard for the principles set out in this Declaration, in order to safeguard respect for human rights, fundamental freedoms and human dignity and to protect public health. They should seek to ensure that research results are not used for non-peaceful purposes.

The requirement for ethical review committees is becoming widely accepted already for neuroscience research, though in some countries psychological and sociological research is not subject to ethical review. Article 16, however, is being widely applied in biomedical research.

16. States should recognize the value of promoting, at various levels as appropriate, the establishment of independent, multidisciplinary and pluralist ethics committees to assess the ethical, legal and social issues raised by research on the human genome and its applications.

The general sharing of neuroscience knowledge is similar to genetic knowledge, despite the potentially more powerful nature of the ways that the mind works and could be influenced. Similar texts appear in most UNESCO documents for all fields of scientific knowledge, as seen in section E:

E. SOLIDARITY AND INTERNATIONAL CO-OPERATION

17. States should respect and promote the practice of solidarity towards individuals, families and population groups who are particularly vulnerable to or affected by disease or

disability of a genetic character. They should foster, inter alia, research on the identification, prevention and treatment of genetically-based and genetically-influenced diseases, in particular rare as well as endemic diseases which affect large numbers of the world's population.

18. States should make every effort, with due and appropriate regard for the principles set out in this Declaration, to continue fostering the international dissemination of scientific knowledge concerning the human genome, human diversity and genetic research and, in that regard, to foster scientific and cultural co-operation, particularly between industrialized and developing countries.

There could be application of the principle of do no harm, in the case where new mental interventions could be expected to harm individuals. The same knowledge could also be used to protect individuals from brainwashing and advertising campaigns, political campaigns, however, like any intervention we can expect it to be misused. The point is international cooperation in mapping the human mind is required, and this needs to be accompanied by full ethical analysis between cultures, as discussed in article 19:

19. a) In the framework of international co-operation with developing countries, States should seek to encourage measures enabling:

1. assessment of the risks and benefits pertaining to research on the human genome to be carried out and abuse to be prevented;
2. the capacity of developing countries to carry out research on human biology and genetics, taking into consideration their specific problems, to be developed and strengthened;
3. developing countries to benefit from the achievements of scientific and technological research so that their use in favour of economic and social progress can be to the benefit of all;
4. the free exchange of scientific knowledge and information in the areas of biology, genetics and medicine to be promoted.

b) Relevant international organizations should support and promote the initiatives taken by States for the abovementioned purposes.

The articles in sections F and G of the Declaration equally apply to neuroscience research, and call for education in understanding of the ethics of neurosciences, and on ethics of neurosciences:

F. PROMOTION OF THE PRINCIPLES SET OUT IN THE DECLARATION

20. States should take appropriate measures to promote the principles set out in the Declaration, through education and relevant means, inter alia through the conduct of research and training in interdisciplinary fields and through the promotion of education in bioethics, at all levels, in particular for those responsible for science policies.

21. States should take appropriate measures to encourage other forms of research, training and information dissemination conducive to raising the awareness of society and all of its members of their responsibilities regarding the fundamental issues relating to the defense of human dignity which may be raised by research in biology, in genetics and in medicine, and its applications. They should also undertake to facilitate on this subject an open international discussion, ensuring the free expression of various socio-cultural, religious and philosophical opinions.

G. IMPLEMENTATION OF THE DECLARATION

22. States should make every effort to promote the principles set out in this Declaration and should, by means of all appropriate measures, promote their implementation.

23. States should take appropriate measures to promote, through education, training and information dissemination, respect for the abovementioned principles and to foster their recognition and effective application. States should also encourage exchanges and networks among independent ethics committees, as they are established, to foster full collaboration.

In article 24, the IBC is called upon to further recommend to UNESCO and member states the ethical conduct of research. This could be applied to neuroscience research, along with other areas of science and technology.

24. The International Bioethics Committee of UNESCO should contribute to the dissemination of the principles set out in this Declaration and to the further examination of issues raised by their applications and by the evolution of the technologies in question. It should organize appropriate consultations with parties concerned, such as vulnerable groups. It should make recommendations, in accordance with UNESCO's statutory procedures, addressed to the General Conference and give advice concerning the follow-up of this Declaration, in particular regarding the identification of practices that could be contrary to human dignity, such as germ-line interventions.

25. Nothing in this Declaration may be interpreted as implying for any State, group or person any claim to engage in any activity or to perform any act contrary to human rights and fundamental freedoms, including the principles set out in this Declaration.

3 Analysis of the Universal Declaration on Bioethics and Human Rights (2005)

Let us examine the insights of the 2005 Universal Declaration on Bioethics and Human Rights for developments of law to govern neurosciences.

The preamble states the character of human beings of reflection, a basic feature of humans that neurosciences will assist in understanding:

Conscious of the unique capacity of human beings to reflect upon their own existence and on their environment, to perceive injustice, to avoid danger, to assume responsibility, to seek cooperation and to exhibit the moral sense that gives expression to ethical principles,

The preamble also makes note of some other relevant international ethics guidelines for biomedical research, which apply to neurosciences:

Also noting international and regional instruments in the field of bioethics, including the Convention for the Protection of Human Rights and Dignity of the Human Being with regard to the Application of Biology and Medicine: Convention on Human Rights and Biomedicine of the Council of Europe, which was adopted in 1997 and entered into force in 1999, together with its Additional Protocols, as well as national legislation and regulations in the field of bioethics and the international and regional codes of conduct and guidelines and other texts in the field of bioethics, such as the Declaration of Helsinki of the World Medical Association on Ethical Principles for Medical Research Involving Human Subjects, adopted in 1964 and amended in 1975, 1983, 1989, 1996 and 2000 and the International Ethical Guidelines for Biomedical Research Involving Human Subjects of the

Council for International Organizations of Medical Sciences, adopted in 1982 and amended in 1993 and 2002,

The preamble of the 2005 Declaration also refers to the prior UNESCO bioethics Declarations. In the same spirit, the analysis here will not repeat what is said above for the 1997 Declaration, but consider new articulations of ethical principles, which are more clearly stated in the 2005 text. To quote the preamble:

> ... and that questions of bioethics, which necessarily have an international dimension, should be treated as a whole, drawing on the principles already stated in the Universal Declaration on the Human Genome and Human Rights and the International Declaration on Human Genetic Data and taking account not only of the current scientific context but also of future developments,

The freedom of research balanced with protection of human dignity and human rights is also noted:

> Recognizing that, based on the freedom of science and research, scientific and technological developments have been, and can be, of great benefit to humankind in increasing, inter alia, life expectancy and improving the quality of life, and emphasizing that such developments should always seek to promote the welfare of individuals, families, groups or communities and humankind as a whole in the recognition of the dignity of the human person and universal respect for, and observance of, human rights and fundamental freedoms,

Psychological aspects of research, and thus also neurosciences, are specifically noted in the preamble:

> Recognizing that health does not depend solely on scientific and technological research developments but also on psychosocial and cultural factors,
> Also recognizing that decisions regarding ethical issues in medicine, life sciences and associated technologies may have an impact on individuals, families, groups or communities and humankind as a whole,
> Bearing in mind that cultural diversity, as a source of exchange, innovation and creativity, is necessary to humankind and, in this sense, is the common heritage of humanity, but emphasizing that it may not be invoked at the expense of human rights and fundamental freedoms,
> Also bearing in mind that a person's identity includes biological, psychological, social, cultural and spiritual dimensions,

In the scope of the 2005 Declaration, neuroscience is clearly included, so that all the principles and content of the Declaration can be applied to neuroscience:

> Article 1 – Scope
> 1. This Declaration addresses ethical issues related to medicine, life sciences and associated technologies as applied to human beings, taking into account their social, legal and environmental dimensions.
> 2. This Declaration is addressed to States. As appropriate and relevant, it also provides guidance to decisions or practices of individuals, groups, communities, institutions and corporations, public and private.

The text should also be read in article 1.2 as applying to all types of moral decision maker, individuals, professionals, nations. This is further explained in the stated aims of the document, article 2 (most relevant text included below):

Article 2 – Aims

The aims of this Declaration are:

(a) to provide a universal framework of principles and procedures to guide States in the formulation of their legislation, policies or other instruments in the field of bioethics;

(b) to guide the actions of individuals, groups, communities, institutions and corporations, public and private;

(c) to promote respect for human dignity and protect human rights, by ensuring respect for the life of human beings, and fundamental freedoms, consistent with international human rights law;

(d) to recognize the importance of freedom of scientific research and the benefits derived from scientific and technological developments, while stressing the need for such research and developments to occur within the framework of ethical principles set out in this Declaration and to respect human dignity, human rights and fundamental freedoms;

(e) to foster multidisciplinary and pluralistic dialogue about bioethical issues between all stakeholders and within society as a whole;

(f) to promote equitable access to medical, scientific and technological developments as well as the greatest possible flow and the rapid sharing of knowledge concerning those developments and the sharing of benefits, with particular attention to the needs of developing countries;

(g) to safeguard and promote the interests of the present and future generations;

The international discussion of neuroscience law, as in this book, is supported by article 2e. We need a broader discussion of these issues at the international level. The subsequent series of ethical principles are all relevant to neuroscience, and many are implicit in the 1997 Declaration. They conveniently provide a framework for discussion between cultures of the nature of international law to regulate neuroscience, and for professional ethical guidelines as well. They are quoted below:

Principles

Within the scope of this Declaration, in decisions or practices taken or carried out by those to whom it is addressed, the following principles are to be respected.

Article 3 – Human dignity and human rights

1. Human dignity, human rights and fundamental freedoms are to be fully respected.

2. The interests and welfare of the individual should have priority over the sole interest of science or society.

Article 4 – Benefit and harm

In applying and advancing scientific knowledge, medical practice and associated technologies, direct and indirect benefits to patients, research participants and other affected individuals should be maximized and any possible harm to such individuals should be minimized.

Article 5 – Autonomy and individual responsibility

The autonomy of persons to make decisions, while taking responsibility for those decisions and respecting the autonomy of others, is to be respected. For persons who are not capable of exercising autonomy, special measures are to be taken to protect their rights and interests.

Article 6 – Consent

1. Any preventive, diagnostic and therapeutic medical intervention is only to be carried out with the prior, free and informed consent of the person concerned, based on adequate information. The consent should, where appropriate, be express and may be withdrawn by the person concerned at any time and for any reason without disadvantage or prejudice.

2. Scientific research should only be carried out with the prior, free, express and informed consent of the person concerned. The information should be adequate,

provided in a comprehensible form and should include modalities for withdrawal of consent. Consent may be withdrawn by the person concerned at any time and for any reason without any disadvantage or prejudice. Exceptions to this principle should be made only in accordance with ethical and legal standards adopted by States, consistent with the principles and provisions set out in this Declaration, in particular in Article 27, and international human rights law.

3. In appropriate cases of research carried out on a group of persons or a community, additional agreement of the legal representatives of the group or community concerned may be sought. In no case should a collective community agreement or the consent of a community leader or other authority substitute for an individual's informed consent.

Article 7 – Persons without the capacity to consent

In accordance with domestic law, special protection is to be given to persons who do not have the capacity to consent:

(a) authorization for research and medical practice should be obtained in accordance with the best interest of the person concerned and in accordance with domestic law. However, the person concerned should be involved to the greatest extent possible in the decision-making process of consent, as well as that of withdrawing consent;

(b) research should only be carried out for his or her direct health benefit, subject to the authorization and the protective conditions prescribed by law, and if there is no research alternative of comparable effectiveness with research participants able to consent. Research which does not have potential direct health benefit should only be undertaken by way of exception, with the utmost restraint, exposing the person only to a minimal risk and minimal burden and if the research is expected to contribute to the health benefit of other persons in the same category, subject to the conditions prescribed by law and compatible with the protection of the individual's human rights. Refusal of such persons to take part in research should be respected.

The articles on consent are very detailed, and challenging especially for persons of early stages of mental maturation, and those with mental disorders. They could also be applied to education, and attempts at social engineering.

Article 8 – Respect for human vulnerability and personal integrity

In applying and advancing scientific knowledge, medical practice and associated technologies, human vulnerability should be taken into account. Individuals and groups of special vulnerability should be protected and the personal integrity of such individuals respected.

The text on vulnerability (article 8) is particularly useful for persons with mental disorders, and that is why it is highlighted. The following article 9 on confidentiality reasserts that which was stated in the 1997 Declaration, and is already in many national laws.

Article 9 – Privacy and confidentiality

The privacy of the persons concerned and the confidentiality of their personal information should be respected. To the greatest extent possible, such information should not be used or disclosed for purposes other than those for which it was collected or consented to, consistent with international law, in particular international human rights law.

Articles 10 and 11 apply the principle of justice and equity, as well as nondiscrimination and nonstigmatization. To actually apply these in society to avoid stereotyping of individuals, and provide equal opportunities for humankind, which is composed of persons with different potentials and abilities, will be a continue challenge to society as knowledge from neuroscience is gained.

Article 10 – Equality, justice and equity
The fundamental equality of all human beings in dignity and rights is to be respected so that they are treated justly and equitably.

Article 11 – Non-discrimination and non-stigmatization
No individual or group should be discriminated against or stigmatized on any grounds, in violation of human dignity, human rights and fundamental freedoms.

Article 12 – Respect for cultural diversity and pluralism
The importance of cultural diversity and pluralism should be given due regard. However, such considerations are not to be invoked to infringe upon human dignity, human rights and fundamental freedoms, nor upon the principles set out in this Declaration, nor to limit their scope.

Article 13 – Solidarity and cooperation
Solidarity among human beings and international cooperation towards that end are to be encouraged.

Article 14 – Social responsibility and health
1. The promotion of health and social development for their people is a central purpose of governments that all sectors of society share.
2. Taking into account that the enjoyment of the highest attainable standard of health is one of the fundamental rights of every human being without distinction of race, religion, political belief, economic or social condition, progress in science and technology should advance:
(a) access to quality health care and essential medicines, especially for the health of women and children, because health is essential to life itself and must be considered to be a social and human good;
(b) access to adequate nutrition and water;
(c) improvement of living conditions and the environment;
(b) elimination of the marginalization and the exclusion of persons on the basis of any grounds;
(e) reduction of poverty and illiteracy.

Article 15 – Sharing of benefits
1. Benefits resulting from any scientific research and its applications should be shared with society as a whole and within the international community, in particular with developing countries.

Article 16 – Protecting future generations
The impact of life sciences on future generations, including on their genetic constitution, should be given due regard.

A holistic approach to the health of individuals, including psychological health, is taken in article 14. This will be very important to consider as we do not reduce psychological traits to simple theories of causation due to genetics or other issues. The broader health of citizens is a responsibility of the national and international community.

The text on the application of principles quoted below is similar to the text discussed in the 1997 Declaration, with some more specificity as seen in articles 18–24:

Application of the principles
Article 18 – Decision-making and addressing bioethical issues

1. Professionalism, honesty, integrity and transparency in decision-making should be promoted, in particular declarations of all conflicts of interest and appropriate sharing of knowledge. Every endeavour should be made to use the best available scientific knowledge and methodology in addressing and periodically reviewing bioethical issues.
2. Persons and professionals concerned and society as a whole should be engaged in dialogue on a regular basis.
3. Opportunities for informed pluralistic public debate, seeking the expression of all relevant opinions, should be promoted.

Article 19 – Ethics committees
Independent, multidisciplinary and pluralist ethics committees should be established, promoted and supported at the appropriate level in order to:

(a) assess the relevant ethical, legal, scientific and social issues related to research projects involving human beings;
(b) provide advice on ethical problems in clinical settings;
(c) assess scientific and technological developments, formulate recommendations and contribute to the preparation of guidelines on issues within the scope of this Declaration;
(d) foster debate, education and public awareness of, and engagement in, bioethics.

Article 20 – Risk assessment and management
Appropriate assessment and adequate management of risk related to medicine, life sciences and associated technologies should be promoted.

Article 21 – Transnational practices

1. States, public and private institutions, and professionals associated with transnational activities should endeavour to ensure that any activity within the scope of this Declaration, undertaken, funded or otherwise pursued in whole or in part in different States, is consistent with the principles set out in this Declaration.
2. When research is undertaken or otherwise pursued in one or more States (the host State(s)) and funded by a source in another State, such research should be the object of an appropriate level of ethical review in the host State(s) and the State in which the funder is located. This review should be based on ethical and legal standards that are consistent with the principles set out in this Declaration.
3. Transnational health research should be responsive to the needs of host countries, and the importance of research contributing to the alleviation of urgent global health problems should be recognized.
4. When negotiating a research agreement, terms for collaboration and agreement on the benefits of research should be established with equal participation by those party to the negotiation.
5. States should take appropriate measures, both at the national and international levels, to combat bioterrorism and illicit traffic in organs, tissues, samples, genetic resources and genetic- related materials.

Promotion of the Declaration
Article 22 – Role of States

1. States should take all appropriate measures, whether of a legislative, administrative or other character, to give effect to the principles set out in this Declaration in accordance with international human rights law. Such measures should be supported by action in the spheres of education, training and public information.
2. States should encourage the establishment of independent, multidisciplinary and pluralist ethics committees, as set out in Article 19.

Article 23 – Bioethics education, training and information

1. In order to promote the principles set out in this Declaration and to achieve a better understanding of the ethical implications of scientific and technological developments, in particular for young people, States should endeavour to foster bioethics education and training at all levels as well as to encourage information and knowledge dissemination programmes about bioethics.
2. States should encourage the participation of international and regional intergovernmental organizations and international, regional and national non-governmental organizations in this endeavour.

Article 24 – International cooperation

1. States should foster international dissemination of scientific information and encourage the free flow and sharing of scientific and technological knowledge.
2. Within the framework of international cooperation, States should promote cultural and scientific cooperation and enter into bilateral and multilateral agreements enabling developing countries to build up their capacity to participate in generating and sharing scientific knowledge, the related know-how and the benefits thereof.
3. States should respect and promote solidarity between and among States, as well as individuals, families, groups and communities, with special regard for those rendered vulnerable by disease or disability or other personal, societal or environmental conditions and those with the most limited resources.

4 A Few Reflections on Universal Declaration on Human Rights

The right to life, liberty and security of person [Universal Declaration on Human Rights (UDHR), article 3] is one of the most fundamental rights. This right encompasses a right of access to all of the necessities of life, including food and shelter. People of different mental states will construct different needs. This will be a challenge to apply when we find further determinants of addiction, desire, need, and mental health. Does a right to life encompass universal access to the benefits of technology, regardless of ability to pay? The Infoethics Survey of Emerging Technologies prepared by the NGO Geneva Net Dialogue at the request of UNESCO looked at the ethical implications of future communication and information technologies (Rundle and Conley 2007). It stated that article 3 may be construed as entailing a right of access to the information, ideas, cultural elements, and communication media that allow people to take part in society, it may also be read to allow an individual to opt out of participating in social and technological systems. For example, under this right a person might be permitted to refuse to have an ICT device implanted into their body, or any other intervention. If such an implant were to become de facto requirement for participation in the Information Society, should the law step in to allow that person to have access to all of the necessities of life, including food and shelter, health care, etc., despite his refusal to receive the implant?

The Universal Declaration on Human Rights, article 18 states:

Everyone has the right to freedom of thought, conscience and religion; this right includes freedom to change his religion or belief, and freedom, either alone or in community with others and in public or private, to manifest his religion or belief in teaching, practice, worship and observance.

Neuroscience interventions may also affect our conscience and constructions of religion. They may also affect the understanding of freedom of opinion, challenging article 19:

> Everyone has the right to freedom of opinion and expression; this right includes freedom to hold opinions without interference and to seek, receive and impart information and ideas through any media and regardless of frontiers.

Like the right of privacy that is articulated in the 1997 and 2005 UNESCO Declarations (and in UDHR article 12), the right to freedom of opinion and expression in the emerging Information Society is closely intertwined with ICTs. Technologies can open channels by which information may be shared and opinions expressed; yet can also be used to restrict the information available, and to identify and interfere with people expressing alternative opinions. This will especially be true when we understand more of our mind. In this sense, there is a link between privacy and the freedom to seek, receive and impart information. ICTs can be used to create a wider social forum where communication can take place, or it can restrict expression by placing limits on a person's ability to communicate with and association. The right to seek, receive, accessing and disseminating ideas and impart information, connecting with others in the broad society that shapes our minds, is also tied to freedom of assembly. We can expect the further enhancement of human ability by implanting of new communication devices into people's minds, as already being used in assisting blind persons in sight, deaf persons in hearing, and in virtual reality games.

Article 26 on education reads:

(1) Everyone has the right to education. Education shall be free, at least in the elementary and fundamental stages. Elementary education shall be compulsory. Technical and professional education shall be made generally available and higher education shall be equally accessible to all on the basis of merit.
(2) Education shall be directed to the full development of the human personality and to the strengthening of respect for human rights and fundamental freedoms. It shall promote understanding, tolerance and friendship among all nations, racial or religious groups, and shall further the activities of the United Nations for the maintenance of peace.
(3) Parents have a prior right to choose the kind of education that shall be given to their children.

Education is reliant on technology and neurosciences. Technology is used to enable education on a vast range of subjects, allowing learners to improve their access to outside sources of information, use multimedia educational materials, interact with teachers and fellow learners in new ways, and in more effective psychological interventions to empower citizens. Although all have the right to education we can see increasing stratification based on income level and access to technology and advanced techniques in educational pedagogy.

5 Reflections on Ethical Challenges of Neurosciences

Although there is a substantive background of international law that will be useful for countries to govern neurosciences, and for all involved to ensure neuroscience research and applications proceeds ethically, there are still challenges ahead. One day we may be able to map most human ideas, and predict moral choices people make (Macer 2002a, b). It is not so much a challenge from the use of technology, but rather a challenge from the growing knowledge of human nature and life itself. There are many opportunities offered by greater understanding of the human mind, but also many challenges to greater individual and cross-cultural understanding of human beings.[3]

One of the most interesting questions before a thinking being is whether we can comprehend the ideas and thoughts of other beings, and conversely whether they can also read our mind. Although the human mind appears to be infinitely complex and the diversity of human kind and culture has been considered vast, in 1994 I made a hypothesis that the number of ideas that human beings have is finite (Macer 1994), and in 2002 I called for a project to map the ideas of the human mind (Macer 2002a). What happens to the field of ethics and moral philosophy if we can map ideas of the mind, and predict moral choices? Should we preempt choices that could harm others? What freedom does it imply to ourselves, and to our society?

The scientific literature comes from a variety of fields including psychology, sociology, ethics, and related behavioural subjects. The methodology used in disciplines of genetics, psychology, animal behavior, sociology, history, public understanding of science, religious studies, to mention just a few, needs to be harvested to design an integrative approach to understand the extent of human ideas, and to construct holistic neuroscience. Some cross-cultural studies suggest idea diversity is above boundaries of culture, religion, age or other demographic factors. In the 1993 International Bioethics survey with 6,000 persons in 10 countries in the Asia Pacific area, the survey results revealed that when faced with a diverse range of bioethics dilemmas, the ideas that respondents in different countries such as New Zealand, India, Thailand and Japan gave were similar and finite in number (Macer 1994). For most dilemmas the number of ideas was about 30 for a given dilemma. The majorities of persons chose between groups of five to ten ideas, and most were independent of culture, religion, age, gender or education.

A mental mapping project of neurosciences would endeavour to analyze the ideas human beings have, and the factors behind these different ideas. One way to understand the ideas and mental processing is to ask a person about the moral dilemmas they remember that they used in practice in the past. In this way, we can map the ideas that led to a particular action as a response to a situation. A second is to ask hypothetical questions about cases and explore how persons think. A third is

---

[3] International Behaviourome web site: http://www.eubios.info/menmap.htm.
Behaviourome listserve <http://groups.yahoo.com/group/Behaviourome/>.

to observe the actions and words of the persons. Practice and theory can differ widely, and ideas might vary even in the same situation based on past experience. All these methods raise questions of ethics of research. If we obtain consent for all observations will this affect the actual way of decision-making?

The individual human mind is a societal creation, formed through a series of interactions with other persons. After an initial response to a dilemma, real or hypothetical, our mind generates an idea. That idea is subject to genetic, environmental and cultural factors as discussed above. Then the process of idea development occurs, subject to the cultural restraints and lessons of the past to that person. The action is taken, but this is not the end of the idea for a normal human mind. The consequences are considered, there may be guilt or self-gratification, through the interplay of the conscience and ego. Although neuroscience may normally consider individuals, since the mind is a social construction this will be more complex than any individual themselves.

The call for such mental mapping can be pitched at both individual and social levels. Sociology has considered societies, and psychology has considered individuals, or influences upon individuals. We should develop neurosciences to explore similarities between cultures and communities not just at individual human level, but also as members responding inside biological communities. Cross-cultural studies can inform this process also.

In modern society, the media plays a significant role in formulating people's ideas, so media studies have traced the way that people's thinking in different countries is converging. As societies seek to educate persons about the implications and ethics of the research the actual minds will be modified, thus we are affecting what we are observing.

There are also implications for cultural identity. How should a culture that tries to maintain its cultural uniqueness by claiming everyone thinks the same, face up to the reality that in every culture the full range of idea diversity is found. This diversity is found in almost all groups, excluding those particularly finite groups that are formed to promote particular political aims, such as those who fight for or against abortion, or euthanasia. Religions which have observed already that humankind is universal will have lesser challenges than religions which claim a special religious status for their "chosen" people.

To compare mental maps between species will also allow comparisons of idea diversity between species. The UNESCO Declarations are worded in terms of Homo sapiens, and will need application in spirit to apply to other biological species, as well as to artificial intelligence systems. The principles could be applied to any moral agent, once the moral agent is recognized. Perhaps, these challenges to the uniqueness of humankind will be accompanied by the develop of new international texts, and one day these will be co-written between species.

The most challenging area is that of construction of artificial intelligence. Although I am a biocentric philosopher, I also argue that once a being can love others, and is able to balance principles to make moral decisions, it is a moral agent, no matter what it is made of (Macer 1998). Therefore if we create an artificial intelligence, or so-called "artificial" moral agent, we have responsibilities to treat he/she/them with respect. This theme has been the subject of numerous movies and books, ranging from "2001 Space Odyssey", "Terminator", through "Matrix" to "I Robot". Our community should be open to all moral agents, and one day might even take on a life of its own!

References

Macer DRJ (1994) Bioethics for the people by the people. Eubios Ethics Institute, Christchurch
Macer DRJ (1998) Bioethics is love of life. Eubios Ethics Institute, Christchurch
Macer DRJ (2002a) The next challenge is to map the human mind. Nature 420:12
Macer DRJ (2002b) Finite or infinite mind?: a proposal for an integrative mental mapping project. Eubios J Asian Int Bioeth 12:203–206
Rundle M, Conley C (2007) Ethical implications of emerging technologies: a survey. UNESCO, Paris
UNESCO (1997) Universal declaration on the human genome and human rights
UNESCO (2005) Universal declaration on bioethics and human rights

Law and Neuroscience in the United States

Owen D. Jones and Francis X. Shen

Abstract Neuroscientific evidence is increasingly reaching United States courtrooms in a number of legal contexts. Just in calendar year 2010, the U.S. legal system saw its first evidentiary hearing in federal court on the admissibility of functional magnetic resonance imaging (fMRI) lie-detection evidence; the first admission of quantitative electroencephalography (qEEG) evidence contributing in part to a reduced sentence in a homicide case; and a U.S. Supreme Court ruling explicitly citing brain development research.

Additional indicators suggest rapid growth. The number of cases in the U.S. involving neuroscientific evidence doubled from 2006 to 2009. And since 2000, the number of English-language law review articles including some mention of neuroscience has increased fourfold. In 2008 and again in 2009, more than 200 published scholarly works mentioned neuroscience. The data clearly suggest that there is growing interest on the part of law professors, and growing demand on the part of law reviews, for scholarship on law and the brain (Shen 2010). In addition, a number of symposia on law and neuroscience have been held in the United States over the past few years, and despite the notable youth of the field, courses in Law and Neuroscience have been taught at a number of U.S. law schools.

Preparation of this chapter was supported by the John D. and Catherine T. MacArthur Foundation (Grant # 07-89249-000 HCD), The Regents of the University of California, and Vanderbilt University. Research assistance was provided by Katherine Kuhn and Tim Mitchell.

O.D. Jones (✉)
New York Alumni Chancellor's Professor of Law and Professor of Biological Sciences, Vanderbilt University; Director, MacArthur Foundation Research Network on Law and Neuroscience, 131 21st Avenue South, Nashville, TN 37203-1181, USA
e-mail: owen.jones@vanderbilt.edu

F.X. Shen (✉)
Visiting Assistant Professor, Tulane University Law School and The Murphy Institute, 6329 Freret Street, New Orleans, LA 70118, USA
e-mail: fshen@tulane.edu

T.M. Spranger (ed.), *International Neurolaw*,
DOI 10.1007/978-3-642-21541-4_19, © Springer-Verlag Berlin Heidelberg 2012

This vivid interest in neurolaw, from both scholars and practitioners, is born of the technological developments that allow noninvasive detection of brain activities. But despite the rapid increase of legal interest in neuroscientific evidence, it remains unclear how the U.S. legal system – at the courtroom, regulatory, and policy levels – will resolve the many challenges that new neuroscience applications raise.

The emerging field of law and neuroscience is being built on a foundation joining: (a) rapidly developing technologies and techniques of neuroscience; (b) quickly expanding legal scholarship on implications of neuroscience; and (c) (more recently) neuroscientific research designed specifically to explore legally relevant topics. With the institutional support of many of the country's top research universities, as well as the support of the John D. and Catherine T. MacArthur Foundation, among other private foundations and public funding agencies, the U.S. is well positioned to continue contributing to international developments in neurolaw.

This chapter provides an overview of notable neurolaw developments in the United States. The chapter proceeds in six parts. Section 1 introduces the development of law and neuroscience in the United States. Section 2 then considers several of the evidentiary contexts in which neuroscience has been, and likely will be, introduced. Sections 3 and 4 discuss the implications of neuroscience for the criminal and civil systems, respectively. Section 5 reviews three special topics: lie detection, memory, and legal decision-making. Section 6 concludes with brief thoughts about the future of law and neuroscience in the United States.

As judges, lawyers, legislators, and the public become more acquainted with neuroscientific evidence, and as neuroscience continues to produce more legally relevant findings, it is likely that we will see continued expansion of law and neuroscience in the United States.

1   Law and Neuroscience in the United States

In recent years, the United States has been home to a number of important developments at the intersection of neuroscience and law. Just in calendar year 2010, the U.S. legal system saw its first evidentiary hearing in federal court on the admissibility of fMRI lie-detection evidence (*United States v. Semrau* 2010), the first admission of quantitative electroencephalography (qEEG) evidence contributing in part to a reduced sentence in a homicide case (*State v. Nelson* 2010); and a U.S. Supreme Court ruling explicitly citing brain development research (*Graham v. Florida* 2010).

These examples make clear that in the United States neuroscientific evidence has already reached the courtroom in at least some important legal contexts. Preliminary assessments by Nita Farahany, for example, indicate a rapid rate of growth, with twice as many reported cases involving neuroscientific evidence in 2009 as in 2006 (Farahany 2011).

Not only is neuroscientific evidence reaching the courts, but it is also – at least in some contexts – directly affecting the administration of justice. For example, in 2010 jurors in a U.S. state court considered whether Grady Nelson, who had earlier been found guilty of murdering his wife and raping a child, should receive the death penalty or life in prison (*State v. Nelson* 2010). When Nelson was spared the death penalty, interviews with jurors after their verdict revealed that, for some, the proffered neuroscientific evidence was a tipping point. As one juror remarked, "the technology really swayed me . . . . After seeing the brain scans, I was convinced this guy had some sort of brain problem" (Ovalle 2010, p. 1). Whether or not this particular use of qEEG evidence was appropriate, and whether or not the Grady Nelson case was rightly decided, similarly situated legal defense teams will likely consider offering similar types of evidence in the future.

Whether, and how, the use of neuroscientific evidence in the legal system will expand is an open and hotly debated question. A large number of commentators have begun to weigh in on how this intersection of different technologies, analytic methods, and legal contexts may ultimately allow for a more effective and fair legal system (for overviews, see Goodenough and Tucker 2010; Jones et al. 2009; Greely and Wagner, Forthcoming; Aronson 2010; Tovino 2007a).

Since 2000, the number of English-language law review articles including some mention of neuroscience has increased fourfold. In 2008 and again in 2009, more than 200 published scholarly works mentioned neuroscience. The data clearly suggest that there is growing interest on the part of law professors, and growing demand on the part of law reviews, for scholarship on law and the brain (Shen 2010).

In addition, a number of symposia on law and neuroscience have been held in the United States over the past few years.[1] This vivid interest in neurolaw, from both scholars and practitioners, is born of the technological developments that allow noninvasive detection of brain activities. But despite the rapid increase in legal interest in neuroscientific evidence, it remains unclear how the legal system – at the

---

[1] As a sampling: in 2008, the Berkman Center for Internet and Society and the Petrie-Flom Center for Health Law Policy, Biotechnology and Bioethics of Harvard Law School hosted a roundtable panel titled *Should Criminal Law be Reconsidered in Light of Advances in Neuroscience?*. In 2008, the Initiative on Neuroscience and the Law at Baylor College of Medicine hosted a conference on *Neuroscience and Law*. In 2008, UC Riverside Extension Law & Science Program and the Gruter Institute for Law and Behavioral Research hosted a *Seminar on Law and Neuroscience*. In 2008, the University of Akron School of Law hosted a law review symposium on *Neuroscience, Law, and Government*. In 2009, the MacArthur Foundation Law and Neuroscience Project sponsored a symposium titled *Psychopathy and the Law*. In 2009, the Stanford Technology Law Review hosted a symposium on *Neuroscience and the Courts: The Implications of Advances in Neurotechnology*. In 2009, the Vermont Law Review published a special issue, *Emotions In Context: Exploring The Interaction Between Emotions And Legal Institutions* (which drew heavily on neuroscience research). In 2009, the Gruter Institute for Law and Behavioral Research ran a conference titled *Law, Biology and the Brain*. In 2010, the American Enterprise Institute for Public Policy Research hosted an event titled *Understanding Humans through Neuroscience*. In 2010, Mercer University School of Law hosted a conference on *The Brain Sciences in the Courtroom*. In 2011, the Denver University Law Review hosted a *Symposium on Law and Neuroscience*; and the Dana Foundation hosted a *Law and Neuroscience* conference in New York. Also, in 2011 a *Neuroscience and the Law* forum was co-sponsored by the National Academy of Sciences and the U.K. Royal Society.

courtroom, regulatory, and policy levels – will resolve the many challenges that new neuroscience applications raise.

To address some of these challenges, the John D. and Catherine T. MacArthur Foundation created the *Law and Neuroscience Project*, in 2007, and subsequently created the *Research Network on Law and Neuroscience*, in 2011.[2] The *Project* and the *Research Network*, now headquartered at Vanderbilt University Law School, in Nashville, Tennessee, have fostered interdisciplinary research among more than 50 scientists, law professors, and judges across the United States.

Both the Project and the Research Network have spurred original empirical research to explore brain-imaging techniques for, among many other things: detecting memory and deception; resting-state functional connectivity analysis of impulsivity in juveniles; risk and information processing in addicts; the effects of neuroimaging evidence on juror decision-making; the cognitive processes supporting third-party legal decision-making; and improved methods for making accurate, individualized assessments of psychopathy.

Members have published on neuroscience and law in the context of responsibility, sentencing, evidence, neuroprediction, addiction, juvenile justice, psychopathy, legal and moral reasoning, neuroethics, incidental findings, limits to neuroimaging techniques, emotions, memory, lie detection, pain detection, risk assessment, behavioral genetics, health law, and many other related topics. The Project has also provided education and outreach on neuroscience to more than 800 judges, and developed the first *Primer on Law and Neuroscience* (Morse and Roskies Forthcoming), as well as the first *Law and Neuroscience* case book for law students (Jones et al. Forthcoming). The research, publications, and outreach of the Project – alongside the work of many other notable scholars in the U.S. and elsewhere – are establishing a firm foundation for the future of this interdisciplinary field.

Despite the notable youth of the field, courses in *Law and Neuroscience* have been taught at a number of U.S. law schools, including Vanderbilt University, the University of Colorado, Georgetown University, Mercer University, the University of San Diego, Temple University, Tulane University, and Yale University – reflecting broad and quickly developing interest across the academy. The University of Pennsylvania Intensive Summer Institute in Neuroscience, which will be offered for the fourth consecutive year in 2012, has similarly introduced a number of lawyers to neuroscience.

These collective efforts, in both the legal and scientific communities, have attracted national attention. The press – print, television, and web – has recognized that "law and the brain" stories are of increasing interest to their readers and viewers. For instance, in the past few years:

---

[2] More information on the Law and Neuroscience Project, and on the Research Network on Law and Neuroscience, is available online at: http://www.lawneuro.org. In addition, two other useful online resources for law and neuroscience information are: (1) the "Neuroethics and Law" blog, maintained by Adam Kolber, at http://kolber.typepad.com; and (2) the Research Network blog at: http://lawneuro.typepad.com.

- *Science* magazine described "neuroscience in court" as one of the seven "Areas to Watch" (2008).
- The *New York Times Magazine* investigated the intersection of law and neuro-science in a cover story, "The Brain on the Stand" (Rosen 2007).
- The NBC Nightly News' *Mind Matters* series explored the neuroscience of psychopaths and mind reading in "Dangerous Minds" (2008).
- The *Wall Street Journal* considered neuroscience evidence in an article "The Brain, your Honor, Will Take the Witness Stand" (2009).
- *Scientific American* ran a piece on "The Legal Brain: How Does the Brain Make Judgments about Crimes" (2009).
- The National Public Radio produced show, *Justice Talking*, ran a week-long series on "Neurolaw: The New Frontier" (2008).

Beyond the headlines of these media stories, of course, are many complex challenges that must be addressed as the U.S. legal system attempts to effectively integrate neuroscience research. It is beyond the scope of this chapter to go in depth on each of these challenges. However, we aim here to introduce readers to many of the most important U.S. neurolaw developments and debates.

The chapter proceeds as follows. Section 1 introduces the development of law and neuroscience in the United States. Section 2 then considers some of the evidentiary contexts in which neuroscience has been, and will continue to be, introduced. Sections 3 and 4 discuss the implications of neuroscience for the criminal and civil systems, respectively. Section 5 reviews three special topics: lie detection, memory, and legal decision-making. Section 6 concludes with brief thoughts about the future of law and neuroscience in the United States.

2   Introduction to Neurolaw in the United States

2.1   The Development of Neuroscience in Law

There are a growing number of criminal cases involving neuroscientific evidence (Snead 2006; Marchant 2008; Farahany 2011). Interest in neuroscience in the U.S. stems generally from the intersection of two things. First, the criminal and civil justice systems rely, critically and fundamentally, on the mental operations of its many participants – judges, jurors, lawyers, defendants, law enforcement officers, court officials, and witnesses. Second, new technologies enable unprecedented investigation and observation of how (and sometimes how well) those mental operations occur.

The rise of neuroscience and law follows the quite rapid and large growth of neuroscience more generally. In 1969, the Society for Neuroscience (SfN) formed with 500 members. Today, it numbers more than 40,000 and hosts an annual conference attended by more than 31,000. This 8,000% membership growth in just four decades speaks to two important facts. First, with more than 40,000 scientists studying the brain and nervous system, and a large number of them in the United States, it is clear that neuroscientific research is now front and center in labs across America. Second, the consistent and rapid *growth* of neuroscience suggests that the field is continuing on a trajectory to become even more important in the years to come.[3]

The advances in cognitive neuroscience are an enormous leap forward in understanding how minds work. Until quite recently, brain structure and function were studied separately, inasmuch as it was hard to study structure without a dead brain – and hard to study function with one. Advances in x-ray technologies opened a window on the structure of living brain tissue. But subsequent advances in techniques, such as fMRI, now enable noninvasive brain-imaging that reveals not only a person's brain structure, but also how a person's brain is (and is capable of) functioning.

The potential implications of neuroscience, for many areas of law and policy, are quite broad (Freeman 2011; Zeki and Goodenough 2006; Freeman and Goodenough 2009; Garland 2004; Annas 2007; Arrigo 2007; Farahany 2009b; Garland and Glimcher 2006; Eagleman 2008; O'Hara 2004; Patel et al. 2007; Chorvat and McCabe 2004). For example, scholars have debated both the theoretical and practical implications of neuroscience for law by addressing issues related to free will, determinism, compatabilism, and the like (see, e.g. Morse 2008b; Pardo and Patterson 2010; Greene and Cohen 2004; Nadel and Sinnott-Armstrong 2010; Erickson 2010).

In the courtroom, neuroimaging evidence has been offered in constitutional, disability benefit, and contract cases, among others. Examples include:

- *Entertainment Software Assn. v. Blagojevich* (2005) (the court considered whether a brain imaging study could be used to show that exposure to violent video games increases aggressive thinking and behavior in adolescents) and *Brown v. Entertainment Merchants Assn* (2011) (Supreme Court Justice Stephen Breyer's dissent cited "cutting edge neuroscience" to support the argument that violent video games are linked to more aggressive thinking and behavior in adolescents);
- *Fini v. General Motors Corp* (2003) (brain scans were proffered to help determine the extent of head injuries from a car accident);
- *Boyd v. Bert Bell/Pete Rozelle NFL Players Retirement Plan* (2005) (a former professional football player proffered brain scans in an effort to prove entitlement to neuro-degenerative disability benefits); and
- *Van Middlesworth v. Century Bank and Trust Co* (2000) (involving a dispute over the sale of land, the defendant introduced brain images to prove mental incompetency, resulting in a voidable contract).

Not surprisingly, neuroscience has also been offered in various criminal contexts. However, it has only been relatively recently that neuroscience has begun to appear there with increasing regularity. Here are several examples.

---

[3] See: Society for Neuroscience, *SfN Milestones: 40 Years of Evolution* (2009), http://www.sfn.org/skins/main/pdf/annual_report/fy2009/milestones.pdf.

Brain images are sometimes offered to help show that a defendant is incompetent to stand trial. In *United States v. Kasim* (2008), for example, Kasim introduced medical testimony and accompanying brain images to argue successfully that he was demented, and therefore incompetent to stand trial for Medicaid fraud (see also, *McMurtey v. Ryan* 2008; *United States v. Gigante* 1997).

Brain images are also increasingly proffered by the defense at the guilt-determination phase, in an effort to negate the *mens rea* element of a crime, and to thereby avoid conviction. For example, in *People v. Weinstein* (1992), the defendant argued that he was not responsible for strangling his wife and throwing her from a twelfth floor window, even if he did so. In support, he offered images of allegedly impaired brain function. Similarly, the defendant in *People v. Goldstein* (2004),[4] – who allegedly pushed a woman in front of a subway train, killing her, sought to introduce a brain image of an abnormality, in an effort to prove an insanity defense.

Brain images are also proffered at the sentencing phase of criminal cases, in furtherance of mitigation. In *Oregon v. Kinkel* (2002), for example, a boy convicted of killing and injuring fellow students in a high school cafeteria sought to introduce brain images of abnormalities, in hopes of supporting and securing a more lenient sentence. In *Coe v. State* (2000), a convicted murderer offered brain images to help prove he was not competent to be executed.

Paralleling the rise of neuroscientific evidence in criminal cases, there has been a rise in defendant's arguments – as in *Ferrell v. State* (2005) and *People v. Morgan* (1999) for instance—that a defense counsel's failure to procure a brain image amounted to ineffective assistance of counsel.

Neuroscientific evidence has also been integrated into civil litigation. For example, the term "neurolaw" was coined at least as early as 1995, when attorney J. Sherrod Taylor (1995) discussed the implications of advances in neurology for civil litigation. Since the early 1990s, a publication called *The Neurolaw Letter* has circulated among personal injury lawyers and medical professionals, and The Brain Injury Association of America has been sponsoring conferences for over two decades to bring lawyers up to speed on developments in brain science.

Across these many legal contexts, efforts to bring neuroscience into courtrooms result in a variety of distinct challenges for the legal system. We now explore some of these challenges.

2.2   The Limitations of Neuroscience in Law

Although promising, there are important methodological limitations with fMRI (Cacioppo et al. 2003; Poldrack et al. 2008; Logothetis 2008; Vul et al. 2009;

---

[4] *Overruled on other grounds*, 6 N.Y.3d 119, 843 N.E.2d 727, 2005 N.Y. LEXIS 3389 (2005).

Bennett et al. 2010). Many have commented on the extent to which these limitations, and those of other brain imaging techniques, may affect the probabilities of garnering legally relevant insights (Mobbs et al. 2007; Morse 2006; Pustilnik 2009; Brown and Murphy 2010; Tancredi and Brodie 2007; Rakoff 2008; Racine et al. 2005; Trout 2008; Gazzaniga 2008; Baskin et al. 2007; Uttal 2008; Uttal 2003).[5] The advance of social neuroscience generally, and legal applications in particular, has also been met with significant ethical concerns (see, e.g., Illes 2006; Farah 2002; Roskies 2008; Moreno 2003; Kennedy 2004; Uttal 2003).[6]

One of the most important critiques raised by these scholars, and recognized in court proceedings, is that there exists a long chain of inference from the fMRI scanner to the courtroom. Functional brain imaging is not mind reading (at least not in the broad sense of that term). While fMRI can accurately measure changes in blood flow and oxygen levels, interpreting those changes as reliable indicators of particular types of thought, or as reliable indicators of what a region of the brain is actually doing, requires a series of inferential steps that are not entirely straightforward.

Because the most legally relevant thoughts are likely to be those that occurred in the past (such as those reflecting the mental state of a defendant at the time of an alleged transgression) brain scans taking place long after the behavior may be of limited diagnostic or forensic use. Even if one accepts a given scanner task as legally relevant, the particular images shown in court may still be problematic. Images can be no better than the manner in which the researcher designed the specific task or experiment, deployed the machine, collected the data, analyzed the results, and generated the images.

In addition, making individualized inferences, as law is typically required to do, from group-averaged neuroscientific data presents a particularly difficult problem for courts to overcome (Faigman 2010). For instance, just because a particular pattern of neural activity is associated, on average at the group level, with impaired decision-making, it does not necessarily follow that a defendant before the court whose brain scans produce the same neural patterns necessarily has such a cognitive deficit. As neuroscientists begin to further explore individual differences in brain activity (Hariri 2009), the "group to individual" inference problem will remain central in applying neuroscience to law.

---

[5] We do not review here the science of fMRI, and its many limitations, but refer interested readers to Jones et al. (2009) for an accessible discussion of the technology. For more general introductions to other cognitive neuroscience methods, see Gazzaniga et al. (2009), Ward (2009), and Purves et al. (2008).

[6] In addition, a number of websites have emerged as forums for discussing neuroethics and related bioethics issues: Dana Foundation (http://www.dana.org/); University of Pennsylvania (http://www.neuroethics.upenn.edu/); President's Council on Bioethics (http://www.bioethics.gov/); Center for Cognitive Liberty & Ethics (http://www.cognitiveliberty.org/); Stanford Center for Biomedical Ethics (http://bioethics.stanford.edu/); National Institutes of Health Bioethics Resources on the Web (http://bioethics.od.nih.gov/).

U.S. courts are still figuring out how to optimally apply evidentiary standards to novel forms of neuroscientific evidence. On the one hand, courts must ask whether there are too many faulty links in the inferential chain that leads from an fMRI scan to a relevant issue of legal responsibility. On the other hand, courts must ask whether jurors are capable of assessing, presumably with the aid of cross-examination and opposing expert witnesses, the inferential chain for themselves.

Methodological cautions, and the subsequent challenge of making appropriate legal inferences, are being acknowledged and addressed by those working at the intersection of law and neuroscience. Through publications produced by the Law and Neuroscience Project, among others, the legal community is being made aware of the many difficulties associated with introducing neuroscientific evidence. At the same time, these cautions and limitations have not, and we believe should not, prevent all use of neuroscience in courtrooms and policymaking. Rather, what is called for is careful, context-specific applications.

3   Evidentiary Context

The methods, goals, and evidentiary standards differ for neuroscience and law (Jones 2004; Sapolsky 2004; Schauer 2010). And, even within law, policymakers see the value in different standards of proof when different interests are at stake (Faigman 2002). Thus, we caution at the outset that how, if at all, the legal system integrates neuroscientific evidence will, and should, vary context by context.

To date, neuroscientific evidence has appeared in the form of expert testimony about the brain, from researchers and clinicians, as well as in the form of graphic images produced through methods such as fMRI, electroencephalography (EEG), qEEG, and others. The novel applications of new brain imaging and brain monitoring technologies have created many practical problems for judges in the U.S., as they consider the admissibility of such evidence, its proper interpretation, its impact on jurors, and the like (Greely and Wagner, Forthcoming; Sinnott-Armstrong et al. 2008; Moriarty 2008; Aharoni et al. 2008; Rogers and DuBois 2009; Pettit 2007).

At present, the admissibility of neuroscientific evidence in U.S. courts remains fluid, and is highly contextual. Even if we limit our focus solely to the federal system, the admissibility of neuroscientific evidence will still vary with the specific legal context in which the brain evidence arises.[7]

---

[7] Given the institutional design of the United States criminal justice system, the admissibility of neuroscientific evidence will not be uniform across the country. This is because the United States has multiple, overlapping criminal jurisdictions (Barkow 2011). Local, state, and federal authorities can all bring criminal charges (Barkow 2011; Stuntz 2008). Of particular note for understanding the admissibility of neuroscientific evidence is that the evidentiary rules that apply in the federal system may be different than those that apply in each of the 50 state systems. While there are many similarities across the 50 states, each state criminal code is unique and each state crafts, within Constitutional limits, its own admissibility standards for scientific evidence. Thus, it

In the civil system, for example, neuroscientific evidence might be introduced to help establish liability, such as in the case of a medical malpractice action; to demonstrate a pre-existing condition, such as in the case of a dispute over insurance coverage; or to help estimate damages, such as in the case of a car accident.

In the criminal system, brain evidence may be offered during the liability phase, the sentencing phase, or both. For example, during the liability phase, the defense may offer brain evidence to support an insanity defense, or to defeat the prosecution's claim that the defendant had (and was therefore capable of having) the mental state requisite for conviction, or to provide evidence of truthfulness. During the sentencing phase, brain evidence may be offered to support a mitigated penalty.

At the liability/guilt stage, the admissibility of neuroscientific evidence in the U.S. is governed by rules that are used to assess scientific evidence generally (Faigman et al. 2011).[8] In the federal system, courts primarily apply Federal Rule of Evidence 702 (allowing an expert witness to testify "if (1) the testimony is based upon sufficient facts or data, (2) the testimony is the product of reliable principles and methods, and (3) the witness has applied the principles and methods reliably to the facts of the case"), and Federal Rule of Evidence 403 (allowing for the exclusion of relevant evidence "if its probative value is substantially outweighed by the danger of unfair prejudice, confusion of the issues, or misleading the jury"). Application of Rule 702 is guided by a trilogy of U.S. Supreme Court cases delivered in the 1990s (Saks 2000; *Daubert v. Merrell Dow Pharmaceuticals, Inc.* 1993; *General Electric Co. v. Joiner* 1997; *Kumho Tire v. Carmichael* 1999).[9]

How will U.S. federal courts apply the *Daubert* standards to neuroscientific evidence? While the answer, as we have stressed, will vary across contexts, we can gain purchase on this question by reviewing the 2010 case *United States v. Semrau*, in which the first *Daubert* hearing was held on the admissibility of fMRI lie-detection evidence (Shen and Jones 2011).[10] In *Semrau*, the federal government charged psychologist Dr. Lorne Semrau with Medicare/Medicaid fraud. Proving fraud required proving that Semrau knowingly violated the law. And Semrau's defense was built, in part, around fMRI scans that allegedly demonstrated he was telling the truth when he claimed (some years after the fact) that even though he had mis-billed for services, he did not *knowingly* defraud the government.

---

should be kept in mind that although we discuss (for brevity) only the Federal rules, in practice neuroscientific evidence will be evaluated by many different standards.

[8] It is noted by commentators that scientific evidence, such as fMRI, may be offered to prove an "adjudicative fact" (e.g., determining mental capacity or for diagnosing a brain injury), or to prove a "legislative fact" (e.g., that there is a general relationship between exposure to violent video games and aggressive behavior) (Feigenson 2006; Davis 1942).

[9] At the state level, some states have adopted the *Daubert* approach; some states still rely primarily on a general acceptance test based on *Frye v. United States* (1923); and some states have blended the two.

[10] Our discussion of the *Semrau* case here is derived, in part, from Shen and Jones (2011).

In assessing the reliability of the proffered fMRI evidence, the Court's analysis applied the previously mentioned *Daubert* test and considered four, non-exclusive factors[11]:

1. Whether the theory or technique can be tested and has been tested;
2. Whether the theory or technique has been subjected to peer review and publication;
3. The known or potential rate of error of the method used and the existence and maintenance of standards controlling the technique's operation; and
4. Whether the theory or method has been generally accepted by the scientific community.

The judge found that factors 1 and 2 were satisfied, while factors 3 and 4 were not. He therefore concluded that the evidence was not admissible under Rule 702.

While the *Semrau* case is illustrative for how U.S. courts might apply evidentiary standards, the case is not necessarily instructive on the future of fMRI (and related brain) evidence in U.S. courts. To begin with, other types of neuroscientific evidence (e.g., brain scans in civil brain injury cases) are often admitted. Moreover, in *Semrau* the evidence was offered at the liability/guilt stage, where the Federal Rules and *Daubert* apply.

In the sentencing phase, however, the evidentiary rules are relaxed and "the court may consider relevant information without regard to its admissibility under the rules of evidence applicable at trial, provided that the information has sufficient indicia of reliability to support its probable accuracy" (Federal Sentencing Guidelines, §6A1.3. Resolution of Disputed Factors). This difference in evidentiary standards is in part, as we will see in the next section of this chapter, why neuroscientific evidence has featured more prominently in the sentencing rather than liability/guilt phase of criminal trials.

Even at the guilt phase, sufficient progress in the underlying science may allow for admissibility. In *Semrau*, the judge wrote in a footnote that "in the future, should fMRI-based lie detection undergo further testing, development, and peer review, improve upon standards controlling the technique's operation, and gain acceptance by the scientific community for use in the real world, this methodology may be found to be admissible even if the error rate is not able to be quantified in a real-world setting."[12] The future admissibility of fMRI evidence in U.S. courts remains very much an open question.

As the use of brain imaging and brain monitoring techniques grows, so too will Constitutional concerns about their use in the legal system. The ability to image the brain while it is thinking raises new questions about what has been variously

---

[11] In evaluating the admissibility of the evidence, the federal judge performed a two-prong gatekeeping role for expert scientific evidence, first evaluating the reliability and then the relevance of the testimony. Because the Court did not find the proffered testimony in Semrau to be reliable, it did not reach the relevance prong.

[12] *United States v. Semrau,* Report & Recommendation, p. 31 (2010).

described as, "cognitive privacy," "cognitive liberty," and "cognitive freedom" (Blitz 2010; Tovino 2007b; Halliburton 2007). At issue are the protections offered by the U.S. Constitution against state use of brain imaging on an unwilling or unaware citizen (Fox 2009; Greely 2004; Tovino 2005).

One crucial question is whether a brain scan is "testimonial" (e.g., forced confession) or "physical" (e.g., fingerprints, handwriting samples, blood tests) evidence (Fox 2009). The 3-prong test to invoke self-incrimination protections are: (1) compulsion, (2) incrimination, and (3) testimony. The first two prongs presumably would be met by a nonvoluntary brain scan, but whether an fMRI scan is testimonial or physical evidence is not yet resolved, and the characterization determines the legal implications. In addition, 4th Amendment protections against search and seizure, and 5th Amendment protections against compelled testimony may also arise in the context of brain fingerprinting (Halliburton 2007; New 2008; Taylor 2006).

4 Neuroscience and Criminal Law

In this section, we very briefly review the role that neuroscientific evidence has played in the U.S. criminal justice system.[13] As earlier, we continue to distinguish between neuroscientific evidence used at the liability/guilt phase and at the sentencing phase. Rarely, it seems, will neuroscientific evidence alone determine culpability. At the sentencing phase, however, neuroscientific evidence is already contributing, and may continue to contribute, to the determination of sentences and treatment.

We also discuss the important implications of neuroscientific evidence for several special populations within the justice system: adolescents, addicts, and trauma victims. For all three populations, there is evidence that – especially at the policy level – the legal system is recognizing that brain differences between these groups and the normal population may recommend differences in sentencing.

4.1 Neuroscience and Culpability

Criminal responsibility in the United States can be summarized in this way:

> Crimes are defined by their "elements," which always include a prohibited act and in most cases a mental state, a *mens rea*, such as intent. The Constitution's Due Process Clause has been construed to require that the prosecution must prove all the elements defining a criminal offense beyond a reasonable doubt. Even if the state can prove all the elements beyond a reasonable doubt, the defendant may avoid criminal liability by establishing an affirmative defense of justification or excuse. (Morse and Hoffman 2007, p. 1074).

---

[13] Because of the United States's federal system, each of the fifty states can, within Constitutional limitations, set its own *mens rea* requirements. As we did in discussing evidentiary standards, we will focus here solely on the federal system.

Culpability of the accused thus depends, in part, on a determination of his/her mental state at the time of the offense. The phrase "mens rea" ("guilty mind") derives from the Latin phrase *"Actus non facit reum nisi rea sit,"* which means "An act is not guilty unless the mind is guilty." While virtually all crimes require a guilty mind, the type of intent required varies. Some crimes simply require "general intent," while others, either expressly or impliedly, contain a "specific intent" *mens rea* requirement.

Although as a practical matter it is now extremely rare to succeed with such a defense, one important avenue by which defendants may avoid liability is by proving "legal insanity." One method for assessing sanity, and the test now used by most state and federal courts, is to examine the defendant's "cognitive" ability, at the time of the crime, to know, appreciate, and understand that the conduct he was engaging in was morally or legally wrong. An alternative, now less common, method is to employ a "control" test, asking whether the defendant could control his conduct in conformity with the law. After John Hinckley was acquitted in 1982 by reason of insanity, following his attempted assassination of President Ronald Reagan, the U.S. Congress and most states reacted by eliminating the control test and relying solely on the cognitive test. (Incidentally, Hinckley's defense introduced computed tomography X-ray evidence in support of its claim of Hinckley's brain abnormalities.)

Some have argued that neuroscientific evidence provides reason to push back against this shrinking insanity defense (Sapolsky 2004; Redding 2006). Neuroscientist Robert Sapolsky has provocatively argued that the legal system should, in light of what has been learned about the effects of damage to the prefrontal cortex (PFC), rebut its presumption of responsibility and instead recognize a continua of individual capacity to regulate self-control. Sapolsky (2004, p. 1794) argues that "although it may seem dehumanizing to medicalize people into being broken cars, it can still be vastly more humane than moralizing them into being sinners." Similarly, law professor Richard Redding (2006) has argued for a new neurojurisprudence that would reform the insanity defense in light of neuroscientific findings.

Thus far, these latter policy suggestions regarding the insanity defense have not materialized, and it remains rare for defendants to mount a successful insanity defense. The introduction of neuroscientific evidence seems unlikely to alter this state of affairs. To illustrate, one of the earliest and most prominent cases of brain imaging evidence used at the liability/guilt phase was the 1992 case of Herbert Weinstein. Weinstein strangled his wife and threw her out the window of their apartment in an effort to make the murder look like a suicide (*People v. Weinstein* 1992; Rojas-Burke 1993). Weinstein admitted to his actions, but mounted an insanity defense that included positron emission tomography (PET) evidence showing the presence of an arachnoid cyst that, Weinstein argued, had impaired his ability to reason (Relkin et al. 1996). Although Weinstein pled out his case, and went on to serve many years in prison, the judge's admission of the PET evidence drew much attention and critique in the legal and scientific communities (Martell 1996; Morse 1996; Weiss 1996; Denno 2002).

The prosecutor in the case predicted that, with *Weinstein*, "the age of scanning has dawned in our courtrooms. This is not a technological genie we are going to be able to put back in the bottle" (Weiss 1996, p. 202). Nonetheless, in the 20 years since the Weinstein case was decided, neuroimaging evidence has rarely been used successfully by defendants to avoid convictions. This is because, as Stephen Morse (2006) has pointed out, the U.S. legal system establishes criminal responsibility based on behavior, not brain states. Put simply, "brains do not commit crimes; people commit crimes" (Morse 2006, p. 397). In the United States, neuroscientific evidence has thus far been, and most likely will continue to be, only minimally useful in exculpating criminal defendants.

4.2   Neuroscience and Sentencing

While the prospects for successful "my brain made me do it" defenses seem slim, neuroscientific evidence is already having a significant mitigating impact in some cases at the sentencing phase. There remains, however, much disagreement over how brain evidence should be interpreted.

We earlier quoted a juror in the *Nelson* case, stating that the qEEG evidence presented was persuasive. But other jurors disagreed. For example, one remarked that "all that testimony, that was a waste of taxpayer money. That's phony. There's nothing wrong with that guy's brain." (Ovalle 2010, p. 1). The net effects of neuroscientific evidence on sentencing decisions remain unknown.

One particularly important sentencing context in which neuroscientific evidence has been used is in death penalty cases. Sentencing procedures for civilian capital cases are governed by federal law and allow the Court to consider both mitigating and aggravating factors. Using neuroscientific evidence in capital sentencing, however, introduces a double-edged sword problem that multiple commentators have recognized (Snead 2007; Farahany and Coleman 2009; Barth 2007). That is, a brain *too* broken may be simply too dangerous to have at large, even *if* it is somehow less culpable.

Neuroscientific evidence may also be used in other types of challenges to the death penalty. For instance, Farahany (2009a) argues that when the U.S. Supreme Court outlawed the death penalty for mentally retarded capital offenders (*Atkins v. Virginia* 2002), the Court created a new type of inequality because it did not protect similarly situated individuals who – by virtue of a traumatic brain injury suffered as an adult – have the same limits in cognitive and behavioral ability as those medically diagnosed as mentally retarded. Thus, Farahany suggests that a challenge may be ripe under the equal protection guarantees of the 14th Amendment of the U.S. Constitution.

One emerging, but not yet fruitful, area in which neuroscience may play a sentencing role is in the assessment of future dangerousness (Nadelhoffer et al. 2010; Beecher-Monas and Garcia-Rill 2003). Neuroscientist Kent Kiehl, with support from the *Law and Neuroscience Project*, is conducting the first study that may provide traction for brain-based neuroprediction. A number of risk assessment tools, based on a battery of behavioral data, are currently used in the criminal justice system (Monahan 2006). If incorporating brain scan data into these future danger-ousness assessments improves the predictive power of actuarial models it may have important implications in at least three sentencing contexts: (a) capital sentencing; (b) civil commitment hearings; and (c) detention hearings for so-called "sexual predators" (Nadelhoffer et al. 2010).

4.3   The Adolescent Brain

Roughly a century ago, Progressive Era reformers in the United States created separate juvenile courts in the hopes that such courts would allow for better youth rehabilitation (Scott and Steinberg 2008). Toward the end of the twentieth century, however, in response to growing juvenile crime rates, juvenile courts became more punitive and state legislatures allowed for juveniles to be more readily transferred to the adult system (Id.). Today, both the states and the U.S. Supreme Court are reexamining juvenile justice policies. This is happening at a time when the devel-opmental neuroscience of adolescent behavior is beginning to offer important legally relevant insights (Baird 2009). And neuroscience appears to be playing some modest role in affecting legislative enactment and Supreme Court deliberations (Haider 2006; Scott and Steinberg 2008; Maroney 2010).[14]

Two U.S. Supreme Court cases are most prominently discussed: *Roper v. Simmons* (2005) and *Graham v. Florida* (2010). In *Roper*, the Court, with Justice Kennedy writing for the majority, ruled 5–4 that the 8th and 14th amendments of the Constitution prohibited the death penalty for those who were under 18 years of age when committing a capital crime. In *Graham*, the Court, with Justice Kennedy again writing for the majority, ruled 6–3 that it is unconstitutional under the 8th and 14th amendments of the Constitution for juveniles to be sentenced to life in prison without parole for nonhomicide crimes.

In both cases, the Court received numerous "amicus briefs." An amicus brief, which gets its name from the Latin *amicus curiae* (meaning "friend of the court"), is a brief submitted to the Court by individuals or organizations who are not parties to the case. In *Roper*, more than 15 amicus briefs were filed, and more than 20 were filed in *Graham*. Several of these briefs, including the ones submitted by the

---

[14] More generally, U.S. society is now being exposed to explicitly brain-based advertisements related to the developing brain. An ad created by the All-State Insurance company features an illustrated brain, sitting on a pedestal labeled "16-year-old brain". One area of the brain is missing, and the ad reads: "Why do most 16-year-olds drive like they're missing a part of their brain? Because they are." The ad, which encourages readers to contact their legislators and support Good Driving Laws, is illustrative of the ways by which brain-based evidence may affect society and policymaking even outside of the court system. See: http://www.allstate.com/content/refresh-attachments/Brain-Ad.pdf.

American Medical Association and American Psychological Association, touched upon the relevance of neuroscience and psychology research on juveniles.

In *Graham*, both the majority and dissenting opinions discussed, in part, the underlying science of adolescent development. The majority opinion explicitly referred to brain science (Section III.B, p. 17):

> No recent data provide reason to reconsider the Court's observations in *Roper* about the nature of juveniles. As petitioner's *amici* point out, developments in psychology and brain science continue to show fundamental differences between juvenile and adult minds. For example, parts of the brain involved in behavior control continue to mature through late adolescence. See Brief for American Medical Association et al. as *Amici Curiae* 16–24; Brief for American Psychological Association et al. as *Amici Curiae* 22–27.

To be sure, the dissents in *Roper* and in *Graham* interpreted the research differently. In *Roper*, for instance, Justice Scalia dissented that "Given the nuances of scientific methodology and conflicting views, courts—which can only consider the limited evidence on the record before them—are ill equipped to determine which view of science is the right one" and that "At most, these studies conclude that, *on average*, or *in most cases*, persons under 18 are unable to take moral responsibility for their actions. Not one of the cited studies opines that all individuals under 18 are unable to appreciate the nature of their crimes."

Scholars continue to actively debate the role that neuroscience research on adolescent brains does, and should, have in these and related cases (Maroney 2010; Morse 2006; Aronson 2007; Aronson 2009; Katner 2006; Gruber and Yurgelun-Todd 2006; Droback 2006). And so, as in other areas of neurolaw, the future of law and the developing brain remains uncertain.

4.4 Addiction, Trauma, and Responsibility

Central to debates about how the criminal justice system should deal with addicted criminals is the extent to which addiction is considered a brain disease (Bonnie 2002). In 1962, the U.S. Supreme Court ruled that a California statute making the status of drug addiction a punishable offense was cruel and unusual punishment under the 8th and 14th Amendments (*Robinson v. State of California* 1962). In arriving at its decision, the court analogized drug addiction to being mentally ill or having a venereal disease. In 1968, however, the Court ruled that states could punish alcoholics for being drunk in public (*Powell v. State of Texas* 1968). In general, addiction is not recognized as a valid defense to criminal behavior (Bonnie 2002). At sentencing, however, addiction may play some role in mitigation.

Two competing visions, echoing the debate over the neuroscience of legal responsibility more generally, present themselves in the face of neuroscience research on addiction:

> As we learn more about ... the neurobiological substrates that underlie voluntary actions, how will society define the boundaries of personal responsibility in those individuals who have impairments in these brain circuits? ... At present, critics of the medical model of addiction argue that this model removes the responsibility of the addicted individual from

his/her behavior. However, the value of the medical model of addiction as a public policy guide is not to excuse the behavior of the addicted individual, but to provide a framework to understand it and to treat it more effectively (Volkow and Li 2004, p. 969).

Neuroscience, to the extent that it can improve treatment programs, may play an increasing role in specialized U.S. "problem solving" courts, which have emerged in the past two decades, and which now include specialized courts for drug treatment and drug reentry for addicts leaving prison (Hora and Stalcup 2008). As of June 2010, more than 2,500 drug courts were in operation, with at least one in every U.S. state and territory.[15] Addressing addiction in the criminal justice system remains a challenge. Substance-involved inmates accounted for 85% of all incarcerated offenders in federal, state, and local jails in 2006.[16] More than 20% of inmates for violent crimes were under the influence of alcohol when acting violently; more than 40% of first-time offenders have a drug use history; and more than 80% of those with five or more convictions have a history of drug use.[17]

Against this backdrop, advances in our understanding of the neurobiology of addiction may allow courts, and legal actors throughout the justice system, to improve upon the folk psychological explanations for behavior that, at best, are incomplete and, at worst, are counter to prevailing scientific consensus.

In addition to drug addicts, military veterans have also received special attention in the legal system. Modeled after the drug courts, new courts have been created in some states, over the past 5 years, to determine sentences for combat veterans (Russell 2009; Hawkins Hon 2010). These courts raise questions about how, if at all, wartime trauma – or indeed non-wartime trauma – should factor into criminal responsibility (Hafemeister and Stockey 2010; Meszaros 2011).

4.5 Psychopathy

Psychopathy is a personality disorder marked by emotional detachment and antisocial behavior (Weber et al. 2008; Kiehl 2008), and is most frequently diagnosed using the Hare Psychopathy Checklist-Revised (PCL-R). Psychopathy is relevant to law because, while estimated to affect just 1% of the adult male population, it is estimated that psychopaths make up 25% of the adult male prison population (Kiehl 2006). Psychopaths account for a disproportionate percentage of the country's violent crime (Kiehl and Hoffman Forthcoming).

The U.S. legal system does not recognize psychopathy as an excusing condition (Morse 2008a). Moreover, the Model Penal Code, which is influential as a model in most states, though not binding, specifically excludes psychopathy as sufficient for

---

[15] See: National Association of Drug Court Professionals, http://www.nadcp.org/learn/what-are-drug-courts/history.

[16] National Center on Addiction and Substance Abuse at Columbia University (2010). *Behind Bars II: Substance Abuse and America's Prison Population.* NCJ 230327.

[17] Petersilia (2003), p. 48.

establishing an insanity defense. That said, in at least one prominent case (reported in Hughes 2010) brain evidence may have given a jury pause in delivering its sentence to a psychopathic killer.

While legal doctrine may or may not ultimately change in light of neuroscientific studies of psychopathic brains, this will not prevent the parallel development of better treatment programs for psychopathy. It is promising that at least some treatment programs have reported and replicated findings of reduced likelihood of recidivism in a population of violent male adolescents (Caldwell et al. 2006, 2007).

5 Neuroscience and Tort Law

Legal scholarship at the intersection of law and neuroscience, with a few notable exceptions (e.g., Kolber 2007; Kolber 2011; Grey 2011; Viens 2007; Shen 2010), has focused primarily on the criminal justice system. There is also good reason, however, to focus on the civil side. In this section, we examine two ways in which neuroscience intersects with important components of tort law: (1) litigation over brain injury, and (2) litigation over emotional harms.

5.1 Brain Injury

Law and the brain sciences have a longstanding, if at times contentious, relationship in civil litigation over brain injuries. Perhaps, the most difficult hurdle to overcome in civil litigation is that of causation (Smith 2009). In order to successfully recover monetary damages, a plaintiff must not only demonstrate an injury, but also that the defendant's actions (or inaction) caused the injury (Young et al. 2006). This is often difficult to do in the case of brain injuries because there is typically no data on the state of the brain prior to an alleged tortious incident. In this way, the causation conundrum is as difficult to resolve as the complex criminal responsibility issues raised earlier.

Despite these challenges, litigation over brain injuries remains common. In recent years, there has been great interest in cases of Traumatic Brain Injury (TBI), and the related mild Traumatic Brain Injury (mTBI). This type of litigation has gained prominence through high-profile investigations into the relationship between concussions and brain damage in American football players (Kluger 2011). Hundreds of individual law suits are already in progress, and the National Football League also faces a class action lawsuit from players who have suffered brain injuries (Schwarz 2010; Borden 2011).

5.2 Pain and Emotional Harm

A second part of civil litigation in which neuroscience may increasingly play a role is in the determination and valuation of pain (Viens 2007). The subjectiveness of pain makes it difficult for the law to determine (a) who is actually feeling pain (as opposed to simply faking it), and (b) how much pain an individual is experiencing (Kolber 2007). Brain imaging, although it is not yet fully capable of doing so, offers at least the promise of providing more objective measures of pain than are presently available (Id.).

Posttraumatic stress disorder (PTSD) provides a useful illustrative case. It has been observed that "No diagnosis in the history of American psychiatry has had a more dramatic and pervasive impact on law and social justice than PTSD."[18] PTSD litigation remains prevalent in the U.S. today. Scientists are beginning to better understand the neural correlates of PTSD, as distinct from other similar mental disorders (Grey 2011). Such advances could, if they materialize as promised, fundamentally change PTSD litigation.

Neuroscience might also change litigation over PTSD, and related mental harms, by changing the way we conceptualize such harms. Traditionally, the U.S. system draws a bright line distinction between "bodily" versus "mental" (i.e., *non*-bodily) harms (Grey 2011; Shen 2010). But in at least one instance – a case that went up the Michigan state supreme court before settling – neuroscience evidence has been advanced to argue that PTSD is in fact a "bodily" injury (*Allen v. Bloomfield Hills School District* 2008).

In the *Allen* case, affidavits submitted to the court on the plaintiff's behalf included neuroscientific evidence, and although the trial court sided with the defense (which argued that proper statutory interpretation did not include PTSD as bodily), the Appellate Court ruled in favor of Allen, reasoning that: "The brain is a part of the human body, so 'harm or damage done or sustained' is injury to the brain and within the common meaning of 'bodily injury' …plaintiff presented objective medical evidence that a mental or emotional trauma can indeed result in physical changes to the brain."[19] The ruling has no precedential weight, but is a stark reminder of the breadth of neuroscience and law litigation that we may see in the coming years.

6 Special Topics

6.1 Lie Detection

Despite the fact that its short-term prospects for admissibility are dim, and its scientific validity remains in doubt, neuroscience-based lie detection has received considerable attention in both scientific and legal outlets (Wagner 2010; Ganis and Keenan 2009; Bizzi et al. 2009; Schauer 2010; Shen and Jones 2011; Appelbaum 2007; Sip et al. 2007; Wolpe et al. 2005; Greely and Illes 2007; Simpson 2008; Moriarty 2009; Alexander 2006; Stoller and Wolpe 2007).

---

[18] Stone (1993), p. 23.
[19] *Allen v. Bloomfield Hills* (2008), p. 57.

At present, there is a consensus in U.S. scientific circles that brain-based lie detection is not ready for legal use. As neuroscientist Anthony Wagner (2010, p. 14) concluded, in a comprehensive 2010 review of the literature, "there are no relevant published data that unambiguously answer whether fMRI-based neuroscience methods can detect lies at the individual-subject level."

Despite the scientific limitations, there are still several instances in which fMRI and EEG-based lie detection evidence have been proffered in U.S. courts. In 2003, an Iowa state court admitted EEG-based "brain fingerprinting" lie-detection evidence (*Harrington v. State* 2003; see Greely and Illes 2007, p. 387–388). The neuroscientific testimony was not considered directly on appeal in the *Harrington* case, but the case nonetheless drew national attention for the very fact that such evidence had been admitted.

In *Wilson v. Corestaff* (2010), a plaintiff in a New York state court sought an evidentiary hearing on the admissibility of fMRI lie-detection evidence to bolster the credibility of a key witness. The judge found that "since credibility is a matter solely for the jury and is clearly within the ken of the jury ... no other inquiry is required." Such a response is consistent with a U.S. Supreme Court decision in which Justice Clarence Thomas wrote that "a fundamental premise of our criminal trial system is that 'the jury is the lie detector'" (*United States v. Scheffer* 1998, p. 313).

When courts encounter neuroscience-based lie-detection evidence in the future, they are likely to arrive, as the court did in *Semrau* (discussed earlier, in Section 2) at the question of whether novel neuroscientific-based lie-detection technologies are analogous to, or distinguishable from, their polygraph predecessors. Although the polygraph is routinely used in police investigations and in employee screening in some federal agencies (National Research Council 2003), the polygraph is almost uniformly inadmissible in state and federal courts (Greely and Illes 2007). Proponents argue that fMRI is a reliable proxy of brain activity and is not readily susceptible to effective counter measures. Opponents contend that fMRI lie detection is just as unreliable as the polygraph, and therefore should be excluded from evidence.

The future of neuroscientific lie detection will hinge not only on legal analogy, of course, but also on scientific progress. In addition to two private U.S. firms – No Lie MRI and Cephos Corporation – scholars are working on novel neuroscientific approaches to detecting deception. Greene and Paxton (2009), for instance, devised an experimental protocol that did not rely, as previous experiments had, on an instructed lie.

6.2 Memory

Neuroscience and psychology have taught us much about how human memory systems function (Milner et al. 1998; Squire 2004). Memory researchers have pointed out the deficiencies and complexity of human memory (see, e.g. Schacter 2002; Schacter and Slotnick 2004). We know, for instance, that we are susceptible to false memories (Loftus 2005; Bernstein and Loftus 2009); that we forget much of what we experience (Wixted 2004); that emotional state affects the quality of our memories (Phelps 2004); and that our personal experience can affect how we remember an event (Sharot et al. 2007; Kensinger and Schacter 2006).

At the same time, we know that memory and law are intimately intertwined. From police lineups and questioning of suspects at the start of a case, to eye witness testimony and jury recollection of trial material at the end of a case, memory is implicated at every stage of legal proceedings. Courts, then, are faced with an intractable problem: human memory is flawed, yet adjudication by nature must typically rely on it.[20] What are courts to do? And can neuroscience help?

Neuroscience research on memory, over and above the general increase in knowledge it offers the legal system, may one day generate tools for courts to distinguish between real and false memories. For instance, work underway in the laboratory of neuroscientist Anthony Wagner is making progress on the detection of real-world, autobiographical memories (Rissman et al. 2010).

Memory detection using neuroscientific tools also raises new constitutional and ethical considerations (Fox 2008; Kolber 2006). For instance, how strong are our rights to privacy with regard to our memories? What constitutional protections exist to prevent the state from taking a "fingerprint" of one's brain? In what contexts should individuals be allowed, or ever forced, to alter (as through existing memory-altering drugs, for instance) their memories? Questions such as these were debated by the President's Council on Bioethics (2003) and continue to be an active topic for debate.

6.3   Legal Decision-Making

The cognitive shortcoming of participants in the legal system has been well researched (see, e.g., Simon 2011). Cognitive neuroscience builds upon this psychology literature to provide us with new insights into the processes by which judges, jurors, and attorneys arrive at the decisions they make (Knabb et al. 2009; Goodenough 2001). Neuroscience research on moral and legal decision-making has begun uncovering the neural correlates of a number of important aspects of legal decision-making (e.g., Young et al. 2010; Koenigs et al. 2007; Schleim et al. 2010; Borg et. al. 2006; Buckholtz et al. 2008), such as the brain activity underlying the decisions of whether to punish someone and, if so, how much (Buckholtz et al. 2008).

---

[20] One option, often rejected by courts, is to allow an expert witness to testify to the limitations of memory. In rejecting this option, courts may point out that it is the purpose of the jury to make its own estimation of the reliability of the witness' memory (*United States v. Rodriguez-Berrios* 2009).

In addition to these neuroimaging studies, a growing body of psychology and neuroscience research suggests that, when making moral judgments, we are guided by our automatic, evolved emotional responses (Greene and Haidt 2002; Haidt 2001). Numerous scholars are therefore exploring how emotions, across a variety of legal contexts, affect moral and legal reasoning (Salerno and Bottoms 2009; Kahan 2008; Posner 2000; Maroney 2006; Jones 1999; Abrams and Keren 2010; Weinstein and Weinstein 2005; Berman 2008).

It is also important to consider the effects of neuroscientific evidence on juror decision-making. While some early empirical work suggested that the "seductive allure" of brain images would unduly persuade jurors (Weisberg et al. 2008; McCabe and Castel 2008), a more recent and much more robust set of studies suggests just the opposite: relative to other scientific evidence that would be admitted in its place, this research suggests that there is no significant relationship between the introduction of brain imaging evidence, per se, and punishment or blame outcomes (Schweitzer et al 2011; Schweitzer and Saks 2011). Other studies find that the effects of fMRI lie detection evidence may be nullified by cross-examination (McCabe et al 2011).

7 Conclusion

Neuroscientific evidence is increasingly reaching U.S. courtrooms in a number of legal contexts, and this trend is likely to continue for the foreseeable future. The emerging field of law and neuroscience is being built on a foundation joining: (a) rapidly developing technologies and techniques of neuroscience; (b) quickly expanding legal scholarship on implications of neuroscience; and (c) (more recently) neuroscientific research designed specifically to explore legally relevant topics. With the institutional support of many of the country's top research universities, as well as the support of the MacArthur Foundation, among other private foundations and public funding agencies, the U.S. is well positioned to continue contributing to international developments in neurolaw.

In this chapter, we have provided a very brief overview of neurolaw developments in the United States. We did not, of course, reach every facet. And topics omitted here include implications of neuroscience for determinations of brain death, mental health law, intellectual property, consumer law, and employment law (Tovino 2007a), as well as issues surrounding appropriate regulation of neuroimaging for legal and national security purposes (Kulynych 2007; Marks 2007). We are also unable to do justice, in these brief pages, to numerous ethical questions that neurolaw can raise (see, e.g. Illes and Sahakian 2011, Illes 2003; Illes 2006; Farah 2005; Moreno 2003; Roskies 2002; Gillet 2009; Greely 2006), such as those related to the possibility of medical findings incidental to research purposes (Wolf et al. 2008; Richardson 2008; Miller et al. 2008), or those sparked by the possibility of cognitive neuro-enhancement (Farah et al. 2004; Greely et al. 2008). Nonetheless, and in summary, several factors are likely to lead to the continued growth of law and neuroscience in the U.S.

First, legal scholars are demonstrating great interest in expanding the dialogue between law and neuroscience. Evidence from the *Law and Neuroscience Bibliography* suggests that there has been incredibly strong growth in 2008 and 2009 in the annual number of articles published per year on law and neuroscience.[21] The 127 publications in 2009 represents a 300% increase over the number published just 5 years earlier, and represents a 2,000% increase over the number published a decade before. Related scholarly communities, such as the Society for Evolutionary Analysis in Law, are similarly strengthening ties between the legal and scientific communities.[22] These trends suggest that future years will bring continued expansion of interdisciplinary scholarly collaboration.

Second, a practical constraint thus far to expanded use of neuroscientific evidence is the prohibitive costs of brain scanning. To the extent that the costs of fMRI and other neuroscientific technologies drop significantly in the coming years, as brain scanning facilities continue quick proliferation, resource limitations will decline as a barrier to entry – both for researchers and for litigants.

Third, practicing lawyers have also shown increasing interest in improving their professional skills through advances in the mind sciences. Books have been published for a practitioner audience (see, e.g., Sousa 2009; Uttal 2008), and Continuing Legal Education classes have been offered as well. Some of these works are optimistic about prospects for legal applications; others sound cautionary notes. Together, they are continuing to capture the attention of key segments of the practicing bar.

If, as we anticipate, the field of law and neuroscience expands, it will require new training for judges. Already a significant number of judges in the United States are being introduced to neuroscience. Since 2007, for instance, the *Law and Neuroscience Project* has partnered with the Gruter Institute for Law and Behavioral Research, the Federal Judicial Center (FJC) and the National Judicial College (NJC) to sponsor major conferences for hundreds of U.S. judges. (The FJC is the research and education agency of the U.S. federal judicial system; The NJC offers an average of 90 courses annually with more than 2,000 judges enrolling from all 50 states.) And the American Association for the Advancement of Science (AAAS) has also sponsored numerous programs for judges. The topics covered at the conferences included, among other things, an introduction to neuroscience; presentations on frontal lobe function including decision-making, behavioral control, and counter-factual thinking; and presentations on measuring individual variation and subjective states including lie detection, pain assessment, and punishment. The indirect effects of these efforts are much larger, as judges who attend the conferences share information with their colleagues on the bench.

---

[21] The bibliography is available online at: http://www.lawneuro.org.

[22] For more information on the Society for Evolutionary Analysis in Law (SEAL), see: http://www.sealsite.org.

U.S. legislators too may play an important role in shaping the future of neurolaw. Some state legislators have already held committee hearings about neuroscience findings. And in the case of early-childhood legislation, for example, Washington State legislator Ruth Kagi (D-WA) credited neuroscience for finally allowing her bills on the issue (which she had been proposing for nearly a decade) to pass. Representative Kagi noted in a 2007 speech that after hearing neuroscientific testimony, a political opponent, "who had stopped every piece of early childhood legislation in the past 5 years, came up to me and said, '*I get it.*'"[23] In addition, a New York state legislator in 2009 proposed a bill that would make certain MRI scans inadmissible in criminal proceedings.

As judges, lawyers, legislators, legal scholars and the public become more acquainted with neuroscientific evidence, and as neuroscience continues to produce more legally relevant findings, it is likely that we will see continued expansion of law and neuroscience in the United States.

References

Abrams K, Keren H (2010) Who's afraid of law and the emotions? Minn Law Rev 94:1997–2074
Aharoni E, Funk C, Sinnott-Armstrong W, Gazzaniga M (2008) Can neurological evidence help courts assess criminal responsibility? Lessons from law and neuroscience. Ann NY Acad Sci 1124:145–160
Alexander A (2006) Functional magnetic resonance imaging lie detection: is a "brainstorm" heading toward the "gatekeeper"? Houston J Health Law Pol 7:1–56
Allen v. Bloomfield Hills School District (2008) 281 Mich App 49; 760 N.W.2d 811
Annas GJ (2007) Foreword: imagining a new era of neuroimaging, neuroethics, and neurolaw. Am J Law Med 33:163–170
Appelbaum PS (2007) The new lie detectors: neuroscience, deception, and the courts. Psychiatr Serv 58:460–462
Aronson JD (2007) Brain imaging, culpability and the juvenile death penalty. Psychol Pub Pol Law 13:115–142
Aronson JD (2009) Neuroscience and juvenile justice. Akron Law Rev 42:917–930
Aronson JD (2010) The law's use of brain evidence. Annu Rev Law Soc Sci 6:93–108
Arrigo BA (2007) Punishment, freedom, and the culture of control: the case of brain imaging and the law. Am J Law Med 33:457–482
Atkins v. Virginia (2002) 536 U.S. 304
Baird AA (2009) The developmental neuroscience of criminal behavior. In: Farahany NA (ed) The impact of behavioral sciences on criminal law. Oxford University Press, New York, p. 81–123
Barkow RE (2011) Federalism and criminal law: what the feds can learn from the states. Mich Law Rev 109:519–580
Barth AS (2007) A double-edged sword: the role of neuroimaging in federal capital sentencing. Am J Law Med 33:501–522
Baskin JH, Edersheim JG, Price BH (2007) Is a picture worth a thousand words? Neuroimaging in the courtroom. Am J Law Med 33:239–269

---

[23] Emphasis added. Quoted in: Gehrman (2007).

Bazan EB (2005) Capital punishment: an overview of federal death penalty statutes. CRS Report for Congress RL30962

Beecher-Monas E, Garcia-Rill E (2003) Danger at the edge of chaos: predicting violent behavior in a post-daubert world. Cardozo Law Rev 24:1845–1901

Bennett CM, Baird AA, Miller MB, Wolford GL (2010) Neural correlates of interspecies perspective taking in the post-mortem atlantic salmon: an argument for multiple comparisons correction. J Serend Unexpect Result 1:1–5

Berman DA, Bibas S (2008) Engaging capital emotions. Nw Univ Law Rev Colloquy 102:355–364

Bernstein DM, Loftus EF (2009) How to tell if a particular memory is true or false. Perspect Psychol Sci 4:370–374

Bizzi E, Hyman SE, Raichle ME, Kanwisher N, Phelps EA, Morse SJ, Sinnott-Armstrong W, Rakoff JS, Greely HT (2009) Using imaging to identify deceit: scientific and ethical questions. Am Acad Art Sci. http://www.amacad.org/pdfs/deceit.pdf.

Blitz MJ (2010) Freedom of thought for the extended mind: cognitive enhancement and the constitution. Wisconsin Law Rev 2010:1049–1117

Blumoff TY (2010) How (some) criminals are made. In: Freeman M (ed) Law and neuroscience: current legal issues. Oxford University Press, New York, p. 171–192

Bonnie RJ (2002) Responsibility for addiction. J Am Acad Psychiatry Law 30:405–413

Borden S (2011) Concussion suit seeks better health monitoring, New York Times, 20 August 2011

Borg JS, Hynes C, Van Horn J, Grafton S, Sinnott-Armstrong W (2006). Consequences, action, and intention as factors in moral judgments: an fMRI investigation. Journal of Cognitive Neuroscience 18:803–817

Boyd v. Bert Bell/Pete Rozelle NFL Players Retirement Plan (2005) 410 F.3d 1173

Brown T, Murphy E (2010) Through a scanner darkly: functional neuroimaging as evidence of a criminal defendant's past mental states. Stan Law Rev 62:1119–1208

Buckholtz JW, Asplund CL, Dux PE, Zald DH, Gore JC, Jones OD, Marois R (2008) The neural correlates of third-party punishment. Neuron 60:930–940

Cacioppo JT, Berntson GG, Lorig TS, Norris CJ, Rickett E, Nusbaum H (2003) Just because you're imaging the brain doesn't mean you can stop using your head: a primer and set of first principles. J Pers Soc Psychol 85:650–661

Caldwell MF, Skeem J, Salekin R, Van Rybroek GJ (2006) Treatment response of adolescent offenders with psychopathy features: a 2-year follow-up. Crim Just Behav 33:571–596

Caldwell MF, McCormick DJ, Umstead D, Van Rybroek GJ (2007) Evidence of treatment progress and therapeutic outcomes among adolescents with psychopathic features. Crim Just Behav 34:573–587

Chorvat T, McCabe K (2004) The brain and the law. Philosophical Transactions of the Royal Society B: Biological Sciences 359:1727–1736

Coe v. State (2000) 17 S.W.3d 193

Daubert v. Merrell Dow Pharmaceuticals, Inc. (1993) 509 U.S. 579

Davis KC (1942) An approach to problems of evidence in the administrative process. Harv Law Rev 55:364–425

Denno DW (2002) Crime and consciousness: science and involuntary acts. Minn Law Rev 87:269–389

Droback JA (2006) "Developing capacity": adolescent "consent" at work, at law, and in the sciences of the mind. UC Davis J Juv Law Pol 10:1–68

Eagleman DM (2008) Neuroscience and the law. Houston Lawyer 16:36–40

Entertainment Software Ass'n. v. Blagojevich (2005), 404 F. Supp. 2d 1051

Erickson SK (2010) Blaming the brain. Minn J Law Sci Tech 11:27–77

Faigman DL (2002) Is science different for lawyers? Science 297:339–340

Faigman DL (2010) Evidentiary incommensurability: a preliminary exploration of the problem of reasoning from general scientific data to individualized legal decision-making. Brooklyn Law Rev 75:1115–1136

Faigman DL, Kaye DH, Saks MJ, Sanders J (2011) Modern scientific evidence: the law and science of expert testimony, 2010–2011 edn. West Publishing Co., New York

Farah MJ (2002) Emerging ethical issues in neuroscience. Nat Neurosci 11:1123–1129

Farah MJ (2005) Neuroethics: the practical and the philosophical. Trends Cognitive Sci 9:34–40

Farah MJ, Illes J, Cook-Deegan R, Gardner H, Kandel E, King P, Parens E, Sahakian B, Wolpe PR (2004) Neurocognitive enhancement: what can we do and what should we do? Neuroscience 5:421–425

Farahany NA, Coleman JE, Jr. (2009) Genetics, neuroscience, and criminal responsibility. In: Farahany NA (ed) The impact of behavioral sciences on criminal law. Oxford University Press, New York, p. 183–240

Farahany NA (2009a) Cruel and unequal punishments. Wash Univ Law Rev 86:859–915

Farahany NA (ed) (2009b) The impact of behavioral sciences on criminal law. Oxford University Press, New York

Farahany NA (2011) An empirical study of brains and genes in U.S. Criminal Law, Vanderbilt University Law School

Feigenson N (2006) Brain imaging and courtroom evidence: on the admissibility and persuasiveness of fMRI. Int J Law Context 2:233–255

Ferrell v. State (2005) 918 So.2d 163

Fini v. General Motors Corp (2003) 2003 WL 1861025 (Mich. App.)

Fox D (2008) Brain imaging and the bill of rights: memory detection technologies and American criminal justice. Am J Bioethics 8:1–4

Fox D (2009) The right to silence as protecting mental control. Akron Law Rev 42:763–801

Freeman M, Goodenough OR (eds) (2009) Law, mind, and brain. Ashgate, Surrey, England

Freeman M (ed) (2011) Law and neuroscience: current legal issues volume 13. Oxford University Press, Oxford

Frye v. United States (1923) 293 F. 1013

Ganis G, Keenan JP (2009) The cognitive neuroscience of deception. Soc Neurosci 4:465–472

Garland B (ed) (2004) Neuroscience and the law: brain, mind, and the scales of justice. Dana Press, New York

Garland B, Glimcher PW (2006) Cognitive neuroscience and the law. Curr Opin Neurobiol 16:130–134

Gazzaniga MS (2008) The law and neuroscience. Neuron 60:412–415

Gazzaniga MS, Ivry RB, Mangum GR (2009) Cognitive neuroscience: the biology of the mind. 3rd edn. W.W. Norton & Company, New York

Gehrman E (2007) From neuroscience to childhood policy. 7 December 2007. http://news.harvard.edu/gazette/story/2007/12/from-neuroscience-to-childhood-policy/

General Electric Co. v. Joiner (1997) 522 U.S. 136

Gillet GR (2009) The subjective brain, identity, and neuroethics. Am J Bioethics 9:5–13

Goodenough OR (2001) Mapping cortical areas associated with legal reasoning and moral intuition. Jurimetrics J 41:429–442

Goodenough OR, Tucker M (2010) Law and cognitive neuroscience. Annu Rev Law Soc Sci 6:61–92

Graham v. Florida (2010), 130 S.Ct. 2011

Greely HT (2004) Prediction, litigation, privacy, and property: some possible legal and social implications of advances in neuroscience. In: Garland B (ed) Neuroscience and the law: brain, mind, and the scales of justice. Dana Press, New York, p. 114–156

Greely HT (2006) Neuroethics and ELSI: similarities and differences. Minn J Law Sci Tech 7:599–614

Greely HT, Illes J (2007) Neuroscience-based lie detection: the urgent need for regulation. Am J Law Med 33:377–431

Greely HT, Sahakian B, Harris J, Kessler RC, Gazzaniga M, Campbell P, Farah MJ (2008) Towards responsible use of cognitive-enhancing drugs by the healthy. Nature 456:702–705

Greely HT, Wagner AD (Forthcoming) Reference guide on neuroscience. In: Federal judicial center reference manual on scientific evidence, 3rd edn., Federal Judicial Center, Washington, DC

Greene J, Cohen J (2004) For the law, neuroscience changes nothing and everything. Phil Trans Royal Soc B Biol Sci 359:1775–1785

Greene J, Haidt J (2002) How (and where) does moral judgment work? Trends Cogn Sci 6:517–523

Greene JD, Paxton JM (2009) Patterns of neural activity associated with honest and dishonest moral decisions. Proc Natl Acad Sci 106:12506–12511

Grey BJ (2011) Neuroscience and emotional harm in tort law: rethinking the American approach to free-standing emotional distress claims. In: Freeman M (ed) Law and neuroscience. Oxford University Press, New York, p. 203–230

Gruber SA, Yurgelun-Todd DA (2006) Neurobiology and the law: a role in juvenile justice? Ohio St J Crim Law 3:321–340

Hafemeister TL, Stockey NA (2010) Last stand? the criminal responsibility of war veterans returning from Iraq and Afghanistan with posttraumatic stress disorder. Ind Law J 85:87–141

Haider A (2006) Roper v Simmons: the role of the science brief. Ohio St J Crim Law 3:369–377

Haidt J (2001) The emotional dog and its rational tail: a social intuitionist approach to moral judgment. Psychol Rev 108:814–834

Halliburton CM (2007) Letting Katz out of the bag: cognitive freedom and fourth amendment fidelity. Hastings Law J 59:309–368

Hariri AR (2009) The neurobiology of individual differences in complex behavioral traits. Annu Rev Neurosci 32:225–247

Harrington v. State (2003) 659 N.W.2d 509

Hawkins Hon MD (2010) Coming home: accommodating the special needs of military veterans to the criminal justice system. Ohio St J Crim Law 7:563–573

Hora PF, Stalcup T (2008) Drug treatment courts in the twenty-first century: the evolution of the revolution in problem-solving courts. Georgia L Rev 42:717–811

Hughes V (2010) Science in court: head case. Nature 464:340–342

Illes J (2003) Neuroethics in a new era of neuroimaging. AJNR 24:1739–1741

Illes J (ed) (2006) Neuroethics: defining the issues in theory, practice, and policy. Oxford University Press, New York

Illes J, Sahakian BJ, eds. (2011) Oxford handbook of neuroethics. Oxford University Press, New York

Jones OD (1999) Law, emotions, and behavioral biology. Jurimetrics 39:283–289

Jones OD (2004) Law, evolution and the brain: applications and open questions. Philosophical Transactions of the Royal Society B: Biological Sciences 359:1697–1707

Jones OD, Buckholtz JW, Schall JD, Marois R (2009) Brain imaging for legal thinkers: a guide for the perplexed. Stan Tech Law Rev 2009:5–53

Jones OD, Schall JD, Shen FX (Forthcoming) Law and neuroscience. Aspen Publishers, New York.

Kahan DM (2008) Two conceptions of emotion in risk regulation. Univ Penn Law Rev 156:741–766

Katner DR (2006) The mental health paradigm and the MacArthur study: emerging issues challenging the competence of juveniles in delinquency systems. American Journal of Law and Medicine 32:503–583

Kennedy D (2004) Neuroscience and neuroethics. Science 306:373

Kensinger EA, Schacter DL (2006) When the Red Sox shocked the Yankees: comparing negative and positive memories. Psychon Bull Rev 13:757–763

Kiehl KA (2006) A cognitive neuroscience perspective on psychopathy: evidence for paralimbic system dysfunction. Psychiatry Res 142:107–128

Kiehl KA (2008) Without morals: the cognitive neuroscience of criminal psychopaths. In: Sinnott-Armstrong W (ed) Moral psychology. Vol. 3. The MIT Press, Cambridge, p. 119–149

Kiehl KA, Hoffman MB (Forthcoming) The criminal psychopath: history, neuroscience, treatment, and economics. Jurimetrics

Kluger J (2011) Football searches for the cause of another tragedy. TIME Magazine. February 23, 2011

Knabb JJ, Welsh RK, Ziebell JG, Reimer KS (2009) Neuroscience, moral reasoning, and the law. Behav Sci Law 27:219–236

Koenigs M, Young L, Adolphs R, Tranel D, Cushman F, Hauser M, Damasio A (2007) Damage to the prefrontal cortex increases utilitarian moral judgements. Nature 446:908–911

Kolber AJ (2006) Therapeutic forgetting: the legal and ethical implications of memory dampening. Vand Law Rev 59:1561–1626

Kolber AJ (2007) Pain detection and the privacy of subjective experience. Am J Law Med 33:433–456

Kolber AJ (2011) The experiential future of the law. Emory Law Journal 60: 585–652

Kulynych JJ (2007) The regulation of MR neuroimaging research: disentangling the Gordian knot. Am J Law Med 33:295–317

Kumho Tire Co. v. Carmichael (1999) 526 U.S. 137

Loftus EF (2005) Planting misinformation in the human mind: a 30-year investigation of the malleability of memory. Learn Mem 12:361–366

Logothetis NK (2008) What we can do and what we cannot do with fMRI. Nature 453: 869–878

Marchant G (2008) Brain scanning and the courts: criminal cases. Presentation to the Research Network on Legal Decision Making, MacArthur Foundation Law and Neuroscience Project, 11 Oct 2008

Marks JH (2007) Interrogational neuroimaging in counterterrorism: a "No-Brainer" or a human rights hazard? Am J Law Med 33:483–500

Maroney TA (2006) Law and emotion: a proposed taxonomy of an emerging field. Law Hum Behav 30:119–142

Maroney TA (2010) The false promise of adolescent brain science in juvenile justice. Notre Dame Law Rev 85:89–176

Martell DA (1996) Causal relation between brain damage and homicide: the prosecution. Semin Clin Neuropsychiatry 1:184–193

McCabe DP, Castel AD (2008) Seeing is believing: the effect of brain images on judgments of scientific reasoning. Cognition 107:343–352

McCabe DP, Castel AD, Rhodes MG (2011) The influence of fMRI lie detection evidence on juror decision making. Behav Sci Law 29:566–577

McMurtey v. Ryan (2008), 539 F.3d 1112 (9th Cir. 2008)

Meilaender G (2003) Why remember? First Things 135:20–24

Meszaros J (2011) Achieving peace of mind: the benefits of neurobiology evidence for battered women defendants. Yale J Law Femin 23:117–177

Miller FG, Mello MM, Jaffe S (2008) Incidental findings in human subjects research: what do investigators owe research participants? J Law Med Ethics 36:271–279

Milner B, Squire LR, Kandel ER (1998) Cognitive neuroscience and the study of memory. Neuron 20:445–453

Mobbs D, Lau HC, Jones OD, Frith CD (2007) Law, responsibility, and the brain. PLoS Biol 5:693–700

Mobley v. The State (1995), 265 Ga. 292

Monahan J (2006) A jurisprudence of risk assessment: forecasting harm among prisoners, patients, and patients. Va Law Rev 92:391–435

Moreno JD (2003) Neuroethics: an agenda for neuroscience and society. Nat Rev Neurosci 2:149–153

Moriarty JC (2008) Flickering admissibility: neuroimaging evidence in the U.S. courts. Behav Sci Law 26:29–49

Moriarty JC (2009) Visions of deception: neuroimages and the search for truth. Akron Law Rev 42:739–761

Morse SJ (1996) Brain and blame. Geo Law J 84:527–546

Morse SJ (2006) Brain overclaim syndrome and criminal responsibility: a diagnostic note. Ohio St J Crim Law 3:397–412

Morse SJ, Hoffman MB (2007) The uneasy entente between legal insanity and *mens rea*: beyond *Clark v. Arizona*

Morse SJ (2008a) Psychopathy and criminal responsbility. Neuroethics 1:205–212

Morse SJ (2008b) Determinism and the death of folk psychology: two challenges to responsibility from neuroscience. Minn. J.L. Sci. & Tech. 9:1–35

Morse SJ, Roskies AL, eds. (Forthcoming) MacArthur primer on law & neuroscience. Oxford University Press, New York

Nadel L, Sinnott-Armstrong W (2010) Conscious will and responsibility. Oxford University Press, New York

Nadelhoffer T, Bibas S, Grafton S, Kiehl KA, Mansfield A, Sinnott-Armstrong W, Gazzaniga MS (2010) Neuroprediction, violence, and the law: setting the stage. Neuroethics. doi:10.1007/s12152-010-9095-z: 1-33

National Research Council (2003) The polygraph and lie detection: executive summary

New JG (2008) If you could read my mind: implications of neurological evidence for twenty-first century criminal jurisprudence. J Legal Med 29:179–198

O'Hara EA (2004) How neuroscience might advance the law. Philosophical Transactions of the Royal Society B: Biological Sciences 359:1677–1684

Oregon v. Kinkel (2002) 56 P.3d 463

Ovalle D (2010) Miami-Dade killer gets life sentence for murder, stabbings, rape. Miami Herald, 2 December 2010

Pardo MS, Patterson D (2010) Philosophical foundations of law and neuroscience. Univ Illinois Law Rev 2010:1211–1250

Patel P, Levine K, Mayberg H, Meltzer C (2007) The role of imaging in United States courtrooms. Neuroimaging Clin N Am 17:557–567

People v. Goldstein (2004) 786 N.Y.S.2d 428

People v. Morgan (1999) 719 N.E.2d 681 (Ill. 1999)

People v. Weinstein (1992) 591 N.Y.S.2d 715

Petersilia J (2003) When prisoners come home: parole and prisoner reentry. Oxford University Press, New York

Pettit Jr, M (2007) fMRI and BF meet FRE: brain imaging and the federal rules of evidence. Am J Law Med 33:319–340

Phelps EA (2004) Human emotion and memory: interactions of the amygdala and hippocampal complex. Curr Opin Neurobiol 14:198–202

Poldrack RA, Fletcher PC, Henson RN, Worsley KJ, Brett M, Nichols TE (2008) Guidelines for reporting an fMRI study. NeuroImage 40:409–414

Posner EA (2000) Law and the emotions. Geo Law J 89:1977–2012

Powell v. State of Texas (1968) 392 U.S. 514

President's Council on Bioethics (2003) Beyond therapy: biotechnology and the pursuit of happiness. Regan Books, New York

Purves D, Brannon EM, Cabeza R, Huettel SA, LaBar KS, Platt ML, and Woldorff MG (2008) Principles of cognitive neuroscience. Sinauer, Sunderland

Pustilnik AC (2009) Violence on the brain: a critique of neuroscience in criminal law. Wake Forest Law Rev 44:183–237

Racine E, Bar-Ilan O, Illes J (2005) fMRI in the public eye. Nat Rev Neurosci 6:159–164

Rakoff J (2008) Science and the law: uncomfortable bedfellows. Seton Hall Law Rev 38:1379–1393

Redding RE (2006) The brain-disordered defendant: neuroscience and legal insanity in the twenty-first century. Am Univ Law Rev 56:51–127

Relkin N, Plum F, Mattis S, Eidelberg D, Tranel D (1996) Impulsive homicide associated with an arachnoid cyst and unilateral frontotemporal cerebral dysfunction. Semin Clin Neuropsychiatry 1:172–183

Richardson HS (2008) Incidental findings and ancillary-care obligations. J Law Med Ethics 36:256–270

Rissman J, Greely H, Wagner AD (2010) Detecting individual memories through the neural decoding of memory states and past experience. Proc Natl Acad Sci 107:9849–9854

Robinson v. State of California (1962) 370 U.S. 660

Rogers KG, DuBois A (2009) The present and future impact of neuroscience evidence on criminal law. APR Champion 33:18–23

Rojas-Burke J (1993) PET scans advance as tools in insanity defense. J Nuclear Med 34:13N–26N

Roper v. Simmons (2005) 543 U.S. 551

Rosen J (2007) The brain on the stand. New York Times Sunday Magazine, 11 March 2007

Roskies A (2002) Neuroethics for the new millennium. Neuron 35:21–23

Roskies A (2008) Neuroimaging and inferential distance. Neuroethics 1:19–30

Russell RT (2009) Veterans treatment court: a proactive approach. N Eng J Crim Civ Confinement 35:357–372

Saks MJ (2000) The aftermath of Daubert: an evolving jurisprudence of expert evidence. Jurimetrics J 40:229–241

Salerno JM, Bottoms BL (2009) Emotional evidence and jurors' judgments: the promise of neuroscience for informing psychology and law. Behav Sci Law 27:273–296

Sapolsky R (2004) The frontal cortex and the criminal justice system. Philosophical Transactions of the Royal Society B: Biological Sciences 359:1787–1796

Schacter DL (2002) The seven sins of memory: how the mind forgets and remembers. Mariner Books, New York

Schacter DL, Slotnick SD (2004) The cognitive neuroscience of memory distortion. Neuron 44:149–160

Schauer F (2010) Can bad science be good evidence? Lie detection, neuroscience and the mistaken conflation of legal and scientific norms. Cornell Law Rev 95:1191–1220

Schleim S, Spranger TM, Erk S, Walter H (2010) From moral to legal judgment: the influence of normative context in lawyers and other academics. Soc Cogn Affect Neurosci 6:48–57

Schwarz A (2010) Case will test NFL teams' liability in dementia. New York Times, 5 April 2010

Schweitzer NJ, Saks M, Murphy E, Roskies A, Sinnott-Armstrong W, Gaudet L (2011) Neuroimages as evidence in a mens rea defense: no impact. Psychology, Public Policy, and Law 17:357–392

Schweitzer NJ, Saks M (2011) Neuroimage evidence and the insanity defense. Behavioral Sciences & the Law 29:592–607

Scott ES, Steinberg L (2008) Rethinking juvenile justice. Harvard University Press, Cambridge

Sharot T, Martorella EA, Delgado MR, Phelps EA (2007) How personal experience modulates the neural circuitry of memories of September 11. Proc Natl Acad Sci 104:389–394

Shen FX (2010) Monetizing memory science: neuroscience and the future of PTSD litigation. Paper presented at the Memory and Law Conference, Tucson Arizona, January 2010

Shen FX (2010) The law and neuroscience bibliography: navigating the emerging field of neurolaw. International Journal of Legal Information 38:352–399

Shen FX, Jones OD (2011) Brain scans as evidence: truths, proofs, lies, and lessons. Mercer L. Rev 62:861–884

Simon D (2011) The limited diagnosticity of criminal trials. Vanderbilt Law Rev 64:143–223

Simpson JR (2008) Functional MRI lie detection: too good to be true? J Am Acad Psychiatry law 36:491–498

Sinnott-Armstrong W, Roskies A, Brown T, Murhpy E (2008) Brain images as legal evidence. Episteme 5:359–373

Sip KE, Roepstorff A, McGregor W, Frith CD (2007) Detecting deception: the scope and limits. Trend Cogn Sci 12:48–53

Smith DM (2009) The disordered and discredited plaintiff: psychiatric evidence in civil litigation. Cardozo Law Rev 31:757–771

Snead OC (2006) Neuroimaging and the courts: standard and illustrative case index. http://www.ncsconline.org/d_research/stl/June06/Snead.doc. Accessed 15 July 2010

Snead OC (2007) Neuroimaging and the "complexity" of capital punishment. NY Univ Law Rev 82:1265–1339

Sousa D (2009) How brain science can make you a better lawyer. American Bar Association, Chicago

Squire LR (2004) Memory systems of the brain: a brief history and current perspective. Neurobiol Learn Mem 82:171–177

State v. Nelson (2010) 11th Fl Cir. Ct., F05-846

Stoller SE, Wolpe PR (2007) Emerging neurotechnologies for lie detection and the fifth amendment. Am J Law Med 33:359–375

Stone AA (1993) Posttraumatic stress disorder and the law: critical review of the new frontier. Bull Am Acad Psychiatry Law 21:23–36

Stuntz WJ (2008) Unequal justice. Harv Law Rev 121:1969–2040

Tancredi LR, Brodie JD (2007) The brain and behavior: limitations in the legal use of functional magnetic resonance imaging. Am J Law Med 33:271–294

Taylor J (1995) Neurolaw: towards a new medical jurisprudence. Brain Inj 9:745–751

Taylor E (2006) A new wave of police interrogation? Brain fingerprinting, the constitutional privilege against self-incrimination, and hearsay jurisprudence. Univ Illinois J Law Technol Pol 2006:287–312

Tovino SA (2005) The confidentiality and privacy implications of functional magnetic resonance imaging. J Law Med Ethics 33:844–848

Tovino SA (2007a) Functional neuroimaging and the law: trends and directions for future scholarship. Am J Bioethics 7:44–56

Tovino SA (2007b) Functional neuroimaging information: a case for neuro exceptionalism? Fla St Univ Law Rev 34:415–489

Trout JD (2008) Seduction without cause: uncovering explanatory neurophilia. Trends Cogn Sci 12:281–282

U.S. v. Scheffer (1998) 523 U.S. 303

United States v. Gigante (1997) 982 F. Supp. 140

United States v. Kasim (2008) 2008 WL 4822291 (N.D.Ind.)

United States v. Rodriguez-Berrios (2009), 573 F.3d 55

United States v. Semrau (2010), U.S. District Court for the Western District of Tennessee, No. 07–10074

Uttal WR (2003) The new phrenology: the limits of localizing cognitive processes in the brain. The MIT Press, Cambridge

Uttal WR (2008) Neuroscience in the courtroom: what every lawyer should know about the mind and the brain. Lawyers & Judges Publishing, Danvers

Van Middlesworth v. Century Bank & Trust Co. (2000) 2000 WL 33421451 (Mich.App.)

Viens AM (2007) The use of functional neuroimaging technology in the assessment of loss and damages in tort law. The American Journal of Bioethics 7:63–65

Volkow ND, Li T (2004) Drug addiction: the neurobiology of behavior gone awry. Nat Rev Neurosci 5:963–970

Vul E, Harris C, Winkielman P, Pashler H (2009) Puzzlingly high correlations in fMRI studies of emotion, personality, and social cognition. Perspect Psychol Sci 4:274–290

Wagner AD (2010) Can neuroscience identify lies? In: Gazzaniga M (ed) A judge's guide to neuroscience, SAGE Center, University of California, Santa Barbara, p. - 13–25

Ward J (2009) The student's guide to cognitive neuroscience, 2nd edn. Psychology Press, Hove

Weber S, Habel U, Amunts K, Schneider F (2008) Structural brain abnormalities in psychopaths: a review. Behav Sci Law 26:7–28

Weinstein J, Weinstein R (2005) "I know better than that": the role of emotions and the brain in family law disputes. J Law Fam Stud 7:351–403

Weisberg DS, Keil FC, Goodstein J, Rawson E, Gray JR (2008) The seductive allure of neuroscience explanations. J Cogn Neurosci 20:470–477

Weiss Z (1996) The legal admissibility of positron emission tomography scans in criminal cases: People v Spyder Cystkopf. Semin Clin Neuropsychiatry 1:202–210

Wilson v. Corestaff (2010) Kings County, Supreme Court of the State of New York, Index No. 32996/07

Wixted JT (2004) The psychology and neuroscience of forgetting. Annu Rev Psychol 2004:235–269

Wolf SM, Paradise J, Caga Anan C (2008) The law of incidental findings in human subjects research: establishing researchers' duties. J Law Med Ethics 26:361–383

Wolpe PR, Foster KR, Langleben DD (2005) Emerging neurotechnologies for lie-detection: promises and perils. Am J Bioeth 5:39–49

Young G, Kane AW, Nicholson K (eds) (2006) Causality: psychological knowledge in court: PTSD, pain, and TBI. Springer, New York

Young L, Bechara A, Tranel D, Damasio H, Hauser M, Damasio A (2010) Damage to ventrome- dial prefrontal cortex impairs judgment of harmful intent. Neuron 65:845–851

Zeki S, Goodenough OR (eds) (2006) Law and the brain. Oxford University Press, Oxford

Summary: Neurolaw in an International Comparison

Henning Wegmann

Abstract After the legal landscape in neurosciences has been described for the individual countries, the last chapter is to give an overview on the most common topics connected to neurolaw. As so far none of the countries has established a genuine neurolaw, it seems to be the major challenge for the near future to try and apply existing regulations on neurosciences.

1 Introduction: Topicality of Neurolaw in the International Discussion

The topic of neurosciences and its diverse connections to the law is a legal challenge with a broad international dimension. Whereas the topics being discussed with respect to neurosciences differ from country to country, one aspect holds true for all countries involved in this comparison: "Neurolaw", which is commonly referred to as a set of legal questions raised by the ongoing and current developments in neuroscientific research and treatment,[1] has not yet been discovered as an independent research discipline. To the contrary, the debate about the legal implications of neuroscientific developments is still at a very early stage and thus shows a wide range of topics, which are still in need of being structured to foster the debate. Thus, the country reports in this book show widely different perspectives in their descriptions of the individual national legal systems. However, one can already see from the various country reports that some of the topics are seen as relevant in many of the countries, although they are discussed in different intensities. Nevertheless, also "less

---

The author works as a research associate in the Research Group "Norm-setting in the Modern Life Sciences" at the Institute of Science and Ethics of the University of Bonn, Germany.

[1] Schleim et al. (2007), p. 8; also referred to by Hilf, Stöger, Country Report Austria.

H. Wegmann (✉)
Institute of Science and Ethics, Bonner Talweg 57, 53113 Bonn, Germany
e-mail: wegmann@iwe.uni-bonn.de

T.M. Spranger (ed.), *International Neurolaw*,
DOI 10.1007/978-3-642-21541-4_20, © Springer-Verlag Berlin Heidelberg 2012

prominent" subjects have to be taken into consideration in order to develop a consistent legal framework approaching neuroscientific developments.

This chapter is first to try and point out the most important legal questions in connection to neurosciences within the different countries, in order to set up a charter of concern, from which one can draw the degree of importance and thus may be able to set up a kind of sequence, in which the different topics have to be dealt with. On the other hand, it is also the pronounced task of a concluding chapter to draw the attention to a greater number of the described legal implications in connection to neurosciences, as this seems to be the only way of getting underway a broad international debate, which is not only restricted to certain fields of interests, but tries to see neurosciences as a phenomenon of Modern Life Sciences, which should provoke a general and consistent legal approach toward areas, which have so far not been discovered.

One of the most interesting and also most difficult fields of neurosciences tends to be the wider field of neuroimaging. In one way or the other, this affects various branches of law, such as intellectual property law, tort law, consumer law, health law, employment law, constitutional and criminal law.[2] Nevertheless, cases of evidence in court are still relatively scarce in this area, which is probably due to the fact that the relevant technologies of 'lie detection' are still in an infant state and their role in the legal systems has yet to be found.[3]

In a first step, this concluding chapter will promote some information on the different states of the national legal frameworks concerning neurosciences. In a second step, the fields of incidental findings, informed consent, criminal law and lie detection, free will and neuroenhancement will be identified as the six major fields of concern in modern neurosciences.

2 Legal Framework

It has to be pointed out right at the beginning that till today there cannot be determined a single country providing specific regulations concerning neuroscientific research on human beings. As a consequence, the general legal framework on research with human beings is commonly applied to the field of neuroscientific research in the individual countries. However, the state of regulation on scientific research differs from country to country and in some cases also shows strong references to the cultural composition of the country. Thus, it seems to be appropriate to give a short overview over the most common legal standards of the different countries.

On an international level, the *UNESCO* plays a decisive role when it comes to setting up common standards for international research, which may serve as an ethical compass for the law, working across the borders of many different countries.

---

[2] Shafi (2009), p. 27.

[3] Morse (2008).

The UNESCO ethics program started in 1993 with the establishment of the International Bioethics Committee (IBC), the first bioethics committee with a global scope and expert membership. As a forum for international reflection on bioethics, the IBC and the resultant bioethics Declarations provide an example of international reflection on these subjects, and the texts were unanimously endorsed by all member countries of UNESCO, making them useful texts for development of national laws, and more specific international guidelines.

Of special importance for the international research landscape are the Universal Declaration on the Human Genome and Human Rights of 1997 and the Universal Declaration on Bioethics and Human Rights of 2005. Besides these two documents, there are useful references in the International Declaration on Human Genetic Data adopted by the General Conference of UNESCO on 16 October 2003.

The Universal Declarations constitute nonbinding instruments in international law. The Declarations gain their importance from the degree to which they have been adopted by the member states and thus develop some moral authority and create a spirit of commitment to ethical principles in the use of applications of science and technology, as all states of the international community are committed to respecting and implementing the basic principles of bioethics. The rather general tone of the Declarations allows the member states to interpret them in an individual manner. Despite this room for individual interpretation left by the Declarations, it is the declared aim of the UNESCO to bring forward proposals for a more harmonized international approach toward the topics discussed in its declarations.

On a more European level, the regulations by the *Council of Europe* are of special interest for neuroscientific research. The European scientific community has, alike the relevant actors of the Council of Europe, recognized that neurosciences have become one of the largest research areas within the entire sphere of modern biology, bearing in mind that highly important findings are coming to light almost daily, which is enhancing the comprehension of the human brain structures and their functions and advancing medical care for patients with serious diseases. In the last few years, a high number of research projects have been supported to investigate the impact of neuroscientific knowledge on different fields of law.

Although working groups, research committees and other institutions have already published several guidelines on the ethical aspects of the topic, there has so far not been enacted some kind of international treaty or document on neuroscientific research. Nevertheless, there is already a respectable number of practical recommendations and international documents on the entire field of biomedical research existing, which could also give some advice for the further approach by the Council of Europe. Among those documents are the Nuremberg Code of 1947, the Declaration of Helsinki, or the UNESCO declarations of 1997 and 2005. Besides those, in particular the Council of Europe's Convention on Human Rights and Biomedicine of 1997 is of interest in this context. Apart from the Convention on Biomedicine, those documents do not have any legally binding force and thus do not show any direct effects on the national legislations. However, those instruments set out some of the most important legal standards such as the informed consent, which have by now been accepted by the international scientific community. On the other hand, they are not able to give detailed

answers on specific problems of neurosciences, such as the handling of incidental
findings.[4] It is suggested that the Council of Europe should indeed use its competence to
enact Additional Protocols to the Convention on Human Rights and Biomedicine in
order to set up a specific additional protocol on matters of neurosciences.[5]

As for *Austria*, regulations on research involving human beings cannot be found
in a specific statute, but rather a great number of single pieces, which are scattered
all over the Austrian legal order have to be taken into account.[6] In addition to this,
several fundamental and constitutional rights are standing in the background and
have to be considered in the application of the single laws. Different parts of this
framework apply according to the type of research performed and to the institution
where it takes place. As for the relevant fundamental rights, particularly the
freedom of research, which is guaranteed in article 17 of the Basic Law of the
Austrian Constitution, shows a very high relevance for the topic, as rules specifically prohibiting research or teaching in certain areas are to be declared unconstitutional. As a kind of restriction to this freedom of research, all activities in the field
of research and teaching have to be carried out in conformity with the (general)
laws of the land, especially those protecting the rights of other persons.[7] But not
only the nationally guaranteed rights of other persons have to be paid attention to,
but also other persons' rights guaranteed by European Law. Among the most
important rights in this context, one must name the right to privacy under Art 8
of the European Charter of Human Rights (ECHR), which has been integrated into
the Austrian constitution. In addition to this, the right of freedom from torture and
inhuman or degrading treatment under Art 3 ECHR and the right to protection of
data are of special importance in the field of scientific research. In contrast to other
states – such as Germany for instance – the Austrian Constitution does not contain a
fundamental right protecting human dignity as such. In order to cut back this lack of
personal protection, the Austrian case-law and the relevant literature in this field are
in agreement about the fact that the elements forming the content of human dignity
in other states can be drawn from some other provisions of the Austrian constitution
and that Art 3 ECHR somehow acts as a security clause in this context. It has to be
noted that neurosciences has already gained a quite high reputation in Austria, as
the practical fields of application reach from the development of certain assistive
technologies, which can support disabled, chronically ill or elderly people and
which are linked with the brain or the nervous system (e.g., stimulators implanted
in the brain) to issues such as brain/neuronal–computer interaction (BNCI) devices.

Like in most of the other countries to come, also the *Brazilian* legal system does
not know any specific provisions on neurosciences. To the contrary, there are lots of
different resolutions existing within the Brazilian legal system, which are dealing

---

[4] See on this below at Sect. 4.

[5] See Rödiger 2011, Report on the Council of Europe.

[6] See Kopetzki (2010a, b).

[7] For an overview see Berka (1999), para. 587.

with the conditions of research on human beings in general. The most important legal aspects in Brazil are considered the supervision by an ethics committee and the prior informed consent of the concerned person. However, those requirements rather deal with the preconditions of getting neuroscientific research projects started in Brazil. This does not give any statement about the legal approach toward neuroscientific developments, which have already become reality. Thus, it can be considered that, in comparison with other countries, neurosciences in Brazil are even at an earlier stage.

In contrast to this, the legal framework in *Canada* shows a very high level of specification, but seems to be both complex and fragmented. It is characteristic for the Canadian framework that it contains a vast number of legally nonbinding, but guiding documents, reaching from a local level up to the national and international level.[8] The roots of Canadian research ethics can be traced back – alike in other countries, especially in Europe – to the Nuremberg Code of 1947 and the Declaration of Helsinki of 1964. Since the sentencing within the Nuremberg Doctor's Trial, the Nuremberg Code belongs to the fundamentals of medical ethics. Still, it is worth mentioning that, apart from any kinds of abuse by the Nazis, there were regulations of Medical Law in Germany and also other countries long before the Nuremberg Trials. Although the Nuremberg Code has never been converted into legal force, it still provides important basic principles which are relevant for current questions of medical ethics. However, the principles have been elaborated ever since and adjusted to the special challenges of the modern sciences. First and most important, the Nuremberg code establishes the standard of informed consent and states that this is crucial and essential for any kind of medical treatment to the individual and that medical treatment must not be carried out without it. Furthermore, only clinical studies shall be allowed which aim at improving the common welfare. In its last points the Nuremberg Code states that every experiment underlies a careful evaluation of benefits and harms and that it has to be secured that the benefit overweighs possible harms.[9] The Declaration of Helsinki was developed by the World Medical Association as a set of ethical principles for the medical community regarding human experimentation in 1964. It has constantly been modified ever since, especially by the Declaration of Tokyo in 1975, the Declaration of Venice in 1983, the Declaration of Somerset-West in 1996, and the Declaration of Edinburgh in 2000.[10] After a small revision with a few amendments to the Declaration in 2004, the last revision is the one of Seoul 2008, which is also the current version of the Declaration. The new version especially contains stricter regulations concerning the principally hostile opinion concerning the usage of placebos, as well as further and more detailed provisions concerning the so-called informed consent. Generally, it embodies three basic principles: respect for all persons, beneficence in the

---

[8] Hadskis (2007), p. 261; Downie and McDonald (2004), p. 174.

[9] Spranger (2009a), p.100.

[10] Spranger (2009a), pp. 49–51.

maximization of benefits over harms, and justice for all those who could benefit from the research. These principles have also found their way into the Canadian regulatory framework of research ethics and shaped it decisively.

To name only one, the most important regulatory tool in Canadian research ethics is the so-called Tri-Council Policy Statement: Ethical Conduct for Research Involving Humans. The original version of 1998 goes back to an initiative by the Medical Research Council, the National Sciences and Engineering Research Council and the Social Sciences and Humanities Research Council. The updated version, TCPS 2, came into effect November 29, 2010. It is the result of a lengthy revision process undertaken by the Interagency Advisory Panel on Research Ethics (PRE), and reflects the commitment of the Agencies to ensure the TCPS remains a living and evolving document.

In *Finland*, it is stressed that the legal landscape of neuroscientific research is tightly connected with the nature of this kind of research. Although neuroscientific research brings together elements from many different disciplines, among them also technology, it is fundamentally based on medical research. As medical research is extensively regulated in the Finnish legal system, it is only consequent to apply the general rules on medical research in the field of neuroscientific research.

According to the Finnish Medical Research Act, the main characteristic of medical research is the intervention into the integrity of a person. With this Act, Finland is one of the few European countries to provide a specific Act on medical research. The Act is also applied if human embryos or fetuses are investigated for special purposes.

A level of high importance is paid to the work of ethics committees in Finland. However, the practical developments make clear that due to the increasing number of research projects and the number of different duties of ethics committees, the time a single ethics committee is able to spend for a single research project becomes shorter and shorter. This is, in particular, challenging with respect to the interdisciplinary character of neuroscientific research projects. Thus some hospital districts have reacted to this by establishing multiple committees or permanent subcommittees to assist with protocols that require expert knowledge. In addition, all ethics committees have the opportunity to request additional ad hoc expert opinions, if necessary. However, it has to be noted that the application of neuroscientific knowledge in the Finnish scientific community is still rather sporadic. The only recorded demands for guidance in the area involve the use of polygraphs in criminal investigation and court proceedings.

France is one of the view countries with a specific national regulation on biomedical research, the so-called Law on Bioethics (Loi de bioéthique), which is also known as Huriet's Law (Loi de Huriet). As a part of the French Public Health Code, the Law on Bioethics brings forward the basic legal conditions for biomedical research with human beings in France. It has strongly been influenced by the European Directive 2001/20/EC on the approximation of laws, regulations and administrative provisions of the Member States relating to the implementation of good clinical practice in the conduct of clinical trials on medicinal products for human use. The inserted dispositions especially aimed at a higher degree of protection for persons involved in research projects. In reaction to the fast-growing knowledge in biomedical research, the French legislator has established a rotational revision of the Law on Bioethics. The implementation of specific provisions on neurosciences has been suggested in January 2010 for the first time. However, such demands have not yet been put into a legally applicable form and thus also France can be mentioned among the countries having no neurospecific legislation to date. From the practical point of view, one of the most important issues in the French debate seems to be possibilities of application for neuroimaging techniques such as the use of fMRI scanners as lie detectors in courts or as communication methods for vegetative state patients. However, such methods are not yet occurring that often in France, but, nevertheless, it is important to discuss the legal approach toward such methods in advance.

Alike the abovementioned countries, also *Germany* does not yet have any legal provisions dealing with neuroscientific research or neurosciences in general. The quite special historical background of the country leads to the fact that the so-called Basic Law, which is formed by the first 19 articles of the German constitution, plays an outstanding part in the legal approach toward not only neuroscientific, but also in general new technical developments. Among the most important provisions, which become relevant in the context of neurosciences, one may name a broad range of protective rights, such as the human dignity of article 1, the equality before the law of article 3, the freedom of faith, conscience, and creed of article 4, or the right to school education of article 7 Basic Law. Any new legislative action in connection with a regulation of neurosciences will have to be measured at the very strict standards of those human rights, which makes clear why Germany is seen to have a rather restrictive and conservative approach toward many new developments in the sector of Modern Life Sciences. Speaking from a practical point of view, it has to be demanded that the already relatively vivid discussion of the legal framework is stretched to further fields of application. Among those are the discussions about the impact on criminal responsibility, the prevention of criminal actions, the future of a neuroscientific lie-detector or assistive technologies for challenged persons. Besides this, there are discussions about the abusive use of thoughts by promotion companies as well as discussions about the fight against terrorism with brain-scanning machines at airports, or the use of neuroscientific methods in the employment sector, in order to find out details about the employee's motivation and character.

The legal approach in *Greece* shows in its origin some similarity to the German legal approach, as it is also strongly influenced by the Greek constitution, which establishes in its article 2 the principle of "respect and protection of the value of the human being" as the primary duty of the state. In other words, this article establishes the general principle of self-determinism.[11] Also, other fundamental

---

[11] See in extenso Dagtoglou, pp. 1133–1141.

principles of research, such as the freedom of will and personal autonomy, in general, originate from this provision. Thus, statutory provisions dealing with the capacity to consent or to form legally valid decisions are related directly to this constitutional article, in the meaning that human value is inevitably violated in case of unjustified loss of capacity. Apart from this high-ranking principle, the most relevant provisions of the Greek constitution in connection to neurosciences are article 5 par. 1 acknowledging the development of personality in social contexts as a general right, article 5 par. 5 concerning the individual right to health and the protection from biomedical experimentation, article 7 par. 2 prohibiting any sort of tortures, including those involving psychological violence, article 9 A on the protection of personal data, article 16 protecting the freedom of science and research and article 21 par. 1 protecting the family and, thus, prioritizing family members as caregivers and legal representatives of persons unable to form and express their will, due to their neurological or mental situation.

These provisions are of special interest for any kind of biomedical experimentation in the field of neurosciences, especially when other individual rights of the persons undergoing research activities may be at a certain risk, and need to be balanced along with this freedom. Some of the mentioned provisions are yet relatively new (so is art. 5 par. 5 of 2001) and are of interest particularly when a person undergoes neurological, psychiatric, or psychological treatment and, furthermore, when the person participates in clinical trials related to experimental methods for such a treatment. Other constitutional provisions have indirect effects in the specific context of neurolaw, since in any case they affect medical acts and health care in general. Apart from the constitutional basis, the Greek legal system also provides some specific provisions within the Civil Code, which can become of interest for neurosciences especially when dealing with the legal capacity of a person in regular life.[12]

In *Japan*, the field of neurosciences has not yet been discussed at length on a legal level, although there can be found a broad discussion on the ethical aspects of the topic. Without providing any detailed legal provisions on neurosciences, the scientific debate in Japan focuses on the aspects of how to explain the concept of free will, the criminal responsibility, and the limitation of intervention into human brain in the field of human experimentation or enhancement for as far as possible.

With the lack of legal provisions in mind, the study of case law gains special importance in Japan. In this connection, one has to take a closer look at the so-called "Lobotomy Case" of 1978. In this case, the accused person was sentenced to undergo an operation of the lobotomy, which was an operation on the brain by cutting a part of the frontal lobe (and in most western states has no longer been used since the late 1950s), without his own consent. As a result, he had an aftereffect of frontal lobe syndrome, and claimed damages from the responsible hospital. In the end, the court decided that this operation had been unlawful medical treatment due to the lack of the

---

[12] See on this Vidalis et al. 2006, Country Report Greece.

concerned person's informed consent. Although the court did not see the operation as an unlawful human experimentation, it was decisive for the Japanese legal system that it declared the operation without the patient's consent as unlawful.

Within the legal system of *Switzerland*, one of the first questions of law is in how far the Swiss constitutional law can or shall adopt the neuroscientific results; furthermore, how namely the basic and human rights constitute a framework which limits the neuroscientific research and its potential implications.[13]

As far as *Turkey* is concerned, it holds true that the approach toward any kind of research is strongly influenced by the cultural background of the individual country, as many of the relevant principle derive from sources being about 1,000 years old. Nevertheless, the principles in the Turkish legal system are not that different from principles in other countries. Generally speaking, research projects involving human beings are only acceptable in cases where there is no other chance to obtain the expected scientific development from such experiment. If then such a research project is carried out, the impact on the legal goods of the concerned person must be kept as low as possible, which corresponds to the general ethical principle of doing no harm.

In Turkey, the experimental researches to be conducted on human beings are defined by laws, bylaws and regulations. One of the earliest regulations in Turkey that monitor the scientific research practices on human beings is the so-called Medical Deontology Regulation of 1960, which sets out some of the most important standards for scientific research. Inter alia, the regulation stresses the general prohibition of any research projects involving human beings and only finds them admissible under strict conditions, especially with no other effective method at hand. In addition to the regulations of this act, article 17 of the Turkish constitution explicitly contains the requirement of informed consent prior to any medical or scientific research project on human beings. Furthermore, Turkey has actively imbedded the Convention on Human Rights and Biomedicine, which the country signed in 1997, into the Turkish legal system by an act of 2003.

In contrast to most of the European legal systems, the legal framework in the *USA* is decisively based on the judgment of individual cases, so-called case law. There can be seen a growing number of criminal cases involving neuroscientific evidence. The general interest in neurosciences in the U.S. seems to originate from mainly two things. First, the criminal and civil justice systems rely, critically, and fundamentally, on the mental operations of its many participants – judges, jurors, lawyers, defendants, law enforcement officers, court officials, and witnesses. Second, new technologies enable unprecedented investigation and observation of how (and sometimes how well) those mental operations occur. Just in calendar year 2010, the U.S. legal system saw its first evidentiary hearing in federal court on the admissibility of fMRI lie-detection evidence,[14] the first admission of quantitative

---

[13] See Schweizer and van Spyk 2007, Country Report Switzerland.

[14] U.S. v. Semrau (2010).

electroencephalography (qEEG) evidence contributing in part to a reduced sentence in a homicide case,[15] and a U.S. Supreme Court ruling explicitly citing brain development research.[16]

> Not only is neuroscientific evidence reaching the courts, it is also – at least in some contexts – directly affecting the administration of justice. For example, in 2010 jurors in a U.S. state court considered whether Grady Nelson, who had earlier been found guilty of murdering his wife and raping a child, should receive the death penalty or life in prison. When Nelson was spared the death penalty, interviews with jurors after their verdict revealed that, for some, the proffered neuroscientific evidence was a tipping point. As one juror remarked, "the technology really swayed me. . . After seeing the brain scans, I was convinced this guy had some sort of brain problem" (Ovalle 2010, 1). Whether or not this particular use of qEEG evidence was appropriate, and whether or not the Grady Nelson case was rightly decided, similarly situated legal defense teams will likely consider offering similar types of evidence in the future.[17]

Although there has not been determined a single provision explicitly addressing the specific problems of neurosciences, most of the mentioned countries provide more general regulations on biomedical research, which might likely be adopted for the field of neurosciences mutatis mutandis. Thus, also despite the lack of specific legal provisions on the topic, certain fields of interest in connection with neurosciences can already today be discussed in a very intense manner, paving the way toward appropriate future legal provisions on the issue.

3   Informed Consent

The issue of informed consent is considered a foundational principle for the ethical practice of research with human participants around the world. This holds true for neurosciences, and in particular for neuroimaging, as well. It is by now commonly accepted that no kind of medical or scientific intervention may be carried out without the prior informed consent of the concerned person. However, the detailed content of the principle of informed consent is not yet that clear. Whereas ethical criteria try to deduce the principles from the general principle of autonomy, legal documents try to pick up this idea and put it into concrete legal terms. However, the Convention on Human Rights and Medicine of 1997 is so far the only legally binding document. All other documents do not develop any legal force. In addition, the Convention has so far been ratified by just 21 European states. That is why still today the detailed standards of informed consent differ from state to state.[18]

---

[15] State of Florida v. Grady Nelson (2010).

[16] Graham v. Florida (2010).

[17] Jones/Shen, Country Report USA.

[18] Wegmann (2009a), p. 20.

The principle of informed consent is of special interest, when it comes to the protection of vulnerable groups. For instance, carrying out research on neuropsychiatric disorders such as schizophrenia or Alzheimer's disease demands the involvement of vulnerable groups. In this context, art. 17 para. 1 of the European Convention on Human Rights and Biomedicine inter alia demands that research projects on persons without the capacity to consent are only allowed if "the results of the research have the potential to produce real and direct benefit" to the health of the concerned person. In the following, art. 17 Para. 2 of the convention points out that research is, nevertheless, admissible in rare exceptions and under certain conditions, which are then listed in the convention. By these exceptions, the article endorses research projects on human beings without the ability to consent, even if such research is of no immediate benefit for the individual him or herself. Due to this, the provision still has to face some severe criticism by the member states and has until today not been signed by all member states, including for instance Germany.

With regard to the content, the informed consent of a participant to involvement in a research study is intended to represent a process of communication between the researcher and participant, and should not be reduced to a legal technicality. As far as biomedical research is concerned, the Council of Europe has set into force an "Additional Protocol to the Convention on Human Rights and Biomedicine" in Strasbourg on January 25th, 2005. This protocol is to specify the most important principles of the Convention in additional standards.[19] It contains a whole chapter dealing with the topic "information and consent". Article 13 makes clear that the necessary information for the participants have to be outlined in a comprehensible form and states certain minimum contents, which have to be covered by the consent, such as the overall plan, possible risks and benefits of the project, the nature and purpose of the research project. This is followed by a detailed catalogue of further contents, such as extent and duration of the project, available preventive, diagnostic, and therapeutic procedures, arrangements of fair compensation in the case of damage, potential further uses of the research results, data or biological materials or the source of funding of the research project.

Whereas the standards and the content of informed consent have by now been worked out quite clearly by means of ethical discourse, accompanied by international declarations and conventions, the consequences of violating the principle of informed consent cannot be determined that clearly. What can indeed be stated is that a violation of the principle of informed consent may lead to claims for damages. In the field of medical treatment, such claims for damages can not only be based on a violation of the personal integrity, but additionally on a breach of contract by the physician.

---

[19] Explanatory Report to the Convention for the Protection of Human Rights and Dignity of the Human Being with regard to the application of Biology and Medicine, ETS No. 164, Structure of the Convention, No. 7, see http://conventions.coe.int/treaty/en/Reports/Html/164.htm (March 17th, 2011).

In the context of neuroimaging research, complying with the requirements of informed consent may be challenging in a number of different respects.

The information of the concerned person about possible risks and harms may include physical, psychological, emotional, and social risks, among others. "While the physical risks associated with neuroimaging research, and particularly fMRI scanning, may be relatively straightforward, other forms of risk may be more difficult to predict. Psychological risks may manifest in the form of discomfort or anxiety from having to remain still within the relatively small magnetic chamber, or may involve unpredictable reactions to study results or incidental findings. Even more challenging are the potential social risks which may be associated with future uses of the technology, for example behavioural diagnosis, lie detection, or insurance eligibility. The relative novelty of techniques and applications of neuroimaging technology also makes it challenging for researchers to adequately disclose the scientific and therapeutic limitations of the study and to address currently unanticipated future uses of the data."[20]

Furthermore, it is also mostly agreed that the risk of possible incidental findings has to be covered by the informed consent.[21] It has to be noted once more that due to the lack of international regulation on the topic there has been developed a broad variety of national regulations in the different countries. For instance, according to the Finnish Medical Research Act, medical research cannot be conducted without the research participant's informed consent in writing, which would then also fulfil the requirements of the Finnish Personal Data Act. However, it seems to be the common understanding in Finland that neuroscientific research may well be conducted on a research participant who is not able to write due to a neurological condition. In this case, provided that the potential research participant has the legal capacity to consent, the research participant can consent orally in the presence of at least one witness who is not dependent on the research.[22]

Although the researchers may not know exactly what kind of information can be extrapolated, e.g., from the scanned images of the research participant's brain, they will be able to explain the nature of the research, what kind of procedures and equipment will be involved, how invasive the procedures will be and the physical risks involved in them. Research, by its nature, is inquisitive, and the purpose of it is to gather new scientific information. It should be explained to the participants that unexpected issues may arise.

Also, the French legal system provides some specific regulations on the content and form of the informed consent in biomedical research projects. But still it is not possible to find a regulation with special focus to neurosciences or neuroscientific research. This makes it seem consequent to call the general regulations being applicable to neuroscientific research, although they have to be concretized and

---

[20] See Caulfield et al. (2010), Country Report Canada.

[21] See Sect. 4.

[22] See Lötjönen (2009a), pp. 161–175.

extended paying attention to the detailed problems of neurosciences, such as the problem of how to cope with so-called incidental findings, which will be discussed in detail in the following chapter.

4   Incidental Findings

As one of the main aspects and thus also one of the main challenges for law in the field of neurosciences, the problem of so-called incidental findings can be determined. An incidental finding is commonly determined as any unanticipated result of (neuroscientific) research which makes it advisable to seek medical assistance to rule out (or to confirm) potential negative consequences of this finding to the mental or physical health of the proband.[23] Other tries of definition state that incidental findings shall be defined as "observations of potential clinical significance unexpectedly discovered in healthy subjects or in patients recruited to brain imaging research studies and unrelated to the purpose or variables of the study".[24] Such slight differences in definitions already show the first aspect of the difficult problem: around the world, it is not yet absolutely clear how to define incidental findings. This may lead to the critical situation that a research team tries to shift its responsibility by negating that it has actually made an incidental finding. Consequently, in a first step it would be helpful to consent on one appropriate definition of incidental findings which the further discussion could be based on. From the country reports in this book, one may see that there are some elements which come back in most of the definitions, so that the way to one common definition should no longer be that far.

Research projects operating with the procedure of neuroimaging have to cope with the problem of a proper approach toward incidental findings. Although most cases lack a contractual relationship between the test person and the principal investigator or the staff carrying out the measurements, this does not allow any kind of failure to act. The duty to act can also derive from other bases than contractual ones.

In a first step, one has to consider a general duty to help other persons, which in many legal systems originate explicitly or implicitly in criminal regulations. Therefore, one can in general stress that there is a possibility of criminal prosecution of a person who incidentally finds out something and does not act, although this dormancy means a concrete danger for the test person. In addition, also aspects of the civil law have to be considered: The failure to act after an incidental finding can cause the duty of compensatory damages, if the test person has to suffer from a physical or financial damage due to the fact of not-acting. The same applies in cases

---

[23] See e.g. Schleim et al. (2007).

[24] Illes et al. (2006), p. 783.

where the incidental finding causes a decline in the insurance or the employment level of the test person.

Whereas in some countries, such as Germany, there does not yet exist a specific regulation on the problem of incidental findings and the discussion rather takes place in the relevant literature,[25] other countries such as Austria are able to draw parallels for the handling of incidental findings in connection with neurosciences from other, already existing legal fields. In Austria, this parallel can be found in the Austrian Gene Technology Act. Genetic research can also yield incidental findings, and this case has been explicitly addressed by this Act, in a way that does not leave too much space for discussion. The Gene Technology Act draws a sharp line between testing for medical purposes and testing for scientific purposes. Sec. 71 of the Gene Technology Act clearly states that "the proband has to be informed of unexpected findings of direct clinical relevance or of results the communication of which he or she has explicitly demanded. The communication, especially in cases where the proband has not asked for the results, has to be arranged in such a manner that it will not upset the proband. In borderline cases, the information may be completely withheld". The wording of this provision does not leave many questions. With only one exception, the tested person has to be informed about possible incidental findings.

For the field of neurosciences, the provision may be applied mutatis mutandis as follows: if a neuroscientific test shows the risk of an actual danger to the concerned person's health, this has to be qualified as a "casualty"[26] and creates an obligation to inform the proband. In contrast to this, a mere predisposition for a certain disease or health risk as such is not sufficient to constitute a "casualty" and therefore does not lead to an obligation to act.

Speaking of the content of the proband's informed consent, this will have to stretch on the will to be informed about incidental findings. Otherwise, it would come to an unfair shift of responsibility toward the researcher in a way that he would be the only person to be held responsible just because of his special knowledge. On the other hand, the researcher would have to face the situation that he is kept away from carrying out indicated measures of treatment on the concerned person.

As with many difficult problems, one also has to consider the problem from the opposite side. A quite intense discussion has spun out about the question, whether or not the concerned person does also have a right to refuse to be informed about incidental findings. In many literature statements, this right is named as a "right not to know". As to some opinions, this right not to know does not exist in cases, where there can be determined "severe" incidental findings, in other words findings which may point toward a serious health risk; in those cases, the research team is under an obligation to inform. On the other hand, findings of minor significance can probably

---

[25] Schleim et al. (2007); Spranger (2009a), pp. 194–197.

[26] Jerabek (2010), para. 4.

be kept secret from the concerned person if he has explicitly expressed this wish. Should, however, the research team erroneously classify findings as being of minor importance while there has been an obligation to inform, the negligent breach of this obligation may well result in a duty to pay damages.

Other approaches rather stress the point that there is a general duty to disclose the knowledge about incidental findings. However, this duty to disclosure shall only come to full power, if the information can be used to result into some kind of benefit for the research participant, which means that his condition shall at least partly be treatable, and the participant has not specifically forbidden the disclosure. Having said this, one can further argue that if the participant has given an informed refusal on the disclosure, this could also be understood as protecting the research participant from unwanted information. A participant claiming his right not to know should consequently be excluded from the research project in the best own interest of the researcher, in order to avoid unwanted legal uncertainty.

However, it has to be clarified to what extent abnormalities can create burdens. The researcher should indicate that – apart from their clinical significance – incidental findings can cause psychic problems and may have serious negative side effects on the state of insurance and employment conditions. From the above-said one can see that there is taking place a quite intense international discussion on the handling of incidental findings in connection with neuroscientific research projects. Whereas the definition lines for an incidental findings can by now be drawn quite clearly, the follow-up problem of how to handle a possible wish of non-disclosure has not yet been subject to a sufficient suggestion.

Consequently, it seems to be possible to set up the following standards in connection with incidental findings. It is essential to gain (written and) informed consent from persons participating in the program. This informed consent has to fulfil special minimum requirements. At first, the test person has to be explicitly informed that he is not participating in any kind of medical or diagnostic examination, so that it is clear there is no specific examination with the goal of finding pathological or other findings needing medical treatment. On the other hand, the possibility is to be stressed that existing aberrances might not be found.

At the same time, the test person has to be informed about the fact that, in case such a finding is made, specialists are asked for advice and that these specialists inform the concerned person about the finding and its consequences and possibly also about possibilities of treatment. If the test person refuses to consent, it is suggested that he is to be excluded from the program. Accordingly, the program should not be carried out, if due to the staff situation it cannot be secured that a specialist is available for the evaluation of incidental findings.

With regard to the researcher's best own will, it is stated by a solid meaning in literature that the researcher should be advised of not explicitly taking responsibility for the evaluation of clinical relevance. To the contrary, the researcher should better make sure the finding is carefully evaluated by a competent professional before the decision about the information. However, the idea of just enabling an expert review of all images is neither feasible with a view to financial burdens nor advisable, because otherwise the participants would participate in the research study with the expectation that if a brain abnormality exists, it will be discovered and reported. The research study would partially obtain a therapeutic facet.

Still, also in this respect, the lack of provisions dealing with neuroscientific incidental findings seems to make it advisable for the individual countries to enact such provisions in order to gain some legal safety. As it can be drawn from the mentioned country reports, there is already some international consensus on the topic of incidental findings, which might very well lead to a fast progress in this field.

5 The Concept of Free Will

The current debate on the question of free will is genuinely based on the experiment carried out by the Californian researcher Benjamin Libet.[27] If one takes the results of the experiment for granted, one could principally criticize the figures of free will and liability of the human being.

Within the *Brazilian* discussion, the researchers discussing the physical substrates of the moral already call into question the principle of free will very clearly:

> The implications for the law are obvious: if there is no free will, all the civil law, based on the idea of autonomy, and the entire criminal law, founded on the notion of guilt (personal responsibility), must be rethought.[28]

But in order to put such statements into a wider sense, one has to mention that the research carried out on the issue of free will in Brazil is still at its beginning. Demands that the law will have to follow the new developments are definitely correct in their core, but they are too vague from today's point of view, due to the fact that it is not yet clear if there really will be new developments to be followed.

The *Austrian* Criminal Code, alike most comparable codifications in other states, such as Germany, clearly points out that only who acts culpably must be punished (Sec. 4 Criminal Code). Taking this initial point into consideration, one has to recognize that the concept of guilt sets as a precondition that the perpetrator can be accused of not having acted otherwise than he or she did. A strongly determinist approach, however, will doubt that a person had this chance as his actions or omissions were predetermined by (neuronal) factors beyond one's control.

Opinions stating that the general concept of guilt in the Austrian criminal law has to be rethought have to bear in mind that the Austrian Constitution possibly may hinder such a construction.[29] On the other hand, the idea of defining guilt as

---

[27] For a short introduction to the experiment, see Sect. 5 above.

[28] Free translation of: http://blog.sbnec.org.br/2010/07/direito-e-neurociencias-neurodireito-o-que-e-isso/ (March 21st, 2011).

[29] For further references on this discussion, see Tipold (2005), paras. 44–48.

deviation from average might also been as enshrined in the Austrian Constitution in its article 142. "Taking all this into account, it seems very improbable that new neuroscientific findings concerning the non-existence of a free will could "undermine the concept of guilt"[30] in Austrian criminal law."[31]

Within the *German* legal system, the option to renew the general understanding of personal debt and thus to abort the principle of free will is definitely not resulting from human dignity as the pivotal principle in the German Constitution.

> It is important to note that this concept would not question the original image of the human being, which is the basis for the definition of human dignity. To the contrary, in such a case the existing understanding of human beings and their dignity would have to be strengthened, because otherwise the thesis of free will would lack a connection to the Basic Rights. Despite these concepts, it is to be pointed out that the principle of debt is at least also based on human dignity and that not the principle of human dignity is based on the principle of debt. Consequently, any recalibration of this constellation must take into account the frame of Art. 1 Basic Law, which is not violated by such modifications. Therefore, the whole discussion is only linked with constitutional matters in that way, that any penalty without debt would indicate an intervention into human dignity in a legal sense. But so far neuroscientific findings do not have the necessary power to overcome the old principle of debt.[32]

Whereas the approach of the *Italian* legal system shares the generally skeptical view toward a reconsideration of the principle of free will, it takes the problem at an earlier stage and questions whether the discussion about the freedom of will is at all the right discussion to be held. The Nordmann Report (2004) outlines the field in the following way:

> Information and communication technology helped produce the profound transformation of daily life in the 20th Century. Biotechnology is transforming agriculture, medical diagnosis and treatment, human and animal reproduction. Most recently, the transformative potential of nanotechnology has captured the imagination. Add to this that cognitive and neurosciences are challenging how we think of ourselves, or that the rise of the social sciences parallels that of bureaucracies and modern forms of governance.[33]

In *Japan*, the issue of free will has so far been discussed from an ethical perspective according to the so-called "Modern School" and the "Classical School". To shortly name the main aspects of those opinions, the Classical School insisted that free will exists from the viewpoint of indeterminism and principle of culpability, whereas the Modern School rather speaks of a so-called soft

---

[30] Spranger (2009b), pp. 42–43 and Spranger (2009d), p. 1035 comes to a similar conclusion for German criminal law. See also Paeffgen (2010), paras. 230i–230j; Streng (2007), pp. 690–691; Hochhuth (2005), p. 753 comes to the general conclusion (not limited to criminal law) that the legal concept of a free will can be upheld.

[31] Hilf/Stöger, Country Report Austria.

[32] Spranger, Country Report Germany.

[33] Nordmann (Rapporteur), Converging Technologies – Shaping the Future of European Societies, Report 2004, reperibile in: http://ec.europa.eu/research/conferences/2004/ntw/pdf/final_report_en.pdf. (March 21st, 2011).

determinism. To date, this viewpoint can be considered as predominant in the Japanese debate on free will.

It is stated in the Japanese literature that determinism tends to accept interventions into the brain as a social treatment or a measure for preserving public health. However, the application of these viewpoints on neurosciences is generally seen very critical, as there is a profound fear that psychiatric patients might base their defense solely on the argument that they were not free in will.

As for *New Zealand*, there are raised similar concerns like in Japan, pointing at the thought that the offender or the accused person in criminal law might no longer be seen as a person, but as a victim of his own brain, as this seems to be the point which decides whether or not a person is able to form a free will in a legal sense. This neuropersonal model promotes the idea that individuals only commit crimes because they possess brains that are unwell.[34]

> If neuroscience and behavioural genetics can go some way to explaining and predicting behaviour, with particular reference to issues like free will, determinism and their effect on the extent to which criminal culpability is likely to be undone by new scientific discoveries, then neuroscience may have an important role to play in assessing offenders[35]. Social agencies may learn how to best handle criminal behaviour accompanying drug addiction, given an apparent genetic predisposition towards the same. It is suggested by Garland and Frankel that by changing the way society views and understands addiction, drug use, and treatment, neuroscience has the potential to reshape our policies on criminalisation and incarceration as they pertain to drug-related offences.[36]

A comparison between the different countries and their approaches toward the debate about free will shows a clear result. Although the impact of neuroscientific developments on the concept of free will is discussed in many of the countries in a very broad sense, it seems to be only the initial point of the general discussion in the field of neurosciences. Most authors agree that the influence of neurosciences on the general understanding of the free will or the picture of human beings will be held at minimum and that there will not be a complete renewal of these general understandings. Thus, it does not at all seem wrong to state the view that the debate about free will has for the moment reached its limits, as there are no serious results at the moment, which would change the general understanding of the human being or the concept of free will.

---

[34] Pockett (2010), pp. 281–293.

[35] Garland and Frankel (2006), pp. 101–103.

[36] Henaghan/Rouch, Country Report New Zealand.

6   Criminal Law and Mind Reading

A broad discussion in the field of neurosciences examines the possible use of neuroscientific methods for evidence matters in the court rooms. The fast developing achievements in neuroscientific examination techniques, especially imaging techniques, has led to a growing interest in their possible use in criminal or civil procedures.

The use of neuroscience in criminal law has largely come from the research carried out by the Californian researcher Benjamin Libet, who in the 1980s had a profound impact on the scientific understanding of the association between the brain and behavior, while challenging the notion of free will. Through a series of experiments, in which Libet had subjects make voluntary hand movements while measuring their brain activity, he concluded the brain was active even before the subject was aware of having made the conscious decision to move their hand. This inferred that choice may be determined in the brain before the mind acts, making free will illusory. According to Libet, this has major implications for individual responsibility and the concept of guilt.

In some countries such as in the USA, Singapore, Israel, and India, the use of lie detectors in court trials is no longer an exception, but has become some kind of daily routine. For instance, the Brain Electrical Oscillations Signature (BEOS) Test, a brain-based lie-detection test, has been used in more than 300 Indian cases since 2003,[37] with the negative peak of this development in 2008, when the test was used to convict the accused person of murder for the first time. In *Australia*, structural neuroimaging methods such as CT scans and diagnostic MRI scans are routinely admitted as evidence in civil and criminal trials. Studies are proffered as evidence in, for example, cases determining the presence of brain injury due to trauma, declaration of brain death due to pathology or injury, diagnosis of brain pathology, testamentary capacity, and dementia and mental illness.

There is a certain expectation that new technologies could in the future be used as a viable alternative to the polygraph. However, already the use of the polygraph is broadly discussed on an international level and the views reach from admission of the procedure over some suspicion towards the polygraph to a complete ban of the polygraph in other countries. This fact makes it even more difficult to examine the admissibility of neuroimaging procedures in court trials, due to the fact that those techniques differ from the polygraph in one way or the other. Nevertheless, the discussions about the admissibility of neuroimaging methods in court trials mostly take their initial points in the discussion about the admissibility of a polygraph test, which is why this way is the one followed in this chapter as well. Apart from the use of neuroimaging methods in court trials, the use of mind-reading techniques can also be discussed in connection with possible uses to the benefit of disabled persons or persons without the capacity to consent.

---

[37] Puranik et al. (2009), S. 817.

The use of a polygraph is currently seen as illegal in *Austria*. Bearing this in mind, the usage of neuroimaging techniques in the context of establishing the facts in a criminal or civil case is at first a constitutional question as it touches on several fundamental rights. According to the European Court of Human Rights,

> the right not to incriminate oneself is primarily concerned, however, with respecting the will of an accused person to remain silent. As commonly understood in the legal systems of the Contracting Parties to the Convention and elsewhere, it does not extend to the use in criminal proceedings of material which may be obtained from the accused through the use of compulsory powers but which has an existence independent of the will of the suspect such as, inter alia, documents acquired pursuant to a warrant, breath, blood and urine samples and bodily tissue for the purpose of DNA testing.[38]

According to this decision and the fact that the European Charter of Human Rights has been incriminated into the Austrian constitution, the forced administration of neuroimaging procedures on the accused would touch the core of the right not to incriminate oneself as described by the European Court of Human Rights as the accused could not remain "silent" but would give away some of his and her thoughts. In other countries, such as Germany, also the voluntary application at least of the polygraph is not admissible according to the Federal Supreme Court, as it is seen as an invalid evidence device according to section 244 of the German Code of Criminal Procedure.

Apart from this, according to the Austrian understanding, the use of mind-reading techniques touches the right to a fair trial according to article 6 ECHR. This is due to the fact that the results cannot yet be seen as reliable in way that would prove the guilt or nonguilt of the accused person. Furthermore, from the perspective of the witness, the forced use of such techniques touches on his or her right to private life under article 8 ECHR.

Alike in the other spheres of neurosciences, the *Finnish* law does not provide any regulations on the usage of a polygraph in court trials. It is commonly respected that all coercive measures against the suspect must follow a certain legal provision. Taking this into account, one can only come to the conclusion that the use of a polygraph test is voluntary. Refusing to undergo the test cannot be used as evidence against the party to the proceedings.[39] According to the Finnish Code of Judicial Procedure, the court can evaluate all evidence freely unless there is a special provision in law on its significance. A consequent application of this provision would likely result in the admission of the polygraph test or other procedures of neuroimaging in court trials. However, there have so far been no cases reported, where the conviction has solely been based on the polygraph.

In *France*, the fMRI scanner has not yet been used as a lie-detection test in court trials. It is commonly regarded as violating the human dignity and therefore not allowed to be used. In addition to this argumentation, the French Supreme Court has decided in a few cases so far that hearings conducted under hypnosis are not to be

---

[38] ECtHR 17.12.1996, 19187/91 Saunders v. United Kingdom, para. 69.

[39] Klami (1996), p. 215.

allowed because they are violating the accused person's free will and oppose to his consent.

As for *Germany*, the admissibility of lie-detection methods decisively depends on the constitutional approach toward such methods. Whereas positive reactions state that the methods are a significant improvement compared to the common lie-detector, but show lighter impacts on the concerned person's rights, it seems to be advisable to compare the methods of neuroimaging and the common lie detector not only in a technical, but also in a legal sense. However, this comparison does not hold true in all aspects and thus has to been quite critical. In contrast to the polygraph, procedures of neuroimaging do not measure physical reactions, which hint at excitement, fear or nervousness, but only show the brain activity indirectly. Thus, the usage of image-guided procedures is not aiming at a conclusion from physical reactions to a feeling to truth or untruth, but from the brain function to the quality of a thought.

However, this distinction does not seem to be forcing in any way, as it holds true that the interpretation of the result is much more important in connection to the usage of image-guided procedures than in connection to a polygraph.[40] According to the German jurisdiction, the human dignity of the concerned person cannot be violated, if the concerned person has consented to the procedure. Of course, this result can only be held by those who state that human dignity does not have to be protected against the will of the concerned person and thus demands a protection against one own. The dogmatic relevance and acceptance of this option show the need for a closer glance at the problem, which has so far not come into action in German literature.

A truly very interesting decision on the usage of neuroscientific measures in court trials can be found in the *Greek* jurisdiction. For the first time in the history of Greek court trials, the Mixed Jury Court of Athens admitted in its decision 93/2002 the use of neuroscientific techniques in a court trial. It is interesting to note that this court trial went its way vice versa to the common court trials mentioned in connection to neurosciences, where it is discussed whether or not to use the new methods in order to convict the accused. In the above-mentioned trial, one of the defendants submitted the request to be subjected in the procedure of lie-detection with the method of Event Related Potentials (ERPs) in order to prove his innocence. The defendant was facing – among others – the charge of man slaughter. For this reason, the defendant asked the court to give the permission to two neuropsychiatrists to visit the prison, in order to examine him with the method of ERPs. In its decision, the court fully accepted the demand of the accused, allowed him to be examined with this scientific method and allowed the use of the findings as evidence. Some months later, the court, with its decision 312/2002, acquitted the accused, without making special reference to the method of lie-detection.

---

[40] Spranger (2007), p. 164 et seq.

By this decision, one can very clearly see the other side of the medal concerning the use of neuroscientific measures in court trials. Whereas the general discussion mainly focuses on the aspect, whether or not an accused can be forced to undergo methods of lie-detection, this argumentation lacks a more positive view of the topic. There are indeed scenarios, where it is the declared wish of the accused person to take methods of neurosciences in order to prove his innocence. In such cases, it could very well be an appropriate option to admit the usage of neuroscientific methods in the court trial. Nevertheless, also according to the Greek discussion, the use of neuroscientific measures is rather seen skeptical. First, according to the wording of the Greek constitution, the assessment of capacity for criminal responsibility is always examined in relation with the specific action committed. As a consequence of this, there is no general irresponsibility or diminished responsibility according to the Greek legal system. In addition to this, the Greek Penal Code adopts a mixed system, composed of biological as well as psychological factors in order to determine criminal responsibility. According to this system, in order to assess a person's criminal responsibility, one has to look first for the possible biological factors and second to ask a psychological question, i.e., what effect these biological factors had on the appreciation of the wrongful character of the specific action committed and the ability to act according to this appreciation, at the time of crime. Therefore, even if neuroscientific techniques succeed in proving a specific anomaly or malfunction in the brain of the accused, it does not follow that this is a reason to exculpate or diminish responsibility. What clearly needs to be established is a causal link between the brain dysfunction (or predispositions that this dysfunction entails for the individual) and the specific action committed by the perpetrator.

All in all, there are two elements which compose the general notion of criminal irresponsibility under the Greek law, namely a cognitive and a volitional element. If neuroscientific techniques manage to show some defect in one of these two elements, then they will have been able to prove or to play a significant role in proving the irresponsibility or diminished responsibility of the perpetrator.[41]

However, the use of neuroscientific techniques in the Greek criminal courts is restricted by some key provisions related to the protection of some rights and freedoms of citizens. One of these provisions is the article 7 par. 2 of the Greek Constitution, which generally prohibits any torture, bodily injury, damage to health or psychological violence, and any violation of human dignity, under form of punishment or under any other form (e.g., interrogation method). Psychological violence includes violation, investigation, and legal assessment of the subconscious world of a person exercised by state institutions. As psychological violence could be considered the use of lie detector, which captures the unconscious mental events by changes in respiration, eyelid movements, etc.

The article 137 PC on tortures considers any affront of human dignity as a torture strongly prohibited by the Greek Constitution. In the 3 rd paragraph of the art. 137, the Penal Code explicitly refers to the use of a lie detector as an affront to human dignity. Thus, the combination of Articles 2 par. 1 Const, 7 par. 2 Const. and 137 A par. 3 PC can infer a

---

[41] Vidalis/Gkotsi, Country Report Greece.

general prohibition of the penetration and the forced investigation of a person's innate mind, especially with the use of techniques such as lie detectors.[42]

In *Italy*, a huge legal and forensic medicine literature is dedicated to the issue and some recent case law tries to deal with the issue of neuroscientific methods of lie-detection in conclusive terms. According to the Italian jurisdiction, which could so far been established on the topic, "moral damage" has to be conceived in its widest meaning as the violation of the individual's personal sphere, even if there are no immediate economic consequences. Italian courts regularly admit imaging evidence (traditional radiography, CT scans, and MRI) to provide insight into the extent of a person's pain and suffering. "A large number of cases involve people unable to work due to serious and painful conditions, that do not have objectively measurable symptoms or tests, and that may therefore face difficult problems when making and supporting a claim for disability insurance benefits. These hurdles are common also to people who suffer from conditions like fibromyalgia, chronic fatigue syndrome, or chronic pain conditions like complex regional pain syndrome (CRPS). Another frequent set of cases regards plaintiffs who sustain injuries in motor vehicle accidents and claim to have chronic pain well beyond the time that the objective injuries have healed."[43]

The Italian point of view seems to be, that despite all reasons to be cautious, there are indeed some strong reasons, why neuroscientific evidence should be admitted in the court rooms.

In the *Netherlands*, responsibility is typically measured by a number of clinical variables, which are determined by means of an anamnesis and standardized behavioral and/or (neuro) psychological tests, as well as by demographic and crime-related variables. Whereas the complete absence of responsibility is a reason for discharge, diminished responsibility is typically a reason for mitigated sentencing and typically leads to an order of detention during Her Majesty's pleasure. The term is used to describe detainment in prison or a psychiatric hospital for an indefinite length of time. A judge may rule that a person be "detained at Her Majesty's pleasure" for serious offences or based on a successful insanity defence.

Although the number of cases is unknown, neuroscientific evidence aimed at the assessment of responsibility has already entered the courtroom in the Netherlands. In some cases, the expert witness – a behavioral neurologist – did not find any brain damage or not sufficient connections between the brain damage and the behavior that constituted the criminal act, and hence the suspect was considered fully responsible. In other cases, however, brain damage and a link between the damage and the behavior was found that did influence the decision about the degree of responsibility.[44]

---

[42] Vidalis/Gkotsi, Country Report Greece.

[43] Santosuosso, Country Report Italy.

[44] Klaming/Koops, Country Report Netherlands.

In the Dutch literature, one of the most important challenges of using neuroscientific methods in court trials seems to be the possibility that neuroscientific evidence is inappropriately persuasive and may therefore unduly affect legal decision-making. Serious concerns are raised about the fact that judges and juries may perceive evidence derived by means of insights from neuroscience without sufficient critical appraisal. Thus, it is seen as important to further empirically explore the effect of neuroscientific evidence on legal decision-making in order to ensure the responsible use of this type of evidence. If this precondition is met, the general use of neuroscientific methods might become legal in the Dutch legal system.

Like in other countries as well, in *New Zealand* most defendants in criminal cases provide neuroimaging evidence to try and show diminished capacity or insanity during trial, and as supporting mitigation during sentencing. In contrast to the Dutch view, the New Zealand jurisdiction rather sees the techniques of neurosciences negatively. The decision R v Dixon[45] explicitly excludes the possibility of admitting neuroimaging evidence to provide an assessment of the accused's brain, or its working order, in establishing its capacity to know right from wrong. From a human rights perspective, especially potential clashes with the New Zealand Bill of Rights Act are important. The use of neuroscientific methods is seen as "being a witness against oneself", meaning a violation of the Bill of Rights.

The efficacy of such evidence is also of concern. Common law courts appear universally suspicious of such things as polygraph evidence, and it is suggested that some standard of accuracy would need to be established before neuroscience evidence can be regularly introduced as supporting evidence. However, in some areas such as accident insurances these methods are already admitted, for instance in Germany.

In *Switzerland*, a great attention is also being paid to neuroscientific methods in the field of police- and security law. In this context, neuroscientific methods could possibly affect on the prognosis of danger of a person by reading his thoughts. On a longer sight, even the search for terrorists or other dangerous perpetrators might be possible via controls of the public area.

Also in the *United Kingdom*, there have been faced some court trials in the recent past, where the court's decision has decisively been based on methods of neurosciences. In 2005, the Court of Appeal revisited four cases where the appellants' convictions which ranged from murder to serious assault were based, in part, upon scan evidence of brain injuries suffered by children.[46] Although the mentioned cases have been very unlike in their content, they all have in common that there was a great disagreement about the quality of evidence methods deriving out of neurosciences.

---

[45] R v Dixon [2008] 2 NZLR 617 (CA).

[46] R v Harris; R v Rock; R v Cherry; R v Faulder [2005] EWCA Crim 1980.

It is of high relevance to recognize the extremely positive view of the court on methods of neurosciences. The court explicitly stated that 'developments in scientific thinking should not be kept from the Court, simply because they remain at the stage of hypothesis.' On the other hand, the court also clearly pointed out that the quality of the evidence given has to be explained in a first step: 'Obviously, it is of the first importance that the true status of the expert's evidence is frankly indicated to the court.'

In the first UK court trial to use methods of neuroimaging ever,[47] an EEG test was carried out on the defendant and the consultant neurologist's report was submitted for consideration by the court. It showed no abnormality which could explain the defendant's loss of consciousness. The appeal court notes its disapproval of the use of this written medical evidence commenting 'medical reports so often used in civil actions have no place in criminal courts'.

Despite the generally friendly view toward neuroscientific evidence, the courts in the UK tend to remain skeptical about the efficacy of such methods. Although in some American advertisements methods such as fMRI testing are declared as the 'first and only direct measure of truth verification and lie detection in human history!',[48] the factual quality of neuroimaging methods are still today not quite clear. But even with this in mind, the UK legal system seems to be at a highly topical level concerning the use of neuroscientific evidence in court trials, which may be underlined by the following:

In England and Wales under a new offender management regime polygraph tests which are designed to test the truthfulness of statements about behaviour of sexual offenders after release on licence from prison are being introduced. The Offender Management Act 2007 enables the imposition on such offenders of a mandatory requirement to undertake polygraph tests to enable the risk they pose to the community to be better managed. Section 29 of the Act controls the manner in which the tests are conducted and Section 30 prohibits using the results of the tests taken by an offender as part of this regime being used in evidence 'against the released person for an offence'. Of course this does not preclude the instigation of an investigation into the offender's activities using information gleaned from the offender during the test interviews. Polygraph evidence has been stated in a recent case from the Court of Appeal to be inadmissible.[49]

The Act makes the testing regime mandatory for certain sexual offenders. The government are reviewing the use of the testing and the effectiveness of the tests as a management tool. What is interesting is the justification given for making such testing mandatory by Baroness Scotland, Minister of State for the Home Office in the House of Lords. She said 'We will commission a research study to run alongside the mandatory testing pilot, with a view to determining whether the polygraph test is efficacious in assisting the collection of useful evidence about offenders' behaviour and whether it genuinely facilitates effective

---

[47] Hill v Baxter.

[48] www.noliemri.com (March 21st, 2011).

[49] R v Malcolm MacMahon [2010] ECWA CRIM 1953.

offender management without disproportionally affecting the rights of those tested. We think it is a sensible way forward.'[50,51]

In reality, however, one cannot yet find signs of the British parliament to be introducing such tests, which seems to hold true particularly if they do not provide any specific advantages in terms of accuracy over polygraph tests. But nonetheless the use of the polygraph in such a manner suggests that where there is a test which appears to distinguish truth from lies then the government may be tempted to use such tests in the management of offenders, particularly where the risk posed by the release of those offenders is seen by the public as high risk.

To conclude with a view to the UK legal system, one may cite a statement by the Law Commission, published on its website: 'The final report will recommend that a clear admissibility test for expert evidence should be created to replace the present application of the 'relevance and reliability' test. To accompany the test, we will also set out proposed guidelines to assist Crown Court trial judges (and magistrates' courts) to determine whether proffered expert evidence is sufficiently reliable to be considered.'

In particular, the decision of the Mixed Jury Court of Athens shows that the general discussion in some way lacks a more positive view of the situation of neuroscientific methods of lie-detection in the court rooms. There are indeed scenarios, where it is the declared wish of the accused person to take methods of neurosciences in order to prove his innocence. In such cases, it seems to be the best option to admit the usage of neuroscientific methods in the court trial. There are no reasonable arguments, why the court should not be able to judge the quality of this method of proof by the same measures it does in cases of common methods of proof. For instance, if the relevant neuroscientific method is still very new, and it is not yet clear whether or not this method does deliver reliable results, the court will have to take this into consideration. This opinion is also stressed by the *Australian* point of view. Although so far there have been no cases in Australia where neuroscientific techniques have been successfully introduced as evidence in both criminal and civil trials, the issue is already discussed in Australia in an intense way. To date, the discussion about the admissibility of neuroscientific techniques in Australian courts rests with the admissibility in criminal hearings. In Australia, the responsibility for evaluating the validity of scientific tests falls on the judiciary via the rules of evidence, in particular the Evidence Acts. Expert testimony based on functional studies is deemed to constitute a scientific technique warranting elucidation through the provision of expert evidence and therefore subject to the strict rules of evidence. In this context, the threshold enquiry when considering the admissibility of expert opinion evidence, as with evidence of any kind, is to identify its relevance.[52] Due to the Australian system of case law, the question of relevance

---

[50] Hansard, HL Vol.692, 5.45 pm (June 12, 2007).

[51] Claydon/Catley, Country Report United Kingdom.

[52] Houston/Vierboom, Country Report Australia.

cannot be answered in general but has to be evaluated with regard to the details of the individual case.

However, the general approach in Australia seems to be rather skeptical. The introduction of neuroscientific evidence regarding mental capacities may be considered an appropriate alternative approach toward the problem of responsibility in Australian courts. While opening the courts to fMRI would open this use of scans to a prosecution, responses to this approach point out that this confuses "capacity responsibility" with "virtue responsibility" (a description of how one has acted based in fMRI evidence) and thus deals with elements of the sentencing process, rather than the guilt assessment process.

In the *USA*, the testing whether or not a method of neuroscientific evidence is seen as reliable for proof in court trials follows a four-step test, which was developed in the so-called Daubert case. In assessing the reliability of the proffered fMRI evidence, the analysis of the court follows four nonexclusive factors:

1. Whether the theory or technique can be tested and has been tested;
2. Whether the theory or technique has been subjected to peer review and publication;
3. The known or potential rate of error of the method used and the existence and maintenance of standards controlling the technique's operation; and
4. Whether the theory or method has been generally accepted by the scientific community.[53]

However, within the U.S. scientific community, there is at the moment a strong consensus that methods of lie detection are not yet ready for use in the court rooms. The neuroscientist Anthony Wagner stated in 2010 that today there are not yet any reliable data which give proof, whether or not methods of neuroimaging can really detect true from false and thus serve as lie-detection measures. Still, and despite the scientific limitations, there are indeed several court trials in which fMRI and EEG-based lie detection evidence have been proffered in U.S. courts. For instance, in 2003 an Iowa state court admitted EEG-based "brain fingerprinting" lie detection evidence.

Following the argumentation lined out above, one could come to the conclusion to generally admit methods of neurosciences in the court room, as long as the accused has consented to the procedure. In a second step, the court itself will have to evaluate, probably with the help of external experts, in which ways the method of proof is qualified to give serious evidence and consequently to which degree the decision of the court might be based upon the neuroscientific evidence. Also in comparison with other methods of proof, it does not seem very likely that it will be able to base the court's decision on a neuroscientific method *alone*. Alike some other methods of proof, for instance the blood withdrawal in the German Criminal Procedure Code, the admission of neuroscientific methods against the will of the

---

[53] See Jones/Shen, Country Report USA.

accused person shall only be admissible in rare exceptional situations. On a level beneath the criminal law level, there can be determined further possible fields of application, which have so far not been discussed in depth.

7   Neuro-Enhancement

In recent publications, the issue of neuroenhancement comes to speech more and more. The term describes any situations in which a person tries to enhance his cognitive abilities by measures of neurosciences. Such measure could simply be drugs which affect on the brain and thus lead to an enhancement. Also covered are interventions directly into the brain which then enhance the abilities of the brain. In this context, so-called assistive technologies do also raise some interesting questions from a legal point of view.

In *Austria*, some of these questions have been shortly addressed in a report on the use of assistive technologies, which has been produced by the Austrian Bioethics Commission, an expert advisory board of the Federal government.[54] Although the report deals with assistive technologies in general and is not restricted in content to neurosciences, some of the problems identified might as well, at least in the future, be of relevance for certain forms of technically advanced neuro-enhancement.

A first problem may arise in cases of malfunction or wrong interpretation of the results of such neuroscientific equipment. Second, assistive technologies will regularly be used to help people who are not in a position to declare an informed consent. Although the Austrian legal order contains provisions in this respect, it will have to be discussed whether the potentially grave consequences of the use of such technologies require specific provisions on consent by third persons or whether the legislation in force can sufficiently deal with this problem.

> However, the law of medical devices is to a large extent dominated by EU legislation. As a consequence, it will sooner or later become necessary to (comprehensively) address the use of such technologies at the level of European legislation. In this respect, it should be borne in mind that clinical trials of new medical devices regularly require the consent of an Ethics Committee irrespective of where this research takes places (in other words, not limited to hospitals or universities – see Sec. 58 of the Medical Devices Act).[55]

Furthermore, a specific problem of neuroenhancement is seen in the fear that persons using neuroscientific enhancers my gain an unfair advantage against the "normal" or "average" person. Although these questions have to be addressed, it seems to be likely that the current legal systems are able to cope with these challenges and that there is no greater need for specific regulations. As for

---

[54] Report by the Bioethics Commission on "Assistive Technologies: Ethical Aspects of the Development and Use of Assistive Technologies", dated 13 July 2009. For details, see footnote 50 above.

[55] Hilf/Stöger, Country Report Austria.

Germany, it can be drawn directly from the Constitution that, according to article 3, no person shall be discriminated against another person. It is clear that to date there is no sharp line determining which neuroscientific methods are indeed discriminations for other persons. However, as it holds true in the above-mentioned fields of neurosciences as well, it is to be expected that the serious results deriving from neuroscientific research will become more reliable and that this will enable the legal orders to set up clear provisions on the use of neuroenhancers.

The *Finnish* legal system already provides at least an initial point for the assessment of neuroenhancers. The main problem is seen in the fact that medical knowledge is not only always used in order to cure illness but also to enhance qualities that are considered normal from the medical point of view. Typical examples of this are the use of pharmaceuticals to enhance memory or concentration or to control anxiety. The pharmaceuticals used for these purposes are not necessarily forbidden substances, but they can have been approved for the treatment of an unrelated illness or a related condition, which is medically relevant.[56] The main legal question deriving from this is whether the physician is allowed to prescribe medical drugs for other than medical purposes.

Legal assistance on this can be found in the Finnish Act on Health Care Professionals, which in section 15 states that "the aim of the professional activities of health care professionals is to promote and maintain health, to prevent illness, to cure those who are ill and to alleviate their suffering. In their professional activities, health care professionals must employ generally accepted, empirically justified methods, in accordance with their training, which should be continually supplemented."

Compared to other fields of neurosciences, the area of neuroenhancement seems to be the one with the greatest need of legal clarification so far. In all the other areas, there have been significant improvements in the legal approach towards neurosciences within the last few years, which are pointing into the right direction, but of course do not free from the necessity to further legally discuss the new neuroscientific developments.

8 Conclusion

Although neurosciences have started to be approached from a legal point of view in many countries so far, there has not yet been a significant breakthrough in the struggle for appropriate legal provisions in this field. In a way, this may also be due to the fact that neurosciences in themselves are not yet that clearly shaped and thus nobody is able to clearly determine which future-scenarios are still to be expected. However, even if the contours of neurosciences become much clearer in the near

---

[56] Silvola, Country Report Finland.

future, it is not to be expected that neuroscientific developments will totally undermine fundamental principles such as the concept of free will. Nevertheless, there are already today fields of application where it would be advisable to quickly find solutions to the already existent neuroscientific measures. This holds true in particular for the application of neuroscientific methods in court trials. According to the country reports, this is a field which is discussed in most of the mentioned countries. What is even more important is the fact that such methods have already entered the court rooms in many of the countries. Consequently, it is of utmost importance to find a consensus, whether or not such methods shall be admissible in court trials and in which context or to what extent. As for the European countries, the result will also have to meet the requirements brought forward in the European Convention on Human Rights with its Additional Protocols. To sum up, it will be very interesting to see, in which ways the different countries will develop their legal systems with regard to neurosciences in the nearest future.

References

Berka W (1999) Die Grundrechte. Springer, Wien

Caulfield T, Rachul C, Zarzeczny A (2010) Neurohype and the name game: who's to blame? AJOB Neurosci 1(2):13–15

Downie J, McDonald F (2004) Revisioning the oversight of research involving humans in Canada. Health Law J 12:159–181

Garland B, Frankel M (2006) Considering convergence: a policy dialogue about behavioural genetics, neuroscience and law. Law Contemp Probl 69:101–104

Hadskis M (2007) The regulation of human biomedical research in Canada. In: Downie J, Caulfield T, Flood C (eds) Canadian health law and policy, 3rd edn. LexisNexis Canada Inc., Markham, pp 257–310

Hochhuth M (2005) Die Bedeutung der neuen Willensfreiheitdebatte für das Recht. (Deutsche) JuristenZeitung 15-16:745–753

Illes J, Kirschen M, Edwards E, Stanford L, Bandettini P, Cho M, Ford P, Glover G, Kulynych J, Macklin R, Michael D, Wolf S (2006) Incidental findings in brain imaging research. Science 311:783–784

Jerabek R (2010) § 95 StGB. In: Höpfel F, Ratz E (eds) Kommentar zum StGB (loose-leaf and online), 2nd edn. Manz, Wien

Klami HT (1996) Valheenpaljastuskoe Suomen oikeudessa. Defensor Legis 2:209–216

Kopetzki C (2010a) Behandlungen auf dem "Stand der Wissenschaft". In: Pfeil (ed) Finanzielle Grenzen des Behandlungsanspruchs. Manz, Wien, pp 9–46

Kopetzki C (2010b) Braucht Österreich eine Kodifikation des biomedizinischen Forschungsrechts? In: Körtner U et al (eds) Ethik und Recht in der Humanforschung. Springer, Wien, pp 56–89

Lötjönen S (2009a) Autonomy and dignity in clinical medical research on adults with cognitive impairment. In: Aasen HS, Halvorsen R, Barbosa da Silva A (eds) Human rights, dignity and autonomy in health care and social services: nordic perspectives. Intersentia, Antwerp, pp 161–175

Morse S (2008) (quoted in Associated Press), Neuroscience increasingly presented as evidence for trials in US courts. Fox News, 3 March 2008

Paeffgen U (2010) Vorbemerkungen zu §§ 32ff StGB. In: Kindhäuser U et al (eds) Strafgesetzbuch, vol 1, 3rd edn. Nomos, Baden-Baden

Pockett S (2010) The concept of free will: philosophy, neuroscience and the law. Behav Sci Law 25:281–293

Puranik DA, Joseph SK, Daundkar BB, Garad MV (2009) Brain signature profiling in India: its status as an aid in investigation and as corroborative evidence – as seen from judgments. Proceedings of XX All India Forensic Science Conference, Jaipur, S.815–822. http://www. axxonet.com/cms-filesystem-action/publications/beos_in_india.pdf. Accessed 19 March 2011

Rödiger C (2011) Das Ende des BEOS-Tests? Zum jüngsten Lügendetektor-Urteil des Supreme Court of India. Nervenheilkunde 30:74–79

Schleim S, Spranger T, Urbach H, Walter H (2007) Zufallsfunde in der bildgebenden Hirnforschung. Nervenheilkunde 11:1041–1045

Schweizer RJ, van Spyk B (2007) Arzt und Forschung. In: Kuhn M, Poledna T (eds) Arztrecht in der Praxis, 2nd edn. Genf, Zürich, pp 535–595

Shafi N (2009) Neuroscience and the law: the evidentary value of brain imaging. Grad Stud J Psychol 11:27–39

Spranger TM (2007) Neurowissenschaften und Recht. In: Jahrbuch für Wissenschaft und Ethik, pp 161 et seq

Spranger TM (2009) Medical law in Germany. International encyclopaedia of laws. Kluwer Law International, Suppl 55

Spranger TM (2009a) Rechtliche Implikationen der Generierung und Verwendung neurowissenschaftlicher Erkenntnisse. In: Schleim et al (eds) Von der Neuroethik zum Neurorecht? Vandenhoeck&Ruprecht, Göttingen, pp 193–213

Spranger TM (2009d) Neuroprothetik und bildgebende Hirnforschung. Neue Impulse für die Praxis des Betreuungsrechts. Betreuungsmanagement 4:206–208

Streng F (2007) Schuldbegriff und Hirnforschung. In: Pawlik M et al (eds) Festschrift für Günther Jakobs. Carl Heymanns Verlag, Köln

Tipold A (2005) § 4 StGB. In: Höpfel F, Ratz E (eds) Kommentar zum StGB (loose-leaf and online), 2nd edn. Manz, Wien

Vidalis T, Mitrou L, Takis A (2006) Constitutional reception of technological developments and "new" rights. In: TI Ant. Sakkoulas (ed) Centre of European constitutional law, five years after the constitutional revision of 2001, Athens, pp 273–312 (in Greek)

Wegmann H (2009a) Informed consent – essential contents and consequences of violation. J Int Biotech Law:20–28